Dietary Supplements

ACS SYMPOSIUM SERIES **987**

Dietary Supplements

Chi-Tang Ho, EDITOR
Rutgers, The State University of New Jersey

James E. Simon, EDITOR
Rutgers, The State University of New Jersey

Fereidoon Shahidi, EDITOR
Memorial University of Newfoundland

Yu Shao, EDITOR
GlaxoSmithKline Consumer Healthcare

Sponsored by the
ACS Division of Agricultural and Food
Chemistry, Inc.

American Chemical Society, Washington, DC

ISBN: 978–0–8412–3992–0

The paper used in this publication meets the minimum requirements of American National Standard for Information Sciences—Permanence of Paper for Printed Library Materials, ANSI Z39.48–1984.

Distributed by Oxford University Press

PRINTED IN THE UNITED STATES OF AMERICA

Foreword

The ACS Symposium Series was first published in 1974 to provide a mechanism for publishing symposia quickly in book form. The purpose of the series is to publish timely, comprehensive books developed from ACS sponsored symposia based on current scientific research. Occasionally, books are developed from symposia sponsored by other organizations when the topic is of keen interest to the chemistry audience.

Before agreeing to publish a book, the proposed table of contents is reviewed for appropriate and comprehensive coverage and for interest to the audience. Some papers may be excluded to better focus the book; others may be added to provide comprehensiveness. When appropriate, overview or introductory chapters are added. Drafts of chapters are peer-reviewed prior to final acceptance or rejection, and manuscripts are prepared in camera-ready format.

As a rule, only original research papers and original review papers are included in the volumes. Verbatim reproductions of previously published papers are not accepted.

ACS Books Department

Contents

Tea and Health

Fruits: Their Bioactives and Health Effects

Safety and Bioactivity

Indexes

Preface

Dietary supplements are used by 50–60% of the American population, and more than 300 million people worldwide. Today, dietary supplements are 'mainstream' and can be purchased not only via the Internet, pharmacies, grocery stores and specialty shops, but in virtually all major supermarkets, drug stores and national outlets. Dietary supplements are occupying increased shelf space in these stores largely due to consumer demand and this has lead in part to an increasing number of dietary supplemental products available that are marketed to address a wider range of health and nutritional applications. Besides traditional multivitamins and minerals, dietary supplements are used to boost energy, to maintain health of the digestive system, to aid in weight loss, reduce anxiety and stress, enhance our immune system, and in support of most health and nutritionally related conditions or issues. Factor in the significant increase of American that turn to and proactively seek and visit Alternative and Complimentary Medicine practitioners and therapist as a routine part of their families medical, nutritional and health counsel the use and reliance on dietary supplements will continue to increase as will the expectations of their efficacy. In recent years, the most rapid growing portion of dietary supplements is in natural botanical supplements. However, due to the complexity of chemical components in plant extracts, our scientific understanding of the efficacy of these botanical supplements remains very limited. We are challenged not only by the complexity of the natural products that may be found in a single plant, but in the common mixtures and blends that contain many botanical ingredients. We are challenged in the quality control and standardization of product preparations and in our basic understanding as to the products actual bioavailability, and chemical fate after consumption, such that the identification of the compound(s) responsible and the underlying mechanism of how it may work are elusive. Coupled to issues of safety and toxicity, product stability and even adulteration, it is clear that there is significant work to be researched as we improve our understanding of botanicals used as dietary supplements. The breadth of scientific research in this continuing emerging field continues to bridge the gap from purported uses to one of evidence-based.

The aim of this book is to provide some of the latest findings and perspectives related to dietary supplements (Chapters 1–3), phytochemistry of botanical dietary supplements and newer products (Chapters 4–8), as well as bioactive compounds of teas and fruits (Chapters 9–16). Finally, the safety and bioactivity of selected dietary supplements are discussed (Chapters 17–21). The book will serve as a useful reference for research workers and students in a variety of disciplines, including food science, nutrition, chemistry and biochemistry, as well as for health professionals. Researchers in the universities, industry, and government laboratories will find the book of particular interest as it provides the most recent findings in this ever-growing area.

We are indebted to all authors who contributed to this book and shared their vast knowledge in preparing a state-of-the-art information package for those interested in the area of dietary supplements.

Chi-Tang Ho
Department of Food Science
Rutgers University
65 Dudley Road
New Brunswick, NJ 08901

James E. Simon
New Use Agriculture and Natural Plant Products Program, Department of Plant Biology & Plant Pathology
Rutgers University
59 Dudley Road
New Brunswick, NJ 08901

Fereidoon Shahidi
Department of Biochemistry
Memorial University of Newfoundland
St. John's, NL
Canada A1B 3X9

Yu Shao
GlaxoSmithKline Consumer Healthcare
1500 Littleton Road
Parsippany, NJ 07054

Dietary Supplements

Overview

Chapter 1

Dietary Supplements: An Overview

Chi-Tang Ho[1], James E. Simon[2], Fereidoon Shahidi[3], and Yu Shao[4]

[1]Department of Food Science, Rutgers, The State University of New Jersey, 65 Dudley Road, New Brunswick, NJ 08901
[2]New Use Agriculture and Natural Plant Products Program, Department of Plant Biology and Plant Pathology, Rutgers, The State University of New Jersey, 59 Dudley Road, New Brunswick, NJ 08901
[3]Department of Biochemistry, Memorial University of Newfoundland, St. John's, NL A1B 3X9, Canada
[4]GlaxoSmithKline Consumer Healthcare, 1500 Littleton Road, Parsippany, NJ 07054

The use of dietary supplements in North America and particularly in the United States is widespread. The chemistry and biological studies related to the ingredients involved, particularly botanical bioactives are rapidly progressing. An overview of some aspects of dietary supplements described in detail in the book is provided.

Introduction

A dietary supplement is defined under the Dietary Supplement Health and Education Act (DSHEA) of 1994 and regulated by Food and Drug Administration, in 2003. A dietary supplement can be a product intended to supplement the diet that bears or contains one of more of the dietary ingredients including vitamins, minerals, amino acids, herbs and other botanical extracts (*1*). The sale of products in the dietary supplement market has approached $20 billion in the United States (*1*).

Top Dietary Supplements

The use of dietary supplements in the United States is widespread and this continues to grow as more people become concerned about the access and quality of health care, and seek complementary and alternative medicines (*2*). Major dietary supplements sold in the United States are still vitamins and minerals such as vitamin B-complex, multiple vitamin-minerals, vitamin A and vitamin C, calcium and iron. Other popular dietary supplements are Co-Q10 for energy and performance, glucosamine and chondroitin for managing osteoarthritis, soy and soy products for lowing cholesterol, among others.

Botanical Dietary Supplements

With the increasing demand for prevention of diseases, herbal products have become popular in the international market, not to mention in those countries with a long history of using herbal medicines such as China, India and Japan. Currently there are about 750,000 plant species, 300,000 registered species, 30,000-75,000 medicinal plants, 20,000 medicinal plants listed by World Health Organization (WHO) and 400 medicinal plants widely traded in the world.

In the U.S., herbal products are mainly regulated as dietary supplements and sold in different forms such as fresh plant products, dried botanical powders, liquid botanical extracts, soft extracts, dry extracts, tinctures and purified natural compounds. The global open market, allows all kinds of botanical products to be sold in the U.S. with ginseng, ginkgo, saw palmetto, echinacea, soy, bilberry, grape seed and green tea extracts as top-selling items. Table I lists some of the popular botanical dietary supplements along with their potential active components and functions (*3*).

Several of the botanical dietary supplements listed in Table I such as black cohosh (Chapter 5) and kava (Chapter 17) are discussed in this book. Phytochemistry of other botanicals such as Gnetum species (Chapter 4), Aframomun melegueta (Chapter 6), Xylopia aethiopia (Chapter 7) and basil (Chapter 8) are also reviewed.

Bioactives from Marine Sources

Chapter 3 discusses in details the bioactives from marine sources. Marine species provide a rich source of food as well as by-products that could be used for production of a wide range of bioactive compounds. These bioactives include omega-3 fatty acids, proteins and biopeptides, carotenoids and carotenoproteins, enzymes, chitinous materials and glucosamine as well as

Table I. Popular Botanical Dietary Supplements, Their Potential Active Components and Functions

Dietary Supplements	*Latin name*	*Potential active components*	*Major potential health benefit*
Astragalus	*Astragalus membranaceus*	Polysaccharides, saponins astragalosides)	Immunomodulatory, hepatoprotective
Black cohosh	*Cimicifuga racemosa*	Fukinolic acid 23-epi-26-deoxyactein	Relief of menopausal symptoms
Cranberry	*Vaccinium macrocarpon*	Proanthocyanidins	Prevention and treatment of urinary tract infections
Dang Gui	*Angelica sinensis*	Ligustilide	Treatment of gynecological conditions
Echinacea	*Echinacea purpurea, E. pallida, E. angustifolia*	Polysaccharides and glycoproteins, cichoric acid, alkamides	Treatment of common cold, cough, and upper respiratory infections
Feverfew	*Tanacetum parthenium*	Parthenolide and other sesquiterpene lactones	Alleviation of fever, headache, and women's ailments
Garlic	*Allium sativum*	Allyl sulfur compounds	Antibacterial, anticarcinogenic, antithrombotic, hypolipidemic
Ginger	*Zingiber officinale Roscoe*	Gingerols	Antiemetic, anti-inflammatory, digestive aid
Ginkgo biloba	*Ginkgo biloba*	Ginkgolids, flavonoids	Treatment of cerebral dysfunction and circulatory disorders
American ginseng	*Panax quinquefolium*	Ginsenosides	Therapeutic effects on immune function, cardiovascular diseases, cancer, sexual function
Asian ginseng	*Panax ginseng*	Ginsenosides	Combat psychophysical tiredness and asthenia

Table I. *Continued*

Dietary Supplements	*Latin name*	*Potential active components*	*Major potential health benefit*
Goldenseal	*Hydrastis Canadensis*	Alakaloid berberine and β-hydrastine	Soothing irritated skin and mucous membranes, easing dyspepsia
Grape seed extract	*Vitis vinfera*	Proanthocyanidins	Antioxidant, anti-inflammatory, immunostimulatory, antiviral and anticancer
Green tea polyphenols	*Camellia sinensis*	Epigallocatechin gallate and catechins	Preventive effects on heart diseases, cancer, neurodegenerative disorders and diabetes
Kava	*Piper methysticum*	Kava lactones	Effects on relaxing and mood calming
Licorice	*Glycyrrhiza glabra*	Triterpene saponins, flavonoids and other phenolics	Possess soothing, anti-inflammatory and antitussive properties
Maca	*Lepidium meyenii*	Aromatic isothiocyanates	Use for aphrodisiac purpose
Milk thistle	*Silybum marianum*	Silymarin	Treatment of liver disorders
Pycnogenol	*Pinus pinaster ssp. atlantica*	Procyanidins	Use for protection of the circulation, and to store capillary healing
Red clover	*Trifolium pretense*	Isoflavones	Treatment for menopausal symptoms
Reishi mushroom	*Ganoderma lucidum*	Triterpenoids, polysaccharides	Antitumor and immunomodulating effects
Saw Palmetto	*Serenoa repens*	Unknown	Use for prostate health
Soy isoflavones	*Glycine max*	Genistein, daidzein	Prevention of menopausal symptoms, osteoporosis, coronary heart disease and cancer

Continued on next page.

Table 1. *Continued*

Dietary Supplements	*Latin name*	*Potential active components*	*Major potential health benefit*
St. John's wort	*Hypericum perforatum*	Hyperforin, hypercin	Treatment of mild depression
Valerian	*Valeriana officinalis L.*	Valepotriates (iridoids)	Use for mild sedative and sleep disturbance
Yohimbe	*Pausinystalia johimbe*	Yohimbine	Use for aphrodisiac purpose

carotenoproteins, enzymes, chitinous materials and glucosamine as well as calcium, among others. The bioactives present in marine resources are effective in rendering beneficial health effects and reducing the risk of a number of chronic diseases. Thus, marine bioactives may serve as important value-added nutraceuticals, natural health products and functional food ingredients that can be used for health promotion and disease risk reduction.

Fruits and Their By-products as Dietary Supplements

Epidemiological studies indicate that the frequently and high intake of fresh fruits and vegetables is associated with lower incidence of some types of cancer and other age-related diseases. Table II lists some bioactive phytochemicals in selected fruits. Citrus is one of the most widely available and popular types of fruits, phytochemicals such as limonene, flavonoids and limonoids are recently gaining recognition for their impressive chemopreventive properties (*4*). Chapters 13 to 16 in this book review extensively the chemistry, biological activities and bioavailability of polymethoxyflavones (PMFs) in citrus peels. PMFs found almost exclusively in *Citrus* genus, particularly in the peels of sweet orange *(Citrus sinensis* (L.) Osbeck) and mandarin *(Citrus reticulate* Blanco). They have been of particular interest due to their broad spectrum of biological activities, including anti-inflammatory anti-carcinogenic, and anti-atherogenic properties (Chapter 13), Chronic inflammation was closely associated with the increased risk of various human cancers (*5*).

Quality Control of Dietary Supplements

For different commercial purposes, there are different quality control and quality assurance standards for botanical dietary supplements. However, usually they follow some general rules, the botanicals must be authenticated, safe to use,

Table II. Bioactive Phytochemicals in Selected Fruits

Fruits	Bioactive Phytochemicals
Citrus	limonene, limonoids, flavonoids, polymethoxyflavones
Apple	quercetin-3-galactoside, quercetin-3-glucoside, epicatechin, phloridzin
Grape	proanthocyanidins, quercetin, ellagic acid
Cranberry	proanthocyanidins, anthocyanins, catechins, triterpenes
Noni	ursolic acid, rutin, scopoletin, saccharide fatty acid esters
Pomegranate	ellagic acid, ellagitannins, cyanidin

with a limit of foreign materials, heavy metals, aflatoxins and pesticides. The pH value, ash contents, moisture contents and particle size should be in a reasonable range. Also microbiological tests for ingredients of interest must be passed.

Usually, herbal dietary supplements can be comparatively screened for quality related to their active ingredient content by instrumental methods (colorimetric, HPLC, LC/MS). The HPLC and LC/MS are suitable methods for quality control including the chemical fingerprinting of plant powders, tinctures and extracts. The inclusion of such instrumental methods of analysis, in addition to the typical review of published literature, should be a standard and critical component in the assessment of quality in botanical standardization for all plant-based dietary supplements. Too often, the literature is used in place of, rather than as a compliment to, actual analytical testing. Companies and even research groups may lack specific (or accurate) standards needed for compound identification and quality assessment in specific botanicals due to their lack of availability and/or expense and instead rely on previously reports which may illustrate elution of compounds pictured in labeled peaks from GC and/or LC analysis; thus false positive compound identifications may result. Many botanicals are "quality challenged" in that: (i) natural products may vary significantly from their own stated label claim (if there is one); (ii) another plant part or plant species is used either on purpose or inadvertently; (iii) adulteration is more common than acknowledged; and (iv) some botanical products can be spiked with synthetic chemicals or compounds from other botanicals (*6*).

To overcome these problems now facing the botanical industry, we recommend that reference materials and reference standards for these botanicals should be available through governmental and research institutes or non-profit organizations. Universal analytical methods should continue to enlarge for development and validation of most popular botanicals, and in particular for those botanicals that are potentially problematic from a safety perspective. Detailed genetic and chemical fingerprinting for every popular herb is needed and software programs to integrate the various chemical and genetic fingerprinting systems are long overdue. Using instrumental analysis as a core and standard part of the quality control process in botanicals, though well recognized, needs further implementation and continued focus (*7*).

References

1. Coates, P.M.; Blackman, MR.; Cragg, G.M.; Levine, M.; Moss, J.; White, J.D. *Encyclopedia of Dietary Supplements,* Marcel Dekker, New York, NY, 2005.
2. Marriott, B.M. In *Dietary Supplements of Plant Origin: A Nutrition and Health Approach,* Maffei, M. ed. Taylor & Francis, London, UK, 2003; pp. 1-17.
3. Hathcock, J. *J. Nutr.* **2001,** *131,* 1114S-1117S.
4. Patil, B.S.; Brodbelt, J.S.; Miller, E.G.; Turner, N.D. In *Potential Health Benefits of Citrus,* Patil, B.S.; Turner, N.D.; Miller, E.G.; Brodbelt, J.S. eds. ACS Symp. Ser. 936, American Chemical Society, Washington, DC, 2006, pp. 1-16.
5. Huang, M.-T.; Ghai, G.; Ho, C.-T. *Comprehensive Rev. Food Sci. Food Safety* **2004,** *3,* 127-139.
6. Wang, M.; Wu, Q.L.; Simon, J.E.; Ho, C.-T. In *Asian Functional Foods,* Shi, J.; Shahidi, F.; Ho, C.-T. eds. Marcel Dekker, Inc., New York, NY, 2005; pp. 73-102.
7. Wang, M.; Wu, Q.L.; Simon, J.E.; Liang, C.P.; Ho, C.-T. In *Herbs: Challenges in Chemistry and Biology,* Wang, M.; Sang, S.; Hwang, L.S.; Ho, C.-T. eds. ACS Symp. Ser. 925, American Chemical Society, Washington, DC, 2006, pp. 39-54.

Chapter 2

Calcium: Chemistry and Biology

Adrianne Bendich and Ronit Zilberboim

GlaxoSmithKline Consumer Healthcare, 1500 Littleton Road, Parsippany, NJ 07054

Calcium is an essential mineral that is required for the formation of bone, and is necessary for the clotting of blood, conduction of nerve impulses and contraction of muscle among other functions. Calcium is found naturally in foods, such as dairy products and can be added to certain foods, such as orange juice. Calcium is also available in dietary supplements. In all cases, calcium is found in these diet sources in the form of various calcium salts. The physical characteristics of the salts affect their actual elemental calcium content, solubility in various solvents and most importantly, their bioavailability. Bioavailability includes absorption, metabolism and excretion and each of these metabolic processes can be affected by the type of salt ingested. With regard to dietary supplements, the concentration of calcium in the salt also affects the size of the pill, the number of doses required to reach the recommended intake level and finally the cost. As an example, calcium carbonate contains 40% elemental calcium whereas calcium citrate contains 21% elemental calcium; both are equally well absorbed, yet the number of pills required to deliver 1000 mg of elemental calcium (100% of the daily value) with the citrate salt is twice the number needed with calcium carbonate, and the costs are greater with calcium citrate.

Calcium is the most common mineral in the human body, where it is present in almost the same relative abundance as in the earth's crust. Calcium's biological functions stem directly from its chemical properties which allow for flexibility in its physical as well as chemical interactions. The skeleton and teeth contain about 99% of the body's total calcium that serves two purposes-structural and as a large calcium reserve that is used to maintain the stability of circulating calcium levels. This reserve is readily available for bodily functions and elaborate regulatory mechanisms keep tight control of blood calcium levels. Interestingly, calcium deficiency does not manifest itself as a shortage of calcium for cellular or physiological processes, but rather as a decrease in the skeletal stores that ultimately can result in osteoporosis. It is not surprising therefore that calcium is the most studied mineral in relationship to human health (*1*).

Chemistry

Calcium (Ca), atomic number 20, is the fifth element and the third most abundant metal in the earth's crust. The distribution of calcium is very wide; it is found in almost every terrestrial area in the world. Calcium has a silvery color, is rather hard, and is prepared by electrolysis. It is one of the alkaline earth elements and it is less chemically reactive than alkaline metals and more than the other alkaline-earth metals. Calcium exists in its oxidized form with a valance of 2^+. It is found in nature in a combined form and occurs abundantly as limestone, gypsum, and fluorite; apatite is the fluorophosphate or chlorophosphate of calcium. It readily forms a white coating of nitride in air, reacts with water, and burns with a yellow-red flame, forming largely the nitride. The metal is trimorphic, harder than sodium, but softer than aluminum.

Cellular Functions of Calcium

Calcium is an essential ion in all organisms, where it plays crucial roles in skeletal functions and non-skeletal functions. The non-skeletal functions range from the maintenance of intracellular structures to the temporal and spatial regulation of neuronal function (*2*). Importantly, it influences neural function directly and indirectly since many of the calcium-regulating hormones affect the central nervous system (*3*).

Calcium is vital to many cellular events. Intracellular calcium functions as a regulator (second messenger) mediating various functions including muscle contractions, mitosis and neurotransmitter secretion. Large calcium gradients exist across membranes. For example, free calcium ions within the cytosol of a cell are about 1000 times lower in concentration than in the surrounding extracellular fluids where calcium also has several critical functions. Extracellular calcium is essential in assisting in blood clotting and in signal transmission for neuromuscular functions (skeletal, heart and smooth muscle contraction). Calcium is central to cellular electrophysiological functions (heart beat regulation) with its ability to carry a charge during an action potential across membranes. Lastly it acts as a cofactor for extracellular enzymes and regulatory proteins (*4,5*).

In summary, calcium participates in building and maintaining bones and teeth; regulates heart rhythm and helps maintain proper nerve and muscle function; helps regulate the passage of nutrients in and out of the cell walls; lowers blood pressure; important to normal kidney function; helps reduce blood cholesterol levels and may reduce the incidence of colon cancer.

Calcium Metabolism

Absorption

In humans, calcium balance is dependent on the interplay between 3 organ systems: the gastrointestinal tract (GI), the kidney and bone. The plasma level of calcium is maintained by the parathyroid hormone (PTH) vitamin D axis, a complex mechanism that controls renal absorption and clearance. When an adult ingests 1000 mg calcium, between 300 and 350 mg is absorbed in the small intestine, primarily controlled by the calciotropic hormones (*6*). Specifically, calcium is absorbed in the intestine via two mechanisms. The active, classical epithelial transcellular transport occurs with the help of the calcium binding protein calbindin, and ATPase pump. In this process, calcium is transferred across membranes into the extracellular space. The amount of calcium absorbed is dependent on the bioavailability of the calcium in the diet and the capacity for absorption in the intestine (which is regulated at least in part by 1, 25 dihydroxy vitamin D). The second absorption mechanism is a passive, paracellular process that allows calcium to permeate across gap junctions. The amount of calcium transferred is dependent on the lumen's permeability which is influenced by 1, 25 dihydroxy vitamin D and calcium concentration (*7*) as well as the concentration of free calcium. The majority of ingested calcium is not absorbed and it is excreted in the feces.

Excretion

Most calcium is lost from the body through digestive secretions and urine. Between 200 and 300 mg of calcium is lost through pancreatic, biliary and intestinal secretions in the feces, and about another 100 mg in found in the urine. In addition, 20-60 mg of calcium is lost through skin (perspiration), nails and hair.

The kidney filters about 10,000 mg of calcium daily. The vast majority of it is re-absorbed, in both transcellular and paracellular processes and about 100 mg are excreted in the urine. The balance between excretion and re-absorption is dependent on plasma calcium concentration via PTH which acts as a messenger. When blood levels of calcium drop, the extracellular calcium-sensing receptor (CaSR) in the parathyroid gland tranduces a signal to release PTH. The PTH increases kidney re-absorption of calcium and also increases calcium absorption proteins in the intestine. The kidney excretes calcium by a combination of filtration of calcium across the glomeruli and subsequent reabsorption of the filtered calcium along the renal tubules (*6*).

Dietary Sources of Calcium

Currently, in North America, calcium requirements can be met through dairy consumption, fortified foods and calcium supplements. In the United States, over 80% of calcium from food is found in dairy items and milk is usually used as reference food i.e. one 8 oz cup of milk contains about 300 mg of calcium. Milk is also used as a benchmark for calcium absorption. Despite the high nutritional value of dairy products, consumption has decreased in the past several decades. Leafy vegetables contain a relatively high amount of calcium, but calcium's absorbability varies as a function of other components. Finally, although cereal grains provide a high proportion of dietary calories, these contain a very small amount of calcium naturally.

Bioavailability of Calcium

Calcium bioavailability is dependent upon the combination of calcium content, intrinsic absorbability as well as other individual components in the meal that influence absorption efficacy (*8*). However, the individual's ability to absorb ingested calcium varies and spans a wide range from 17% to 58% for healthy women (*9*). Overall, calcium's absorption efficiency is inversely dependent on its intake and with increasing intake at any one time point, absorption efficiency decreases.

Dietary Supplements

The amount of calcium that a supplement provides is dependent on the actual salt and the matrix used to embed the salt as well as the compressibility forces used during production. Calcium carbonate contains the most elemental calcium as 40% of this salt is the calcium. When absorption of calcium carbonate was compared with absorption form various dairy items including imitation dairy, chocolate milk and yogurt, no significant differences were observed (*8*). In a randomized, controlled crossover study involving 16 men and 29 women, Green et al. assessed differences between calcium from milk and from calcium carbonate. Both were mixed with skim milk powder. Postprandial physiological responses showed significantly higher serum calcium after consumption of the mixture containing the calcium carbonate compared with the milk calcium ($P < 0.0001$). The serum C-telopeptide level, a marker of bone resorption, was significantly lower after calcium carbonate, compared milk calcium ($P < 0.05$). Based on the higher increase in serum calcium levels after consumption of calcium carbonate, the associated lower serum PTH concentrations and a more prolonged postprandial decrease in bone resorption, calcium carbonate appears to be an excellent source of elemental calcium for human consumption (*10*).

Calcium and Bone

Bone mass is influenced by both genetic (~75%) and environmental (~25%) factors including gender, race, hormonal status, nutrition, physical activity and lifestyle (*11*). Bone mass is one factor used in predicting fractures; bone density and bone turnover are other critical factors (*12*). Other factors that reflect bone's potential to fracture include the determination of bone geometry and bone strength (*13*). Critical environmental factors that affect bone include estrogen status, weight-bearing exercise, and calcium and vitamin D (*14*).

Peak Bone Mass

Calcium is a threshold nutrient with regards to bone, and increasing intake up to the threshold level is associated with increased bone mass. However, further increase above the threshold level is not associated with further increase in bone mass (*15*). In a comprehensive literature review, Heaney showed that the vast majority of calcium intervention trials that increased calcium intake had numerous positive outcomes in relation to bone gain or retention and reduced fracture risk (*16*). Furthermore, in most of the intervention trials conducted with children (*17*) and adolescents, regardless of the method of increasing calcium intake, there were positive effects. Unfortunately, once calcium supplementation ceased, the beneficial bone effects were not sustained. Likewise, calcium deficiency in children can limit skeletal development preventing them from reaching their genetically programmed bone mass level. Continued consumption of lower than optimal calcium intake leads to achieving sub-optimal peak bone mass as adults (*6*).

Calcium Homeostasis during Pregnancy and Lactation: Bone Effects

Calcium needs increase during pregnancy and lactation results in physiological adjustments. The usual three organ axis (bone, GI tract and kidney) becomes a four axis system that includes the placenta. Further disruption of the normal coordination also occurs due to the higher levels of hormones. For exam serum levels of 1,25-dihydroxy vitamin D increases due to production by the placenta, but this increase is not accompanied by an increase in maternal PTH levels. Consequently, absorption of calcium from the maternal GI tract increases approximately 2-fold during pregnancy. Specifically, the fractional calcium absorption may increase from about 30% to higher than 80% during pregnancy.

The effect of calcium intake on bone turnover markers during pregnancy was evaluated in a randomized, crossover trial of calcium supplementation (1200 mg/day) in the last trimester. Biomarkers of bone resorption indicated that there was a significant beneficial reduction of skeletal-bone turnover among the supplemented women (*18*). However, it is not clear if the markers are maternal, placental or of fetal origin (*19*).

Elevated calcium absorption occurs prior to the fetal calcium demand and thus allows an overall positive net maternal calcium balance. Overall, most data support almost no maternal bone loss during pregnancy despite the deposition of approximately 33 g of calcium in the fetus (*20,21*). Because of the increase in calcium absorption during pregnancy, there is currently no increase in the recommended calcium intake during pregnancy (*22*). However, data from recent studies challenge this recommendation. In a small longitudinal study using stable calcium isotopes with pregnant women consuming on average about 500 mg/day calcium net bone balance was measured from early in pregnancy through lactation. Overall, negative net bone balance was measured despite periods of positive balance (*23*). It could very well be that calcium supplementation above current recommendations may help to maintain maternal calcium balance as her stores are transferred to the fetus.

In contrast to pregnancy, maternal adaptation to calcium homeostasis is very different during lactation. Calcium absorption drops back to the values seen in non-pregnant women, and renal calcium conservation combined with bone resorption controls calcium needs (*21*). During this period, both a negative balance (*23*) and maternal bone losses in the range of 5-10% have been reported (*24*). Although this bone loss is considered transient (*19*), recent data suggest that bone remodeling change is as important if not more important than mass change. High rate of bone turnover is considered an independent risk factor for fragility (*24,25*).

Calcium and Osteoporosis Risk

Several intervention studies involving osteoporotic patients have shown that calcium supplementation, with or without vitamin D supplementation, reduced the risk of osteoporotic fractures (*26*). Recently, the results of the Women's Health Initiative (WHI) clinical intervention trial involving calcium and vitamin D supplementation (CaD) for a minimum of 7 years were published (*27*). The CaD arm was placebo-controlled and enrolled over 36,000 postmenopausal women without osteoporosis. The primary outcome tested was the potential for CaD to reduce the risk of hip fracture and secondarily to test whether women in the CaD arm would have lower risks of all fractures and/or colorectal cancer. Key findings comparing all participants in the CaD arm with placebo include a significant 1% increase in hip bone mineral density, a 12% non-significant reduction in hip fracture as well as a non-significant reduction in all fractures. Additionally, in subset analyses, adherent women (59%) who took 80% or more of the Ca/D supplements had a significant 29% reduction in hip fracture. The study showed no difference between the active CaD and placebo arms in the rate of adverse gastrointestinal events, but a statistically significant 17% increased risk of kidney stones was observed. Colorectal cancer incidence did not significantly differ in participants assigned to the CaD arm compared to placebo (*28*).

It is important to note that the positive findings occurred despite the fact that the women in the trial had baseline total calcium intake that was already close to current national recommendations of 1000-1500 mg/day. The women in the study had an average daily calcium intake of 1150 mg before the study began, which is nearly double the national average. Moreover, the study population was younger than anticipated, heavier, over 16% were taking anti-osteoporosis drugs by the end of the study and only 59% had taken all of the CaD intended doses.

Finally, all women in the study were allowed to continue to take their own multivitamins as well as their own calcium and vitamin D supplements during the trial which further reduced the expected differences between the active and placebo groups (there was effectively no true placebo group).

With regard to the kidney stone finding, the incidence of kidney stones in this study was self-reported (*27*) and not related to the level of total calcium intake (personal communication), raising questions about the association of kidney stones with the calcium and vitamin D intervention arm of the study. Survey studies have not consistently found a direct association between calcium supplements and kidney stone risk. In the Nurses' Health Study II (*29*) in contrast to the findings in the NHS study published in 1997 (*30*), supplemental calcium intake was not significantly associated with the risk of stone formation. Of note, in a case-control study using baseline information from WHI participants, use of supplemental calcium was associated with a significant reduced risk of kidney stones (*31*). Since this finding was unexpected and there is currently no consensus in the literature about a consistent role of calcium supplements in kidney stone risk, further information is needed before any conclusions can be made regarding this unexpected finding.

The Non-bone Roles of Calcium

Calcium and Blood Pressure

It is well recognized that an inverse relationship between intake of dietary calcium and blood pressure exists (*32*). However, because calcium has many biological roles and interacts with other nutrients, it has been difficult to isolate and fully understand its effects on blood pressure. The association between higher calcium intakes and modestly lower blood pressure was first reported in the late 1970s. Since the link was first suggested, numerous, epidemiological findings associate reduced blood pressure with increased mineral intake (calcium, potassium, and magnesium). Supportive data exists from the first National Health and Nutrition Examination Survey (NHANES I) (*33*), and more recent NHANES III and NHANES IV (*34*). The latest epidemiological analysis found an inverse relationship between mineral intake and systolic blood pressure.

In a randomized double blinded study, Pan et al. reported the effect of calcium supplementation (800 mg) on blood pressure. A significant reduction in the systolic blood pressure in the treatment group was measured and the higher the baseline blood pressure, the greater the reduction (*35*). Putting together all the data available to date confirms the need for and benefit of regular consumption of the recommended daily intake level of dietary calcium (*32*).

Mechanistically, it has been hypothesized that an increase in dietary calcium modulates the parathyroid vitamin D axis and reduces intracellular ionized calcium resulting in relaxation and dilation of smooth muscles (*36*). Recently, McCarron, (*37*) suggested that physiological effects that influence multiple organ systems rather than arterial pressure alone are working together to lower blood pressure.

Calcium, Preeclampsia and Hypertension During Pregnancy

Preeclampsia is a pregnancy-specific condition that increases maternal and infant mortality and morbidity. This syndrome is multisystemic; it involves many organs, the activation of coagulation and increased sensitivity to pressor agents.

Endothelial dysfunction (*38*), inflammatory activation, oxidative stress (*39,40*) and predisposing maternal factors are involved in the genesis of the syndrome (*41*). It has been suggested that the sensitivity of vascular smooth muscle to calcium is increased in preeclampsia (*42*) and furthermore, abnormalities observed in calcium metabolism in preeclampsia were linked to 1, 25-dihydroxy vitamin D and the placenta (*43*).

In the late 70s, epidemiological observations revealed an inverse relationship between calcium intake and the incidence of preeclampsia in populations where dietary calcium intake was low and the incidence of hypertension and preeclampsia high (*44*). Using a case control study with significant differences in the level of milk consumption but without differences in calcium intake by tablets, also showed increased risk of developing preeclampsia associated with lower milk intake (*45*).

Bucher (*46*) reported the results of a meta-analysis of randomized controlled trials reviewing the effect of calcium supplementation during pregnancy on blood pressure, preeclampsia, and adverse outcomes of pregnancy. The meta-analysis found that calcium supplementation during pregnancy leads to reduction in blood pressure and incidence of preeclampsia (*46*). Accumulating and conflicting evidence led to the trial of Calcium for Preeclampsia Prevention (CPEP), carried out in multiple centers in the U.S. (*47*). In this large placebo-controlled trial (over 4500 pregnancies), women who were 13 to 21 weeks pregnant were assigned to receive 2000 mg calcium per day or placebo. Calcium supplementation did not significantly reduce the incidence or severity of preeclampsia or delay its onset. Furthermore, the prevalence of pregnancy-associated hypertension without preeclampsia was not different between the groups (*48*). In contrast, in a similar design but smaller trial (>450 pregnancies) carried out in Australia (with 1800 mg/day of elemental calcium), Crowther showed a significant reduction of both risk of preeclampsia and preterm birth (*49*). This study failed to show an effect on hypertension, however. In both the CPEP in the US and the study in Australia, the women had a dietary intake of calcium above 1000 mg, the recommended intake level, prior to getting the over 1800 mg additional calcium in the active group. Also, in the CPEP study, the women had no other risk factors for pre-eclampsia.

A later rigorous meta-analysis, Atallah et al showed that overall calcium supplementation is beneficial especially for women at high risk of gestational hypertension; the findings are more significant in communities with low dietary calcium intake (*50-53*).

Most recently, the largest (over 8000 women) randomized placebo-controlled, double-blinded trial in nulliparous normotensive women from populations with dietary calcium < 600 mg/day was reported. In this multinational trial, women received a 1500 mg calcium supplement or placebo. Primary outcomes were preeclampsia and preterm delivery; secondary outcomes focused on severe morbidity and maternal and neonatal mortality rates. Calcium supplementation was associated with a small reduction in preeclampsia but with a significant reduction in eclampsia and severe gestational hypertension. Other complications including severe maternal morbidity and mortality index were also reduced in the supplementation group. The authors concluded that calcium did not prevent preeclampsia but did reduce its severity, maternal morbidity, and neonatal mortality.

In summary, despite a lack of consistency in all clinical trials results, it appears that during pregnancy, calcium supplementation has beneficial effects especially with at risk populations and that this effect is biologically plausible.

Maternal Calcium Intake and Offspring's Risk Factors

Several studies show that maternal calcium intake during pregnancy may be associated with the offspring's blood pressure later in life. Belizan *et al* (*54*), based on a follow-up of the children born to mothers who participated in a calcium intervention study during their pregnancy, reported reduced systolic blood pressure in the children of calcium supplemented mothers at mean age 7 years, particularly in overweight children. In a partial cohort from the Calcium for Preeclampsia Prevention intervention study, the blood pressure of infants was measured at 3 month and 2 years. Systolic blood pressure of infants born to the calcium-supplemented mothers was significantly lower than in the placebo group at 2 years (but not 3 month) (*55*). In 2004, Gillman et al reported an association between blood pressure of 936 infants and reported maternal calcium intake during mid pregnancy based on observational data generated during Project Viva also in the U.S. The data suggests that supplementing pregnant women with calcium may lower offspring blood pressure (*56*). The influence of calcium intake during pregnancy on offspring's cardiovascular risk factors in Tasmanian infants followed for 9 years has been reported by Morley (*57*). Supplemented mother's children had lower total cholesterol and low density cholesterol, and was also associated with higher BMI (>17.5). Overall, they hypothesized that calcium availability could permanently program lipid metabolism during fetal life directly, or by influencing maternal lipid profile. These findings have important implications for population health and may also be relevant for establishing guidelines for calcium intake recommendations for during pregnancy based on fetal programming role (*58*).

Premenstrual Syndrome

The term premenstrual syndrome (PMS) refers to a cluster of mood, physical, and cognitive symptoms that occur one to two weeks prior to the start of menstrual bleeding and subside with the onset of menstruation. As many as

80% of women of reproductive age may experience premenstrual emotional and physical changes (*59,60*). Up to 40% of women of reproductive age experience premenstrual symptoms sufficient to affect their daily lives, and 5% experience severe impairment (*61*). Symptoms vary among individuals; the most common symptoms include fatigue, irritability, abdominal bloating, and breast tenderness, labile mood with alternating sadness and anger, and depression (*59*).

There is a long history of a link between calcium status and the menstrual cycle. A 1930 study (*62*) showed that plasma calcium levels were lower in the premenstrual period compared to those seen in the week following menstruation. A recent case-control epidemiological study using a cohort from the Nurse's Health Study II found that participants with the highest intake of calcium from food sources had a significantly lower risk of ever having PMS. Women with the highest intake of calcium from foods (quintile median = 1260 mg/day) had a RR of 0.71 (95% CI = 0.50 – 0.99; P for trend = 0.02) compared to women with the lowest intake (*63*).

A preliminary diet intervention study (*64*) demonstrated decreased symptoms of premenstrual and menstrual distress when women received diets containing 1,336 mg/day of calcium, as opposed to 587 mg. A large U.S.-based, multicenter clinical trial (*65*) involving 466 women with diagnosed PMS received 1,200 mg/day of calcium or placebo for three menstrual cycles. The calcium supplemented cohort showed a significant overall 48% reduction in symptoms as compared to a 30% reduction in the placebo group. All of four symptom factor composite scores (negative affect, water retention, food cravings, and pain) were significantly improved in the calcium supplemented group.

There is evidence that abnormalities in calcium and vitamin D regulation may contribute to the causation of PMS and that PMS may be linked to other disorders associated with inadequate calcium intake, such as osteoporosis. In a study that compared women with established vertebral osteoporosis to controls (*66*), it was found that the risk of osteoporosis was higher among those with a history of PMS. Another study found evidence of reduced bone mass in women with PMS as compared to asymptomatic controls (*67*). PMS may serve as a clinical marker of low calcium status, perhaps reflecting an underlying abnormality in calcium metabolism, and it may serve as an early warning sign to young women of a possible increased risk of osteoporosis.

Conclusions

Calcium is an essential nutrient for humans and its chemical structure defines its biological functions. Calcium is the most studied mineral critical for human health. In fact, a recent whole book is devoted to the complexity of calcium's role in human physiology (*68*). This chapter provides an overview and extensive reference list that concentrates on the bone and non-bone functions of calcium and examines the changing needs of calcium throughout the lifespan especially in women. As an example, epidemiological data indicate a lowered risk of ever suffering from PMS in women with the highest calcium intakes. For women who do have PMS, well controlled studies have shown that calcium supplementation can reduce the severity of PMS symptoms. In addition to helping maintain the health of women of childbearing potential, calcium is essential for fertilization. The higher concentrations of estrogen seen during pregnancy results in the enhanced absorption of calcium during pregnancy and helps to assure that there is calcium available for fetal bone growth. In contrast, the hormonal changes seen during pregnancy are reduced during lactation, and therefore calcium intake during breastfeeding is particularly important for maternal bone health.

Calcium intake during childhood, adolescence and young adulthood is predictive of osteoporosis risk in the senior years because peak bone mass is achieved by about age 30. The risk of developing osteoporosis for U.S. adults over age 50 is very high: approximately 50% of women and 25% of men will suffer from osteoporosis. The publication of the results from the WHI provides strong evidence that post-menopausal women who take about 1000 mg/day of supplemental calcium can reduce their risk of hip fracture by 30% compared to non-supplementing women. This is an enormous public health finding.

Calcium's importance to health is not limited to bone. Rather, there is growing evidence form epidemiological survey as well as intervention studies that consistently find that higher calcium intakes are associated with lower systolic and diastolic blood pressure. During pregnancy, low calcium intake has many adverse effects aside form the potential negative effects on maternal and fetal bone. Although the recent WHO trial did not find a significant decreased risk in pre-eclampsia, there was a significant reduction in infant mortality and maternal complications. Moreover, the dose of calcium was 1500 mg/day in the WHO study and in other intervention studies using 2000 mg/day in similar populations of women with low initial calcium intakes, there were significant findings of reduction in pre-eclampsia.

In conclusion, the chemistry and biology of calcium continue to be active areas of research. Clinical studies are ongoing to better understand the role of calcium in human health in both men and women during different life stages. Both the bone as well as the non-bone functions of calcium continue to provide a strong rationale to consume the optimal intake of calcium daily.

References

1. Weaver, C.M. Introduction. In *Calcium in Human Health.* Weaver, C.M., Ed. Humana Press, Inc.: Totowa, NJ, 2006, pp. 1-3.

2. Hoenderop, J.G.; Nilius, B.; Bindels, R.J. *Physiol. Rev.* **2005**, *85*, 373-422.
3. Hatton, D.C.; McCarron, D.A. *Hypertension* **1994**, *23*, 513-530.
4. Weaver, C.M.; Heaney, R.P. In *Modern Nutrition in Health and Disease.* Shils, M.E.; Olson, J.A.; Shihe, M.; Ross, A.C., Eds. Lippincott Williams & Wilkins: New York, NY, 2006, pp. 194-210.
5. Theobald, H.E. *British Nutrition Foundation Bulletin* **2005**, *30*, 237-277.
6. Heaney, R.P. *OBG Managment* **1998**, 2-5.
7. Awumey, E.M.; Bukoski, R.D. In *Calcium in Human Health.* Weaver, C.M., Ed. Humana Press, Inc.: Totowa, NJ, 2006, pp. 13-35.
8. Recker, R.R.; Bammi, A.; Barger-Lux, M.J.; Heaney, R.P. *Am. J. Clin. Nutr.* **1988**, *47*, 93-95.
9. Wolf, R.L.; Cauley, J.A.; Baker, C.E. et al. *Am. J. Clin. Nutr.* **2000**, *72*, 466-471.
10. Green, J.H.; Booth, C.; Bunning, R. *Asia Pac. J. Clin. Nutr.* **2003**, *12*, 109-119.
11. NIH Fact Sheets. Osteoporosis: Peak Bone Mass in Women. National Institute of Health Osteoporosis and Related Bone Diseases. 2005.
12. Wilkin, T.J.; Devendra, D. *BMJ* **2001**, *323*, 795-797.
13. Schonau, E. *Pediatr. Nephrol.* **2004**, *19*, 825-831.
14. Mora, S.; Gilsanz, V. *Endocrinol. Metab. Clin. North Am.* **2003**, *32*, 39-63.
15. Heaney, R.P.; Bachmann, G.A. *J. Womens Health (Larchmt.)* **2005**, *14*, 889-897.
16. Heaney, R.P. *J. Am. Coll. Nutr.* **2000**, *19*, 83S-99S.
17. Specker, B.; Binkley, T. *J. Bone Miner. Res.* **2003**, *18*, 885-892.
18. Janakiraman, V.; Ettinger, A.; Mercado-Garcia, A.; Hu, H.; Hernandez-Avila, M. *Am. J. Prev. Med.* **2003**, *24*, 260-264.
19. Kalkwarf, H.J. In *Calcium in Human Health.* Weaver, C.M., Ed. Humana Press, Inc.: Totowa, NJ, 2006, pp. 297-309.
20. Fleet, J.C. In *Calcium in Human Health.* Weaver, C.M., Ed. Humana Press, Inc.: Totowa, NJ, 2006, pp. 163-189.
21. King, J.C. *J. Nutr.* **2001**, *131*, 1355S-1358S.
22. Standing Committee on the Scientific Evaluation of Dietary Reference Intakes Food and Nutrition Board Institute of Medicine. In *Dietary Reference Intakes for Calcium, Phosphorus, Magnesium, Vitamin D and Fluoride.* National Academy Press: Washington, DC, 1998, pp. 71-145.
23. O'Brien, K.O.; Donangelo, C.M.; Zapata, C.L.; Abrams, S.A.; Spencer, E.M.; King, J.C. *Am. J. Clin. Nutr.* **2006**, *83*, 317-323.
24. Kovacs, C.S. *J. Mammary. Gland Biol. Neoplasia* **2005**, *10*, 105-118.
25. Heaney, R.P.; Weaver, C.M. *J. Am. Coll. Nutr.* **2005**, *24*, 574S-581S.
26. Bendich, A.; Leader, S.; Muhuri, P. *Clin. Ther.* **1999**, *21*, 1058-1072.
27. Jackson, R.D.; LaCroix, A.Z.; Gass, M. et al. *N. Engl. J. Med.* **2006**, *354*, 669-683.
28. Wactawski-Wende, J.; Kotchen, J.M.; Anderson, G.L. et al. *N. Engl. J. Med.* **2006**, *354*, 684-696.

29. Curhan, G.C.; Willett, W.C.; Knight, E.L.; Stampfer, M.J. *Arch. Intern. Med.* **2004**, *164*, 885-891.
30. Curhan, G.C.; Willett, W.C.; Speizer, F.E.; Spiegelman, D.; Stampfer, M.J. *Ann. Intern. Med* **1997**, *126*, 497-504.
31. Hall, W.D.; Pettinger, M.; Oberman, A. et al. *Am. J. Med Sci.* **2001**, *322*, 12-18.
32. McCarron, D.A.; Reusser, M.E. *J. Am. Coll. Nutr.* **1999**, *18*, 398S-405S.
33. McCarron, D.A. *N. Engl. J. Med* **1982**, *307*, 226-228.
34. Townsend, M.S.; Fulgoni, V.L. III; Stern, J.S.; Adu-Afarwuah, S.; McCarron, D.A. *Am. J. Hypertens* **2005**, *18*, 261-269.
35. Pan, Z.; Zhao, L.; Guo, D.; Yang, R.; Xu, C.; Wu, X. *Zhonghua Yu Fang Yi Xue Za Zhi* **2000**, *34*, 109-112.
36. Repke, J. *OBG Management* **1998**, *10*, 6-11.
37. McCarron, D.A. In *Calcium in Human Health.* Weaver, C.M., Ed. Humana Press, Inc.: Totowa, NJ, 2006, pp. 421-429.
38. Khan, F.; Belch, J.J.; MacLeod, M.; Mires, G. *Hypertension* **2005**, *46*, 1123-1128.
39. Spinnato, J.A.; Livingston, J.C. *Clin. Obstet. Gynecol.* **2005**, *48*, 416-429.
40. Scholl, T.O. In *Preventive Nutrition: The Comprehensive Guide for Health Professionals,* Bendich, A.; Deckelbaum, R.J.; Eds., Humana Press: Totowa, NJ, 2005, pp. 629-663.
41. Roberts, J.M.; Balk, J.L.; Bodnar, L.M.; Belizan, J.M.; Bergel, E.; Martinez, A. *J. Nutr.* **2003**, *133*, 1684S-1692S.
42. VanWijk, M.J.; Boer, K.; van der Meulen, E.T.; Bleker, O.P.; Spaan, J.A.; VanBavel, E. *Am. J. Obstet. Gynecol.* **2002**, *186*, 148-154.
43. Diaz, L.; Arranz, C.; Avila, E.; Halhali, A.; Vilchis, F.; Larrea, F. *J. Clin. Endocrinol. Metab.* **2002**, *87*, 3876-3882.
44. Villar, J.; Belizan, J.M.; Fischer, P.J. *Int. J. Gynaecol. Obstet.* **1983**, *21*, 271-278.
45. Duvekot, E.J.; de Groot, C.J.; Bloemenkamp, K.W.; Oei, S.G. *Eur. J. Obstet. Gynecol. Reprod. Biol.* **2002**, *105*, 11-14.
46. Bucher, H.C.; Guyatt, G.H.; Cook, R.J. et al. *JAMA* **1996**, *275*, 1113-1117.
47. Levine, R.J.; Esterlitz, J.R.; Raymond, E.G. et al. *Control Clin. Trials* **1996**, *17*, 442-469.
48. Levine. R.J.; Hauth, J.C.; Curet, L.B. et al. *N. Engl. J. Med* **1997**, *337*, 69-76.
49. Crowther, C.A.; Hiller, J.E.; Pridmore, B. et al. *Aust. N. Z J. Obstet. Gynaecol.* **1999**, *39*, 12-18.
50. Atallah, A.N.; Hofmeyr, G.J.; Duley, L. *Cochrane. Database. Syst. Rev.* **2002**, CD001059.
51. Kulier, R.; de Onis, M.; Gulmezoglu, A.M.; Villar, J. *Int. J. Gynaecol. Obstet.* **1998**, *63*, 231-246.
52. Hofmeyr, G.J.; Roodt, A.; Atallah, A.N.; Duley, L. *S. Afr. Med J.* **2003**, *93*, 224-228.

53. Vitlar, J.; Belizan, J.M. *Am. J. Clin. Nutr.* **2000**, *71*, 1375S-1379S.
54. Belizan, J.M.; Villar, J.; Bergel, E. et al. *BMJ* **1997**, *315*, 281-285.
55. Hatton, D.C.; Harrison-Hohner, J.; Coste, S.; Reller, M.; McCarron, D. Am. *J. Hypertens* **2003**, *16*, 801-805.
56. Gillman, M.W.; Rifas-Shiman, S.L.; Kleinman, K.P.; Rich-Edwards, J.W.; Lipshultz, S.E. *Circulation* **2004**, *110*, 1990-1995.
57. Morley, R.; Carlin, J.B.; Dwyer, T. *Int. J. Epidemiol.* **2004**, *33*, 1304-1309.
58. Bergel, E.; Belizan, J.M. *Int. J. Epidemiol.* **2004**, *33*, 1309-1310.
59. ACOG Committee Opinion. *Int. J. Gynaecol. Obstet.* **1995**, *50*, 80-84.
60. Alvir, J.; Endicott, J.; Linden, A.; Thomas, M.A.; Thys-Jacobs, S.; Zimmermann, R. *The Premenstrual Syndrome Challenge: Is it a disease? What is the Best Course of Treatment?* 2000, Medical Crossfire, Liberty Publishing.
61. Daugherty, J.E. *Am. Fam. Physician* **1998**, *58*, 183-188.
62. Okey, R.; Stewart, J.A.; Greenwood, M.L. *J. Biol. Chem.* **1930**, *87*, 91-102.
63. Bertone-Johnson, E.R.; Hankinson, S.E.; Bendich, A.; Johnson, S.R.; Willett, W.C.; Manson, J.E. *Arch. Intern. Med.* **2005**, *165*, 1246-1252.
64. Penland, J.G.; Johnson, P.E. *Am. J. Obstet. Gynecol.* **1993**, *168*, 1417-1423.
65. Thys-Jacobs, S.; Starkey, P.; Bernstein, D.; Tian, J. *Am. J. Obstet. Gynecol.* **1998**, *179*, 444-452.
66. Lee, S.J.; Kanis, J.A. *Bone Miner* **1994**, *24*, 127-134.
67. Thys-Jacobs, S.; Alvir, J.M.; Fratarcangelo, P. *Psychopharmacol. Bull.* **1995**, *31*, 389-396.
68. Weaver, C.M.; Heaney, R.P. *Calcium in Human Health.* Humana Press: Totowa, NJ, 2006.

Chapter 3

Bioactives from Marine Resources

Fereidoon Shahidi

Department of Biochemistry, Memorial University of Newfoundland, St. John's, NL A1B 3X9, Canada

Marine and cultured fish, shellfish and other species provide a rich source of food as well as by-products that could be used for production of a wide range of bioactive compounds. These bioactives include omega-3 fatty acids, proteins and biopeptides, carotenoids and carotenoproteins, enzymes, chitinous materials and glucosamine as well as calcium, among others. The bioactives present in seafoods and marine resources are effective in rendering beneficial health effects and reducing the risk of a number of chronic diseases. Thus, marine bioactives may serve as important value-added nutraceuticals, natural health products and functional food ingredients that can be used for health promotion and disease risk reduction.

Introduction

Seafoods have traditionally been used because of their variety of flavor, color, texture and accessibility in the coastal areas. More recently, seafoods are appreciated because of the emergence of new evidence about their role in health promotion and disease risk reduction, primarily arising from their bioactive omega-3 fatty acids, among others. Bioactive components and nutraceuticals from marine resources belong to several classes of compounds and are used in foods and as nutraceuticals or natural health products (see Table I). When processing seafoods or during catch, a large proportion of by-products or low-value fish is produced. The components of interest that could be extracted or produced from marine resources include lipids, proteins and biopeptides, minerals, flavorants, carotenoids and carotenoproteins, enzymes, chitinous materials and other specialty products. The importance of omega-3 fatty acids, bioactive peptides, glucosamine, chitosan and chitosan oligomers and other bioactives in food and non-food applications as well as for health promotion purposes has been well recognized. This chapter provides a cursory account of selected bioactives from marine resources.

Table I. Nutraceuticals and Natural Health Products for Marine Resources

Components	*Application Area*
Chitin, chitosin, chitosan, oligomer, glucosamine	Food, water and juice clarification, agriculture, supplements
Omega-3 oils	Nutraceuticals, immune enhancement, etc.
Chonodostin sulfate	Dietary supplement, arthritic pain
Squalene	Skin care
Biopeptides, collagen and protein	Nutraceuticals, immune enhancement, etc.
Carotenoids and Carotenoprotein	Nutraceuticals, etc.
Minerals (calcium)	Nutraceuticals
Enzymes	Food and speciality application, etc.
Other specialty chemicals	Miscellaneous

Specialty and Nutraceutical Lipids

The occurrence and health benefits of long-chain omega-3 polyunsaturated fatty acids (PUFA) in seafoods and other marine organisms is a well established fact (*1-3*). These fatty acids are produced in phytoplanktons in the oceans and are then consumed by fish and other marine species. Thus, omega-3 PUFA may be procured from marine algae, body of fatty fish, liver of white lean fish and the blubber of marine mammals. The constituent fatty acids present in such oils include eicosapentaenoic acid (EPA, C20:5n-3), docosahexaenoic acid (DHA, C22: 6n-3), and to a lesser extent docosapentaenoic acid (DPA, C22: 5n-3) in different proportions, depending on the species involved. In addition, liver oil from white lean fish serves as an excellent source of vitamin A while that of shark has a high content of squalene and other bioactives.

Table II summarizes the fatty acid composition of selected marine oils and an algal oil produced commercially. As can be seen, the contents of EPA, DHA and DPA in each oil is dependent on the source material. Thus the ratio of EPA to DHA in menhaden, cod liver and seal blubber oils varies considerably, but the algal oil tested almost exclusively contained DHA. Furthermore, positional distribution of omega-3 fatty acids in such oils, again, depends on source material. The omega-3 fatty acids are primarily located in the sn-2 position of triacylglycerols while they are present mainly in the sn-1 and sn-3 positions of seal blubber oil.

Marine oils, similar to other edible oils, are subjected to different processing steps of refining, bleaching and deodorization. As protective components of oils are generally removed to a large extent during processing, it is important to treat the resultant refined, bleached and deodorized (RBD) oils with appropriate antioxidants in order to enhance their oxidative stability. Microencapsulation provides another means for extending the shelf-life of highly unsaturated oils. Regardless, such oils increase the body demand for vitamin E. Therefore, addition of vitamin E, usually in the form of mixed tocopherols, to highly unsaturated oils is necessary for enhancing their oxidative stability and also to augment the body's need for vitamin E.

The use of marine oils containing omega-3 fatty acids is recommended for foods that are used within a short period of time in order to avoid possible off-flavor development during their expected shelf-life. While it is possible to mask some of the off-flavors generated due to production of flavor-active secondary oxidation products, presence of primary products of oxidation remains to be a concern. Another way of introducing omega-3 fatty acids into food products is to employ adequately microencapsulated products that may remain intact until they reach the gastrointestinal tract. In this way, there is no flavor effect on the product even if the oil used initially contained some oxidation products. Table III summarizes a number of food products that are usually selected to enrich them with omega-3 oils.

The beneficial health effects of marine oils are manifold, but include coronary heart disease (CHD), visual and cognitive development, psychiatric conditions, inflammatory diseases, Crohn's disease and type 2 diabetes, among others. The omega-3 fatty acids, especially DHA, are known to dominate the fatty acid profile of brain and retina lipids and play a major role in the development of the fetus and infants as well as health status and body requirement of pregnant and lactating women.

For therapeutic purposes, the natural sources of omega-3 fatty acids as such may not provide the necessary amount of these fatty acids and hence production and use of omega-3 concentrates may be required (*4*). The omega-3 concentrates may be produced in the free fatty acid, simple alkyl ester, and acylglycerol forms. To achieve this, physical, chemical and enzymatic processes may be employed for concentrate production. The available methods suitable for this purpose, on an industrial scale, are low-temperature crystallization, fractional or molecular distillation, urea complexation, chromatography,

Table II. Major Fatty Acids of Omega-3 Rich Marine and Algal oils[a]

Fatty acid	*Menhaden*	*Cod liver*	*Seal blubber*	*Algal*
14:0	8.32	3.33	3.73	14.9
16:0	17.4	11.0	5.58	9.05
16:1 n-7	11.4	7.85	18.0	2.20
18:0	3.33	3.89	0.88	0.20
18:1 n-9, n-11	12.1	21.2	26.0	18.9
20: n-9	1.44	10.4	12.2	-
20:5 n-3	13.2	11.2	6.41	-
22:1 n-11	0.12	9.07	2.01	-
22:5 n-3	2.40	1.14	4.66	0.51
22:6 n-3	10.1	14.8	7.58	47.4

[a] Units are weight percentage of total fatty acids; algal oil is DHASCO (docosahexaenoic acid single cell oil).

supercritical fluid extraction, and enzymatic splitting, among others (*5*). These procedures have been used, albeit to different extent, by the industry to prepare concentrates that are often sold in the ethyl ester form or re-esterified with glycerol to be offered as triacylglycerols to the market. However, it has been demonstrated that acylglycerols are more stable than their corresponding ethyl esters. Regardless, the modified oils need to be stabilized using synthetic or preferably natural antioxidants.

In preparation of modified lipids containing omega-3 fatty acids, structured lipids (SL) may be produced. SL are triacylglycerols (TAG) or phospholipids (PL) containing combinations of short-chain, medium-chain and long-chain fatty acids (SCFA, MCFA and LCFA, respectively) located in the same molecule and may be produced by chemical or enzymatic means (*6,7*). Structured lipids are developed to fully optimize the benefits of their fatty acid constituents in order to affect metabolic parameters such as immune function, nitrogen balance, and lipid clearance from the bloodstream. These specialty lipids may be produced via direct esterification, acidolysis and hydrolysis or interesterification. We have used acidolysis process to incorporate capric acid or lauric acid into seal blubber oil (*8,9*). In addition, we produced 91% gamma-linolenic acid (GLA) concentrate from borage oil (*10*) which was subsequently used in acidolysis of menhaden and seal blubber oils (*11*). Such structured lipids that include GLA, EPA and DHA were also prepared using borage and evening primrose as a source of GLA and acidolysis with EPA and/or DHA (*12,13*). The products so obtained, while similar to those produced by incorporation of GLA into marine oils, differ in the composition and distribution of fatty acids involved.

Bioactive Peptides and Proteins From Marine Resources

Hydrolysis of the amide linkage in the protein chain leads to the formation of peptides with different numbers of amino acids as well as free amino acids. While enzymes with endopeptidase activity provide peptides with different chain lengths, exopeptidases liberate amino acids from the terminal positions of the protein molecules. Depending on reaction variables as well as the type of enzyme, the degree of hydrolysis of proteins may differ considerably. The peptides produced from the action of a specific enzyme may be subjected to further hydrolysis by other enzymes. Thus, use of an enzyme mixture or several enzymes in a sequential manner may be advantageous. The peptides so obtained may be subjected to chromatographic separation and then evaluated for their amino acid sequence as well as their antioxidant and other activities.

In a study on capelin protein hydrolysates, four peptide fractions were separated using Sephadex G-10. While one fraction exerted a strong antioxidant activity in a β-carotene/ starch linoleate model system, two fractions possessed a weak antioxidant activity and the fourth one had a prooxidant effect. Two dimensional HPLC separation showed spots with both pro- and antioxidant effects (*14*). Meanwhile, protein hydrolysates prepared from seal meat were found to serve as phosphate alternatives in processed meat applications and reduced the cooking loss considerably (*15*). Furthermore, Alaska pollock skin hydrolysate was prepared using a multienzyme system in a sequential manner. The enzymes used were in the order of Alcalase, Pronase E, and collagenase. The fraction from the second step, which was hydrolyzed by Pronase E, was composed of peptides ranging from 1.5 to 4.5 kDa and showed a high antioxidant activity. Two peptides were isolated (Table V), using a combination of chromatographic procedures, and these were composed of 13 and 16 amino acid residues (*16*). The sequence of the peptides involved is given in Table IV and compared with those of soy 7S protein hydrolysates (*17*). These peptides exert their antioxidant activity via free radical scavenging as well as chelation effects. Recently, proteases from shrimp processing discards were characterized (*18*) and application of salt-fermented shrimp byproduct sauce as a meat tenderizer was reported (*19*).

Chitin, Chitosan, Chitosan Oligomers and Glucosamine

Chitin is recovered from processing discards of shrimp, crab, lobster, and crayfish following deproteinization and demineralization (*20*). The chitin so obtained may then be deacetylated to afford chitosan (*20*). Depending on the duration of the deacylation process, the chitosan produced may assume different viscosities and molecular weights. The chitosans produced are soluble in weak

Table III. Food Application of Omega-3 Oils

Bread/bread products	Margarine and spreads
Cereals, crackers, noodles	Eggs
Pasta and cakes	Bars and candies
Milk and dairy products, ice cream	Infant formula
Juices	Fabricated seafoods, burgers, etc.
Mayonnaise and salad dressings	Others

acid solutions, thus chitosan ascorbate, chitosan acetate, chitosan lactate, and chitosan malate, among others, may be obtained and these are all soluble in water. Chitosan has a variety of health benefits and may be employed in a number of nutraceutical and health-related applications. Chitosan derivatives may also be produced in order to obtain more effective products for certain applications.

Chitosans with different viscosities were prepared and used in an experiment designed to protect both raw and cooked fish against oxidation as well as microbial spoilage (*21-23*). The content of propanal, an indicator of oxidation of omega-3 fatty acids, was decreased when chitosan was used as an edible invisible film in herring. Furthermore, the effects were more pronounced as the molecular weight of the chitosan increased. In addition, inhibitory effects of chitosan coatings in the total microbial counts for cod and herring showed an approximately 1.5 and 2.0 log cycles difference between coated and uncoated samples, respectively, after 10 days of refrigerated storage (results not shown). However, to have the products solubilized in water without the use of acids, enzymatic processes may be carried out to produce chitosan oligomers. Due to their solubility in water, chitosan oligomers serve best in rendering their benefits under normal physiological conditions and in foods with neutral pH. Furthermore, depending on the type of enzyme employed, chitosan oligomers with specific chain lengths may be produced for certain applications (*24*).

The low-molecular chitin and chitosan oligomers (also known as chitin/chitosan oligosaccharides (COSs) have received considerable attention as physiologically functional materials having antitumor, immuno-enhancing and antibacterial activities (*25-27*). Production of COSs via the hydrolysis of chitosan may be achieved chemically or enzymatically. The enzymatic production is preferred, but cost of enzyme may be prohibitive. Therefore, a continuous low cost production method to produce COSs with desired molecular size has been developed (*28*). The COSs (Table V) with low-molecular weight and hetero-chitosan oligosaccharides, have been reported to have antibacterial,

Table IV. Antioxidative Peptides from Alaska Pollock Skin Hydrolysate and Soy 7S Protein[a]

Peptide	*Amino acid sequence*
Alaska Pollock	
P_1	Gly-Glu-Hyp (Gly-Pro-Hyp)$_3$-Gly
P_2	(Gly-Pro-Hyp)$_4$-Gly
Soy 75 Protein	
P_1	Val-Asn-Pro-His-Asp-His-Glu-Asn
P_2	Leu-Val-Asn-Pro-His-Asn-His-Glu-Asn
P_3	Leu-Leu-Pro-His-His
P_4	Leu-Leu-Pro-His-His-Ala-Asp-Ala-Asp-Tyr
P_5	Val-Ile-Pro-Ala-Gly-Tyr-Pro
P_6	Leu-Gly-Ser-Gly-Asp-Ala-Leu-Arg-Val-Pro-Ser-Gly-Thr-Tyr-Tyr

[a]Adapted from Ref 16,17

Table V. Antibacterial Activity of Different Molecular Weight COS Fractions

Bacteria	*Antibacterial activity (%)[a]*		
	HMWCOSs[b]	*MMWCOSs[c]*	*LMWCOSs[d]*
Escheria coli[e]	98 ± 0	62 ± 6	51 ± 7
Escheria coli O-157[e]	71 ± 3	56 ± 4	60 ± 2
Salmonella typhi[e]	91 ± 2	88 ± 0	89 ± 0
Pseudomonas aeruginosa[e]	47 ± 5	35 ± 5	22 ± 8
Streptococcus mutans[f]	100 ± 0	99 ± 0	99 ± 0
Straphylococcus aureus[f]	97 ± 3	95 ± 0	93 ± 9
Straphylococcus epidermidis[f]	82 ± 0	57 ± 3	23 ± 1
Bacillus subtilis[f]	63 ± 5	60 ± 5	63 ± 7
Micrococcus luteus[f]	70 ± 0	67 ± 3	63 ± 7

[a] Following the incubation of bacterial culture with 0.1% different COSs fractions, the number of colonies formed on the medium was calculated as a percentage compared to the control. [b] High molecular weight chitosan oligosaccharides (molecular weight range 10-5 kDa). [c] Medium molecular weight chitosan oligosaccharides (molecular weight range 5-1 kDa). [d] Low molecular weight chitosan oligosaccharides (molecular weight below than 1 kDa). [e] Gram-negative. [f] Gram-positive.

radical scavenging, ACE inhibitory and anticoagulant activities (*29*). The monomer of chitin, N-acetylglucosamine (NAG), has been shown to possess anti-inflammatory properties. Meanwhile, glucosamine, the monomer of chitosan, prepared via HCl hydrolysis, is marketed as glucosamine sulfate. This formulation is prepared by addition of ferrous sulfate to the preparation. Glucosamine products may also be sold in formulation containing chondroitin 4- and chondroitin 6-sulfates. While glucosamine helps to form proteoglycans that sit within the space in the cartilage, chondroitin sulfate acts like a liquid magnet. Thus glucosamine and chondroitin work in a complementary manner to improve the health of the joint cartilage.

The byproducts in chitin extraction process from shellfish include carotenoids/carotenoproteins, and enzymes (*30-33*). These components may also be isolated for further utilization in a variety of applications.

Enzymes

The aquatic environment contains a wide variety of genetic material and hence represents exciting potential for discovering different enzymes (*34*). Therefore, much effort has been made to recover and characterize enzymes from fish and aquatic invertebrates (*30*). Digestive proteolytic enzymes from stomachless marine fish such as conner, crayfish and puffer appear to inactivate polyphenol oxidize and/or pectin esterase in fruit juices. Successful application of such enzymes has also allowed inactivation of polyphenol oxidase in shrimp processing as an alternate to sulfiting (*31*). Alkaline phosphates from shrimp may be used in different kits and some enzymes may be recovered and used in deskinning of fish and squid or cleaning of fish roe for caviar production, among others.

Carotenoids

Carotenoids and carotenoprotiens are present in salmonoid fish as well as in shellfish. These carotenoids may be recovered from processing by-products and used in a variety of applications (*33*). In addition, certain carotenoids, such as Fucoxanthin occur naturally in seaweeds (*35,36*). Fucoxanthin has been shown to have anti-proliferative activity on tumor cells and has also been implicated in having anti obesity and anti-inflammatory effects. Fucoxanthinol is a known metabolite of fucoxanthinol.

Minerals and Calcium

Among fish processing by-products, fish bone or skeleton serves as a potential source of minerals and calcium which is an essential element for human health. Calcium from fish would be easily absorbed by the body (*37*). However, to incorporate fish bone into calcium-fortified foods, it is necessary to first convert it into an edible form by softening its structure. This could be achieved by hot water treatment and heat treatment in an acetic acid solution. Pepsin-assisted degradation of Alaska Pollock bone in acetic acid solution led to highest degree of hydrolysis and dissolution of both mineral and organic parts of fish bone (*38,39*). As reported by Larsen (*37*), the intake of small fish with bones could increase calcium bioavailability. Fish bone contains hydroxyapatite which unlike other calcium phosphates does not break under physiological conditions and takes part in bone bonding. This property has been exploited for rapid bone repair after major trauma or surgery.

References

1. Simopoulos, A.P. *Am. J. Clin. Nutr.* **1991**, *54*, 438-463.
2. Abeywardena, M.Y.; Head, R.J. *Cardiovascular Res.* **2001**, *52*, 361-371.
3. Shahidi, F.; Kim, S.K. In *Bioactive Compounds in Foods: Effects of Processing and Storage*, Lee, T.C. and Ho, C.T., eds. ACS Symposium Series 816, American Chemical Society, Washington, DC, 2002; pp 1-13.
4. Wanasundara, U.N.; Wanasundara, J.; Shahidi, F. In *Seafoods: Quality, Technology and Nutraceutical Applications*, Alasalvar, C. and Taylor T., eds. Springer, New York, NY. 2002; pp 157-174.
5. Wanasundara, U.N.; Shahidi, F. In: *Flavor and Lipid Chemistry of Seafoods*. Shahidi, F. and Cadwallader, K.R. eds. ACS Symposium Series 674, American Chemical Society, Washington, DC, 1997; pp 240-254.
6. Lee, K.; Akah, C.C. *Food Rev. Int.* **1998**, *14*, 17-34.
7. Senenayake, S.P.J.N.; Shahidi, F. In: *Seafoods in Health and Nutrition in Transformation in Fisheries and Aquaculture: Global perspective.* ScienceTech. Publishing Co., St. John's, Canada, 2000; pp 25-44.
8. Senenayake, S.P.J.N.; Shahidi, F. *Food Chem.* **2002**, *35*, 745-752.
9. Senenayake, S.P.J.N.; Shahidi, F. *J. Food Lipids* **2007**, *14*, 28-96.
10. Spurvey, S.A.; Shahidi, F. *J. Food Lipids* **2000**, *7*, 163-174.
11. Spurvey, S.A.; Senenayake, S.P.J.N.; Shahidi, F. *J. Am. Oil Chem. Soc.* **2001**, *78*, 1105-1112.

12. Senenayake, S.P.J.N.; Shahidi, F. *J. Agric. Food Chem.* **1999**, *47*, 3105-3112.
13. Senenayake, S.P.J.N.; Shahidi, F. *J. Am. Oil Chem. Soc.* **1999**, *76*, 1009-1015.
14. Amarowicz, R.; Shahidi, F. *Food Chem.* **1997**, *58*, 355-359.
15. Shahidi, F., Synowiecki, *J. Food Chem.* **1997**, *60*, 29-32.
16. Kim, S.K.; Kim, Y-T; Byun, H-G.; Nam, K-S.; Joo, D-S; Shahidi, F. *J. Agric. Food Chem.* **2001**, *49*, 1984-1989.
17. Chen, H-M.; Muramato, K.; Yamauchi, F. *J. Agric. Food Chem.* **1994**, *43*, 574-578.
18. Heu, M.S.; Kim, J-S.; Shahidi, F.; Jeong, Y.; Jeon, Y-J. *J. Food Biochem.* **2003**, *27*, 221-236.
19. Kim, J-S.; Shahidi, F.; Heu, M-S. *Food Chem.* **2005**, *93*, 243-250.
20. Shahidi, F.; Synowiecki, J. *J. Agric. Food Chem.* **1991**, *39*, 1527-1532.
21. Jeon, Y-J.; Kamil, J.Y. V.A.; Shahidi, F. *J. Agric. Food Chem.* **2002**, *50*, 5167-5178.
22. Kamil, J.Y.V.A.; Jeon, Y-J.; Shahidi, F. *Food Chem.* **2002**, *79*, 69-77.
23. Shahidi, F.; Kamil, J.; Jeon, Y.J.; Kim, S-K., *J. Food Lipids*, **2002**, *9*, 57-64.
24. Jeon, Y-J.; Shahidi, F; Kim, S-K. *Food Rev. Int.* **2000**, *16*, 159-176.
25. Jeon, Y.J.; Kim, S.K. *J Microbiol. Biotechnol.* **2002**, *12*, 503-307.
26. Tsukada, K.; Matsumoto, T.; Aizawa, K.; Tokoro, A.; Nareuse, R.; Suzuki, S.; Suzuki, M. *Jpn. J. Canar Res.* **1990**, *81*, 259-265.
27. Jeon, Y.J.; Park, P.J.; Kim, S.K. *Carbolyd. Polym.* **2001**, *461*, 71-76.
28. Jeon, Y.J.; Kim, S.K. *Process Biochem.* **2000**, *35*, 623-632.
29. Park, P.J., Je, J.Y.; Bgun, H.G.; Moon, S.H.; Ku, S.K. *J. Microbiol. Biotechnol.* **2004**, *14*, 317-323.
30. Shahidi, F.; Kamil, Y.U.A.J. *Trends Food Sci. Technol.* **2001**, *12*, 435-464.
31. Simpson, B.K. In: *Maximising the value of Marine By-Products.* F. Shahidi, Ed., Woodhead Publishing, Cambridge, UK, 2007; pp 413-432.
32. Simpson, B.K.; Smith, J.P.; Haard, N.F. In *Encyclopedic of Food Science and Technology*, Hui, Y.H., ed. John Wiley and Sons, New York, N.Y.; pp 1645-1653.
33. Shahidi, F.; Metusalach; Brown, J.A. *Crit. Rev. Food Sci. Nutr.* **1998**, *38*, 1-67.
34. Raa, J. In *advances in Fisheries Technology for Increased Profitability.* Voight, M.N. and Botta, J.R. Eds. Technomic Publication Co., Lancester, PA; pp 509-524.
35. Czeczuga, B.; Taylor, F.J. *Biochem. Syst. Ecol.* **1987**, *15*, 5-8.
36. Haugen, J.; Liaeen-Jensen, S. *Biochem. Syst. Ecol.* **1994**, *22*, 31-41.

37. Larsen, T., Thilsted, S.H., Kongsbak, K.; Hansen, M. *Br. J. Nutr*. **2000**, *83*, 191-196.
38. Jung, W.K.; Park, P.J.; Dyun, H.G.; Moon, S.H.; Kim, S.K. *Food Chem.* **2005**, *91*, 333-340.
39. Ishikawa, M.; Kato, M., Mihori, T.; Watanabe, H.; Sakai, Y. *Nippon Suisan Gakkaishi*, **1990**, *56*, 1687-1691.

Phytochemistry

Chapter 4

Oligostilbenes from *Gnetum* Species and Anticarcinogenic and Antiinflammatory Activities of Oligostilbenes

Ka-Wing Cheng[1], Mingfu Wang[1], Feng Chen[1], and Chi-Tang Ho[2]

[1]Department of Botany, The University of Hong Kong, Pokfulam Road, Hong Kong, People's Republic of China
[2]Department of Food Science, Rutgers, The State University of New Jersey, 65 Dudley Road, New Brunswick, NJ 08901

Interest in the physiological roles of phytochemicals has increased dramatically over the last twenty years. Oligostilbenes, which are widely distributed in several plant families and liverworts, have attracted much attention due to emerging evidence suggesting them to be potential disease preventive and curing agents against several chronic diseases, particularly through their anti-inflammatory, antioxidant and anti-carcinogenic activities. The genus *Gnetum* is of particular interest in search of this group of polyphenols owing to its large genus size and wide distribution of its species in both hemispheres. In this paper we summarize the oligostilbenes that have been isolated from *Gnetum* species and we also discuss the anti-carcinogenic and anti-inflammatory activities of oligostilbenes in general.

Stilbenes are a group of naturally occurring phenolic compounds characterized by the presence of diphenylethylene backbone and have been isolated mainly from angiosperms and gymnosperms (*1*). In plants, stilbenes are synthesized through specific stilbene synthase with *p*-coumaroyl-CoA and malonyl-CoA as major precursor molecules (*2*). During the course of identifying the paramount active compounds in prevention and treatment of the age-related diseases, stilbenes have emerged to hold important positions stemming from various *in vitro* and *in vivo* studies which strongly support their potential benefits in a wide range of health disorders through their antioxidant, anti-inflammatory and anticarcinogenic activities, among others (*3-7*). The most remarkable monomeric stilbene compound discovered so far is resveratrol (3,5,4′-trihydroxystilbene), which exhibits a wide range of pharmacological effects, including cardiovascular protection, antioxidation and cancer chemoprevention (*8-16*).

Stilbenes, which exist in oligomeric forms, are known as oligostilbenes. They are believed to arise from oxidative condensation of stilbene nucleus, and are known to be present mainly in nine plant families, *Dipterocarpaceae, Vitaceae, Cyperaceae, Gnetaceae, Leguminosae, Celastraceae, Paeoniaceae, Moraceae* and *Iridaceae*, in woody tissues, as constitutive components, though various other tissues have also been reported to contain such diphenyl ethene-bridged compounds (*17-21*). Oligostilbenes have also been shown to possess strong health-promoting effects (*22-27*) and a number of them have demonstrated more potent anticarcinogenic and anti-inflammatory activities than the corresponding monomers. Most oligostilbenes are not regular dietary components in human populations (*28*), while plants from *Gnetum* are an exception. *Gnetum* is a perennial plant belonging to the Gymnosperms (*29*). It embodies over 40 species that can be monoecious or dioecious (*30*). Most of them are trees or shrubs and some are woody climbers (*29*). They are widely distributed in both hemispheres, particularly inhabiting the tropics of Asia, Africa, South America and certain islands between Asia and Australia (*29,31*). Some species of this perennial plant have a long history in the diets of some populations in under-developed countries in central Africa, where they are a significant source of dietary protein and mineral elements (*29,32*), though other biological roles have not been evaluated in these populations. In addition, various species of the genus have long been used in medical prescriptions as a blood invigorating agent, for treatment of chronic diseases like arthritis, bronchitis and asthma and have been shown to contain oligostilbenes (*33-37*). In recent years, the genus *Gnetum* has attracted considerable attention due to emerging evidence suggesting them to be potential rich sources of oligostilbenes. Myriad oligostilbenes have been isolated and structurally characterized from *Gnetum* species, ranging from homo-, hetero-dimers, to trimers and tetramers, along with their derivatives, but more extensive *in vitro* and *in vivo* studies are demanded to fill the current gap in our understanding of their pharmacological potentials. This review will focus on the chemistry of oligostilbenes from *Gnetum* and the anti-inflammatory and anticarcinogenic activities of oligostilbenes in general.

Chemistry of *Gnetum* Oligostilbenes

Structural Features and Classification

Since the isolation of gnetin A from *G. leyboldii* in 1982 (*36*), more than 80 oligostilbenes have been obtained from this genus, mainly from their liana and root, but also from wood, bark and fruit. Among the *Gnetum* species, *G. hainanense* and *G. gnemon* have been most extensively evaluated for their oligostilbene content. In Table 1, selected *Gnetum* species, their corresponding oligostilbenes together with order of polymerization, molecular formula and constitutional monomers were presented in order of their isolation from the species. These species exhibit wide variation in their oligostilbene constituents on the basis of the types of oligostilbenes that have been isolated and characterized. This thus confirms the need to systematically screen and identify oligostilbenes from this genus. It appears that geographical variation also significantly affects the stilbene contents. As an example, several studies have reported the isolation of stilbene dimmers from *G. montanum*, Xiang *et al.* obtained no dimeric stilbenes from the same species of a different geographical origin (*38*). These geographical factors complicate species-based profiling of oligostilbene contents.

Similar to oligostilbenes isolated from other plant genus, most of the oligostilbenes of *Gnetum* origin contain resveratrol as the fundamental unit. The simplest are those derived from direct oxidative condensation of resveratrol, such as gnetins and (-)-ε-viniferin. More complicated structures result from homo- or heteromeric condensation of resveratrol and its derivatives like oxyresveratrol, piceatannol, isorhapontigenin, gnetol and resveratrolosides (Figure 1). These mainly include dimmers, trimers and tetramers, which are substituted by varying number and pattern of hydroxyl groups. The number of hydroxyl group can be confirmed by the results of acetylation and methylation experiments (*39*). Acetylation is generally performed in a mixture of pyridine and acetic anhydride for 24 hours under room temperature, while methylation usually carried out with MeI and K_2CO_3 in dry acetone under reflux for 3 hours (*40*). Different connectivity between monomers also gives rise to different oligostilbenes. Based on the structural characteristics of their connection, these oligomers can be divided into five groups (A-E). Group A contains at least one oxygen heterocyclic ring, which may be five-membered or six-membered. The former, usually in the form of *trans*-2-aryl-2,3-dihydrobenzofuran moiety,

Table I. Oligostilbenes from Selected Gnetum Species

Source	*Part examined*	*Compound isolated*	*Oligomer type (Mol formula)*	*Constitutional monomers*	*Ref*
Gnetum leyboldii	wood	gnetin A	dimer ($C_{28}H_{22}O_6$)	Res	*36*
	wood	gnetin B	dimer ($C_{28}H_{24}O_6$)	Res	*36*
	wood	gnetin C	dimer ($C_{28}H_{22}O_6$)	Res	*36*
	wood	gnetin D	dimer ($C_{28}H_{22}O_7$)	Res & oxyRes	*36*
	wood	gnetin E	trimer ($C_{42}H_{32}O_9$)	Res	*36*
Gnetum schwa-ckeanum	fruits	gnetin C	dimer ($C_{28}H_{22}O_6$)	?	*36*
	fruits	gnetin E	trimer ($C_{42}H_{32}O_9$)	?	*36*
Gnetum parvi-folium	lianas	gnetifolin C	dimer ($C_{30}H_{26}O_8$)	?	*48*
	lianas	gnetifolin D	dimer ($C_{30}H_{28}O_8$)	?	*48*
	lianas	ε-viniferin	dimer ($C_{28}H_{22}O_6$)	Res	*48*
	bark	parvifolol A	dimer ($C_{28}H_{22}O_7$)	Res & oxyRes	*40*
	bark	parvifolol B	dimer ($C_{28}H_{22}O_7$)	Res & oxyRes	*40*
	bark	parvifolol C	dimer ($C_{28}H_{22}O_8$)	oxyRes	*40*
	bark	parvifolol D	dimer ($C_{30}H_{26}O_8$)	Isorhapontigenin	*40*
	bark	2b-hydroxy-ampelopsin F	dimer ($C_{28}H_{22}O_7$)	Res & oxyRes	*40*
Gnetum venosum	fruit kernels	gnetin C	dimer ($C_{28}H_{22}O_6$)	Res	*37*
	fruit kernels	gnetin E	trimer ($C_{42}H_{32}O_9$)	Res	*37*
	fruit kernels	gnetin J	trimer ($C_{42}H_{32}O_{10}$)	Res & piceatannol	*37*
	fruit kernels	gnetin K	trimer ($C_{43}H_{34}O_{10}$)	Res & isorhapon-tigenin	*37*
Gnetum ula		gnetulin	dimer ($C_{30}H_{26}O_8$)	Isorhapontigenin	*94*

Continued on next page.

Table I. *Continued.*

Source	*Part examined*	*Compound isolated*	*Oligomer type (Mol formula)*	*Constitutional monomers*	*Ref*
Gnetum montanum	lianas	gnetifolin M	dimer ($C_{30}H_{28}O_9$)	isorhapontigenin	*46*
	lianas	gnetifolin N	dimer ($C_{30}H_{28}O_9$)	isorhapontigenin	*46*
	lianas	gnetifolin L	dimer ($C_{30}H_{28}O_8$)	isorhapontigenin	*47*
	lianas	gnetifolin O	dimer ($C_{30}H_{28}O_9$)	isorhapontigenin	*47*
Gnetum hainanense	lianas	gnetuhainin A	dimer ($C_{28}H_{22}O_7$)	Res & oxyRes	*95*
	lianas	gnetuhainin B	dimer ($C_{28}H_{20}O_7$)	Res & oxyRes	*95*
	lianas	gnetuhainin C	dimer ($C_{28}H_{22}O_7$)	Res & oxyRes	*95*
	lianas	gnetuhainin D	dimer ($C_{28}H_{24}O_8$)	Res & oxyRes	*95*
	lianas	gnetuhainin E	dimer ($C_{28}H_{24}O_8$)	Res & oxyRes	*95*
	lianas	gnetuhainin F	dimer ($C_{30}H_{24}O_8$)	isorhapontigenin	*35*
	lianas	gnetuhainin G	dimer ($C_{30}H_{22}O_9$)	isorhapontigenin	*35*
	lianas	gnetuhainin H	dimer ($C_{30}H_{24}O_9$	isorhapontigenin	*35*
	lianas	gnetuhainin I	dimer ($C_{30}H_{28}O_9$)	isorhapontigenin	*35*
	lianas	gnetuhainin J	dimer ($C_{29}H_{24}O_8$)	isorhapontigenin & oxyRes	*35*
	lianas	gnetuhainin K	dimer ($C_{29}H_{24}O_8$)	Isorhapontigenin & gnetol	*96*
	lianas	gnetuhainin L	dimer ($C_{29}H_{24}O_8$)	Isorhapontigen-in & oxyRes	*96*
	lianas	gnetuhainin M	trimer ($C_{42}H_{32}O_{11}$)	Res & oxyRes	*34*
	lianas	gnetuhainin N	trimer ($C_{45}H_{38}O_{12}$)	Isorhapontigenin	*34*
	lianas	gnetuhainin O	trimer ($C_{45}H_{38}O_{12}$)	Isorhaponti-genin	*34*
	lianas	gnetuhainin P	dimer ($C_{30}H_{28}O_9$)	Isorhapontigenin	*34*

Table I. *Continued.*

Source	*Part examined*	*Compound isolated*	*Oligomer type (Mol formula)*	*Constitutional monomers*	*Ref*
Gnetum hainanense (continued)	lianas	gnetuhainin Q	dimer ($C_{29}H_{24}O_7$)	Res & isorhapontigenin	*96*
	lianas	gnetuhainin R	tetramer ($C_{60}H_{51}O_{16}$)	Isorhapontigenin	*51*
	lianas	gnetuhainin S	dimer ($C_{28}H_{23}O_7$)	Res & oxyRes	*51*
Gnetum gnemonoides	stem	Gnemonoside A	dimer ($C_{40}H_{42}O_{16}$)	Resveratroloside	*97*
	stem	Gnemonoside B	dimer ($C_{40}H_{42}O_{16}$)	Resveratroloside	*97*
	stem	Gnemonoside C	dimer ($C_{34}H_{32}O_{11}$)	Res & resveratroloside	*97*
	stem	Gnemonoside D	dimer ($C_{34}H_{32}O_{11}$)	Res & resveratroloside	*97*
	stem lianas	Gnemonoside E	dimer ($C_{40}H_{42}O_{16}$)	Resveratroloside	*39*
	stem lianas	Gnemonoside F	trimer ($C_{60}H_{62}O_{24}$)	Resveratroloside	*44*
	stem lianas	Gnemonoside G	trimer ($C_{54}H_{52}O_{19}$)	Res & resveratroloside	*44*
	stem lianas	Gnemonoside H	trimer ($C_{46}H_{52}O_{21}$)	Res & Resdiglucopyranoside	*44*
	stem lianas	Gnemonoside I	dimer ($C_{40}H_{42}O_{17}$)	Resveratroloside & oxyRes	*44*
	stem lianas	Gnemonoside J	dimer ($C_{40}H_{42}O_{17}$)	NA	*44*
	stem lianas	gnemonol C	Tetramer ($C_{56}H_{42}O_{13}$)	Res & oxyRes	*39*
	stem lianas	gnetin E	trimer ($C_{42}H_{32}O_9$)	Res	*39, 36*
	stem lianas	2b-hydroxyampelopsin F	dimer ($C_{28}H_{22}O_7$)	Res & oxyRes	*39, 40*
Gnetum pendulum	lianas	gnetupendin C	dimer ($C_{28}H_{22}O_7$)	Res & oxyRes	*50*
	lianas	gnetupendin D	dimer $C_{36}H_{36}O_{13}$)	isorhapontigenin	*92*
	lianas	gnetin D	dimer ($C_{28}H_{22}O_7$	Res	*92*

Continued on next page.

Table I. *Continued.*

Source	*Part examined*	*Compound isolated*	*Oligomer type (Mol formula)*	*Constitutional monomers*	*Ref*
	lianas	shegansu B	dimer ($C_{36}H_{26}O_8$)	isorhapontigenin	*92, 98*
Gnetum africanum	stem lianas	gneafricanins A	dimer ($C_{29}H_{24}O_8$)	oxyRes & isorhapontigenin	*99*
	stem lianas	gneafricanins B	dimer ($C_{29}H_{24}O_8$)	isorhapontigenin & piceatannol	*99*
	stem lianas	gneafricanins C	dimer ($C_{28}H_{22}O_8$)	piceatannol	*100*
	stem lianas	gneafricanins D	dimer ($C_{28}H_{24}O_6$)	Res	*100*
	stem lianas	gneafricanins E	dimer ($C_{28}H_{24}O_7$)	Res & oxyRes	*100*
	stem lianas	gneafricanins F	dimer ($C_{30}H_{26}O_8$)	isorhapontigenin	*100*
Gnetum gnemon	root	gnemonol A	trimer ($C_{42}H_{32}O_{16}$)	Res & oxyRes	*39*
	root	gnemonol B	tetramer ($C_{56}H_{42}O_{12}$)	Res	*39*
	root	gnemonol G	dimer ($C_{28}H_{20}O_7$)	Res	*30*
	root	gnemonol H	trimer ($C_{42}H_{32}O_{10}$)	Res	*30*
	root	gnemonol I	trimer ($C_{42}H_{32}O_{16}$)	Res & oxyRes	*30*
	root	gnemonol J	trimer ($C_{42}H_{32}O_{16}$)	Res & oxyRes	*30*
	root	gnetin C	dimer ($C_{28}H_{22}O_6$)	Res	*30*
	root	gnetin D	dimer ($C_{28}H_{22}O_7$)	Res	*30*
	root	gnetin E	trimer ($C_{42}H_{32}O_9$)	Res	*30*
	root	ampelopsin E	trimer ($C_{42}H_{32}O_9$)	Res	*30*
	root	*cis*-ampelopsin E	trimer ($C_{42}H_{32}O_9$)	Res	*30, 101*
	root	gnemonol D	trimer ($C_{42}H_{32}O_{10}$)	Res & oxyRes	*102*

Table I. *Continued.*

Source	*Part examined*	*Compound isolated*	*Oligomer type (Mol formula)*	*Constitutional monomers*	*Ref*
Gnetum gnemon (continued)	root	gnemonol E	trimer ($C_{42}H_{32}O_{10}$)	Res & oxyRes	*102*
	root	gnemonol F	trimer ($C_{42}H_{32}O_{10}$)	Res	*102*
	root	gnemonol K	trimer ($C_{42}H_{32}O_{9}$)	Res	*49*
	root	gnemonol L	trimer ($C_{42}H_{32}O_{9}$)	Res	*49*
	root	gnemonol M	dimer ($C_{30}H_{26}O_{8}$)	isorhapontigenin	*49*
	root	Gnemonoside K	trimer ($C_{60}H_{62}O_{24}$)	Res	*49*
	root	gnemonoside A	dimer ($C_{40}H_{42}O_{16}$)	resveratroloside	*49, 97*
	root	gnemonoside B	dimer ($C_{40}H_{42}O_{16}$)	resveratroloside	*49, 97*
	root	gnemonoside F	trimer ($C_{60}H_{62}O_{24}$)	resveratroloside	*97, 49*
	root	latifolol	trimer ($C_{42}H_{32}O_{10}$)	Res & oxyRes	*49, 43*
	Root	(-)-ε-viniferin	dimer ($C_{28}H_{22}O_{6}$)	Res	*49*
Gnetum latifolium	stem	latifolol	trimer ($C_{42}H_{32}O_{10}$)	Res	*43*
	stem	gnetin C	dimer ($C_{28}H_{22}O_{6}$)	Res	*43, 36*
	stem	gnetin D	dimer ($C_{28}H_{22}O_{7}$)	Res	*43, 36*
	stem	gnetin E	trimer ($C_{42}H_{32}O_{9}$)	Res	*43, 36*
	stem	(-)-ε-viniferin	dimer ($C_{28}H_{22}O_{6}$)	Res	*43, 18*
Gnetum cleistos-tachyum	lianas	bisisorhapon-tigenin A	dimer ($C_{30}H_{26}O_{8}$)	isorhapontigenin	*42*
	lianas	*cis*-shegansu B	dimer ($C_{30}H_{26}O_{8}$)	isorhapontigenin	*42*
	lianas	gnetuhainin P	dimer ($C_{30}H_{28}O_{9}$)	isorhapontigenin	*96*

Continued on next page.

Table I. *Continued.*

Source	*Part examined*	*Compound isolated*	*Oligomer type (Mol formula)*	*Constitutional monomers*	*Ref*
	lianas	gnetulin	dimer ($C_{30}H_{26}O_8$)	isorhapontigenin	*94*
Gnetum klossii	stem	(-)-ε-viniferin	dimer ($C_{28}H_{22}O_6$)	Res	*103, 18*
	stem	gnetulin	dimer ($C_{30}H_{26}O_8$)	isorhapontigenin	*103, 94*
	stem	gnetin C	dimer ($C_{28}H_{22}O_6$)	Res	*103*
	stem	gnetin E	trimer ($C_{42}H_{32}O_9$)	Res	*103*
	stem	latifolol	trimer ($C_{42}H_{32}O_{10}$)	Res & oxyRes	*103, 43*

Res = resveratrol

predominates among oligostilbenes isolated from *Gnetum* (*41*). Group B contains both oxygen heterocyclic ring and homocyclic ring. Group C contains only homocyclic structure in its connective part. Group D has a carbonyl group in its cyclic connective part (bicyclo-octanoid) and its presence can be easily revealed in the IR spectrum with characteristic absorption band at around 1700 cm^{-1} (*36*). Group E does not possess any cyclic structure in connecting the monomeric units. Moreover, some oligostilbenes exist as isomers. Whilst most *Gnetum* oligostilbenes have *trans*-configuration, a few *cis*-oligostilbenes have also been identified (*1,17,30,42*). These structural differences may affect their bioavailability, affinity for receptor, and thus their biological activity. There has not been a clear pattern in the distribution of these oligostilbenes among *Gnetum* species except for group B which has mainly been isolated from *G. parvifolium* and *G. gnemon*; and group D which has so far been identified only in *G. leyboldii*.

Extraction, isolation and identification

Oligostilbenes are usually soluble in moderately polar to polar solvents such as ethyl acetate, ethanol and methanol. A variety of solvents have been used in extraction of oligostilbenes from *Gnetum* species. Ethanol (60-95%) is commonly employed for obtaining the crude extract. The typical extraction method has been refluxing (*35,42*) or percolation with aqueous ethanol (*37*). Alternatively, successive extraction with acetone, MeOH and MeOH-H_2O mixture has also been adopted to obtain crude extract for further separation (*30, 43*). The first separation is usually carried out on silica gel column and eluted

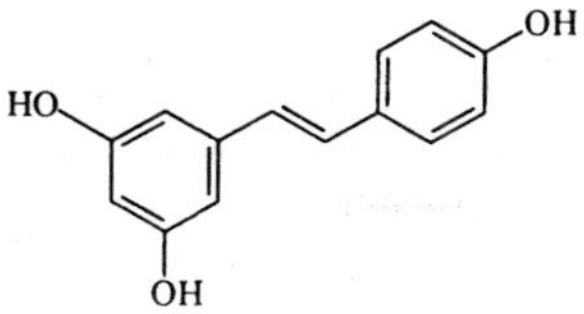

Resveratrol

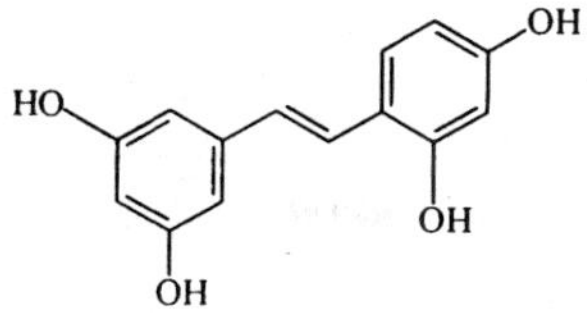

Oxyresveratrol

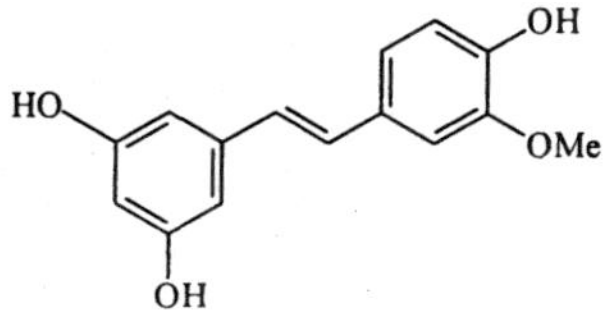

Isorhapontigenin / 3-methoxyresveratrol

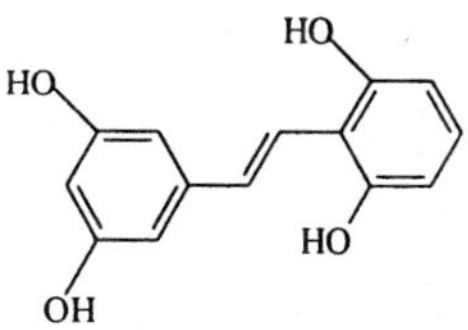

Gnetol

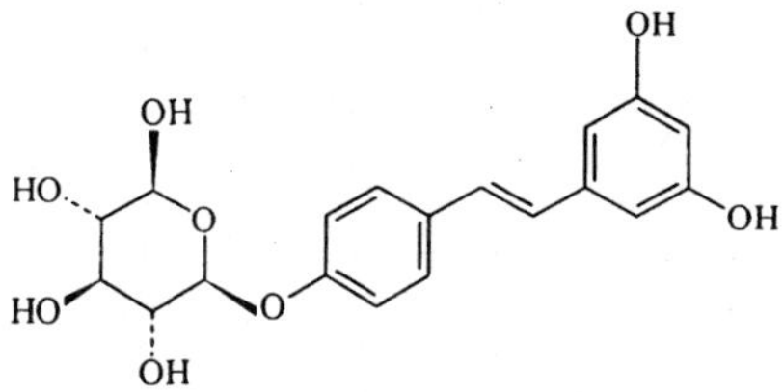

Resveratroloside

Figure 1. Monomeric units constituting oligostilbenes from Gnetum spp.

with solvent system of increased polarity, in particular, by gradually increasing proportion of MeOH in $CHCl_3$-MeOH mixture. A preliminary partition step may also be conducted prior to the above column chromatography. Separation of individual oligostilbenes usually requires further chromatographic procedures, including repeating the column chromatographic process on silica gel, with similar or different elution solvent, such as cyclohexane-acetone mixture (*35*); column chromatography over Sephadex LH-20 often eluted with MeOH; and over reversed phase silica gel. The final purification process is usually by preparative TLC and HPLC (*40,43,44*). In some cases, especially for the more highly hydroxylated oligostilbenes, acetylation may be carried out prior to isolation to avoid oxidative modification or polymerization (*45-47*). The oligostilbenes obtained have a wide range of colors: brown (*39*), yellowish (*48*), greenish (*44*), white (*35*) and colorless (*49*). Many of them show a positive reaction to Gibbs reagent. Structural elucidation is usually accomplished with a combination of spectroscopic techniques. UV spectra usually provide evidence for presence of conjugation or furan ring in the molecule, while IR spectra reveal presence of characteristic functional groups such as hydroxyl, aromatic, olefinic and carbonyl groups. Full elucidation of these oligostilbenes can be best accomplished with MS and various NMR techniques including ^{1}H-^{1}H COSY, HBQC, HMBC and NOESY. Fast atom bombardment (FAB) and electron ionization (EI) are currently the most common modes in molecular mass determination for oligostilbenes (*35,50*). Very few oligostilbenes obtainable from *Gnetum* have been determined for its absolute configuration. A previous circular dichroism study conducted by Kurihara and coworkers proposed a 7R, 8R absolute configuration for (-)-ε-viniferin which was obtained from *Carex pumila* (*18*). There have also been rare studies, which used X-ray crystallography for analysis of the absolute configuration of some oligostilbenes (*51*).

Anticarcinogenic Activities of Oligostilbenes

Studies on the anticarcinogenic activity of oligostilbenes are limited, especially those from *Gnetum* species, yet among those tested, appreciable number of them have shown potent activities in different cancer cell lines. These studies mainly focused on their abilities to prohibit tumor progression rather than prevent its initiation. The plausible anticarcinogenic mechanisms have been ascribed as induction of apoptosis (*52,72,106*), cytotoxic and antiproliferative, antimutagenic activity (*53-55*), and inhibitory effects on the key enzymes that have been implicated in carcinogenesis.

Cytotoxicity

Several studies have demonstrated the cytotoxic activities of oligostilbenes against different cancer cell lines (*54,56,57,104,105*). With the use of three human colon cancer cell lines (SW480, DLD-1 and COLO201), Ito and coworkers evaluated the *in vitro* cytotoxicity of two resveratrol oligomers, vaticanols B and C with resveratrol as the reference (*56*). All these three stilbenoids exhibited cytotoxicity, with vaticanol C being the most potent and the growth inhibition induced by vaticanol C was concentration-dependent. It was suggested that presence of a dibenzobicyclo[3.2.1]octadien framework may contribute to its higher cytotoxic activity (*56*). Seo *et al.* showed that vatdiospyroidol, a resveratrol tetramer, was cytotoxic against human oral epidermoid carcinoma (KB), colon cancer, and breast cancer cell lines with EC_{50} values of 1.0 μg/mL, 1.9 μg/mL, and 3.8 μg/mL, respectively (*57*). Ohyama and coworkers elaborately tested the cytotoxic effect of 11 oligostilbenes against seven cancer cell lines including those of nasopharynx, lung, ileocecal, melanoma, renal, breast and ovary (*54*). Among these oligostilbenes, hopeaphenol, a resveratrol tetramer, showed most potent cytotoxicity against KB cells with an ED_{50} value of 1.2 μg/mL and moderate cytotoxicity against several other cancer cell lines. In contrast, the remaining oligomers (dimmers and trimers) did not exhibit significant cytotoxicity. These findings are in accordance with the hypothesis that oligostilbenes with more than four basic units are endowed with higher cytotoxicity (*52*). Although it was observed that resveratrol oligostilbenes constituting of more than four monomeric units are generally more potent cytotoxic agents, with very high molecular weight (from many monomeric units), it is probable that other factors like steric hindrance may also affect their bioactivities. Oligostilbenes also exhibited high selectivity against different cancer cells. Using five different cancer cell lines, Kim and coworkers found that *trans*-ε-viniferin, *cis*-ε-viniferin and gnetin H possessed marked cytotoxicity against Hela (cervicse) and HCF-7 (breast) human cancer cell lines with IC_{50} values of 20.4, 21.5 and 12.9 μg/mL, respectively, but not effective against HepG2 (liver hepatoma) and HT-29 (colon) (*53*). Pronounced cytotoxicity of viniferin and gnetin H was also observed with HL 60 cells and in a dose dependent manner (*58*). The potential anticarcinogenic activities of oligostibenes are further exemplified by the fact that some oligostilbenes demonstrate good tumor specificity (relative sensitivity between the tumor cell lines and normal cells). In a comparative study engaging both stilbenes and flavonoids, the stilbene trimers demonstrated higher tumor-specific cytotoxicity, especially sophorastilbene A and (+)-α-viniferin, which yielded tumor-specific indices of 3.6 and 4.7, respectively (*59*). However, tumor specificity did not correlate with their ability to induce apoptosis as it was observed that (+)-α-viniferin exhibited the highest tumor specificity, yet induced internucleosomal DNA fragmentation and activation of caspases –3, -8 and –9 in HL-60 cells at concentration much higher than other stilbenes tested (>200 μM) (*59*). Structural modification can significantly affect cytotoxic activity of oligostilbenes. As mentioned above, hopeaphenol was a strong cytotoxic agent against KB cells. However, its oxidative derivative stenophyllol A was ineffective in the same study and it was speculated that the 4-hydroxyphenyl moiety on the seven-membered rings of hopeaphenol may account for its enhanced cytotoxicity (*54*). Vatdiospyroidol (I) was potently cytotoxic against a panel of cancer cell lines, but conversion of the hydroxyl groups to methoxyl groups abrogated its activity, suggesting the importance of the hydroxyl groups in mediating cytotoxicity (*57*). Acetylation could be another strategy to alter cytotoxic activity. By adding an acetate group, Ohyama and coworkers significantly increased cytotoxicity of the dimmers, which displayed negligible activity in their native state (*54*). They suggested that the presence of a lipophilic acetate group could aid in cellular penetration and thus access to the targeted point. However this enhancement seems to be restricted to dimmers as the trimers and tetramers remained inactive after acetylation. Information concerning the effect of structural modification is scarce, especially for oligomers of higher orders. More systematic structure-activity relationship (SAR) studies are demanded to characterize this group of phenolic compounds.

Anti-proliferation

Horoshi *et al.* demonstrated that some oligostilbenes possessed potent inhibitory activity against the isomerase at concentration of <1 μg/mL (*60*). Topoisomerase II inhibitors are also important targets of anticancer research. Inhibition of the enzyme may help arrest cell proliferation as topoisomerase II may be involved in proliferative processes such as DNA replication, chromosome condensation, and chromosome segregation by relaxing supercoiled DNA and resolving knotted or catenated DNA rings (*60*). (+)-α-viniferin and miyabenol C were shown to exhibit inhibitory activity against Protein kinase C (PKC) at micromolar concentrations (*61*) and miyabenol C has also been found to suppress PKC activity of A549 and NCI-H446 lung carcinoma cell lines at 20 μM (*62*). PKC is a family of serine/threonine kinases involved in the signal transduction pathways for cell proliferation and differentiation (*63*). Its activation has been linked to promote tumor formation (*64*). These oligostilbenes may therefore help prevent carcinogenesis and inhibit tumor proliferation (*64*). Some oligostilbenes have been found to exert differential biphasic effects on cell proliferation. At concentrations above 25 μM, Kim and coworkers found that *trans*-(+)-ε-viniferin and gnetin H exhibited strong antiproliferative effects on MCF-7 cells, whereas for resveratrol, similar effects were observed only at concentration of 200 μM (*55*), while within the concentration range of 1 to 10 μM, all the three stilbenes exerted weak stimulatory effect on proliferation of MCF-7 cells. Another oligostilbene, suffruticosol B, also exhibited biphasic effects, but in a different manner. Its pronounced antiproliferative effect on MCF-7 cells occurred at doses below 25 μM, but became stimulatory at concentrations of 50 and 100 μM. More studies are required to evaluate their dose-dependent activity and define the optimal concentration range.

Apoptosis

Apoptosis, which serves to eliminate cells damaged by mutagenic chemicals or irradiation and counterbalance mitosis without perturbing the homeostatic balance of its environment (*65,66*), has become one of the major focuses of phytochemical research for chemopreventive agents. In light of preliminary finding that some oligostibenes may exert their cytotoxic effect via induction of apoptosis, with characteristics of apoptosis (nuclear condensation, fragmentations and DNA ladder formation) after treatment with low dose of vaticanol C (5 μM) (*56*), further investigations carried out using SW 480 cell lines with mutant p53 gene found that vaticanol C activated caspase-3, a protease responsible for cleaving poly(ADP-ribose) polymerase (PARP) involved in DNA repair (*67, 68*). On the other hand, caspase-8, an initiator of the apoptotic cascade, was not recruited (*52, 69*). Several lines of evidence supported a direct action of vaticanol C on the mitochondrial membrane (*52*). First was a marked decrease in mitochondrial membrane potential and time-dependent increased release of cytochrome c (*69*). Second was concurrent activation of caspase-9, a member of the caspase-3 subfamily associated with the cytochrome c-Apaf-1 complex. Third was the markedly reduced growth suppression in cells with over-expressed bcl-2, which probably resulted from the upstream action of bcl-2 to block release of cytochrome c and thus activation of caspase-3 (*70*). These mechanisms differ from traditional apoptotic anticancer drugs such as adriamycin and camptothecin (*52*). However, use of colon cancer cells with mutant p53 may be relevant only to late stages of colon cancer as TP53 mutation would not be detectable in early stages during which tumor development was able to evade the cell cycle inhibition and apoptosis induced by wild-type TP53 (*71*). Subsequent study into the molecular mechanism on HL-60 leukemia cell line showed that vaticanol C (10 μM) strongly prevented phosphorylation of proapoptotic Bcl-2 family member Bad, thus facilitating its interaction with antiapoptotic protein Bcl-xL and/or Bcl-2 to abrogate the effect of the latter and allowing apoptotic activity to dominate (*72,73*). This was mainly brought about by the potent inhibitory effect of vaticanol C on phosphorylation of two upstream regulators, extracellular signal regulated kinase (ERK) and Akt.

Inhibition of Cytochrome P450 1B1

Cytochrome P 450 (CYP) may activate many environmental mutagens and has been found to be overexpressed in different tumor cells (*74,75*). It also catalyzes the 4-hydroxylation of estrogens, which has been considered as an important step in hormonal carcinogenesis (*74*). Suffruticosol B was found to suppress CYP1B1 gene expression in HL-60 cells, thus have pharmacological potential against different cancer cells (*58*).

Anti-inflammatory Activities

Anti-inflammatory activity of oligostilbenes is mediated via multiple mechanisms. These mainly include inhibition of lipoxygenase (LOX) (*76*), cyclooxygenase (COX), inducible nitric oxide synthase (iNOS) (*77*), platelet aggregation (*78*) and release of various inflammatory mediators (*79*). Some have also been shown to possess *in vivo* anti-inflammatory effect in treatment of allergy by skin application (*26*). These imply them to be potential pharmaceuticals in peroxidation and inflammation associated diseases. Oligostilbenes demonstrating strong inhibition of COX and/or LOX activities may also be interpreted as possessing good cancer chemopreventive potential as the major pathways of AA metabolism, PG and LT synthesis are strongly associated with cell proliferation, angiogenesis and invasiveness (*80*).

Inhibition of COX and LOX

Eicosanoids, which arise from regulated oxygenation of arachidonic acid (AA) via the prostaglandin (PG) and leukotriene (LT) pathways, play a central role in mediating inflammatory reactions (*81*). Observation of the inhibitory effect of oligostilbenes on COX and LOX, the two key enzymes regulating eicosanoid synthesis, suggests their potential roles in therapy of inflammatory conditions. A trimer of resveratrol, α-viniferin has been well studied for its COXs inhibitory activity and anti-inflammatory effects. In a study conducted on the microsomal fraction prepared from sheep seminal vesicles, Lee and coworkers demonstrated that (+)-α-viniferin inhibited COX activity of prostaglandin H_2 synthase in a dose-dependent manner with IC50 value of 7 μM (*82*). The inhibitory potency was 3- to 4-fold stronger than that observed with resveratrol. In another study, Chung and coworkers evaluated the anti-inflammatory activity of α-viniferin on carrageenin-induced edema model in mice (*77*). Significant anti-inflammatory effect was observed at 30 mg/kg body weight via the oral route while that via intravenous injection occurred at 3 mg/kg. Although α-viniferin displayed weaker anti-inflammatory effect than the positive control, ibuprofen, a NSAID, chemiluminescence assay uncovered the time-dependent inhibitory effect of α-viniferin on COX-2 activity with IC_{50} value of 4.9 μM and very weak inhibition on COX-1. In addition, an inhibitory effect at transcriptional level was proposed as reflected by the reduced synthesis of COX-2 transcripts after α-viniferin treatment at concentrations of 3 to 10 μM (*77*). α-viniferin thus demonstrated good potential in anti-inflammatory therapy. They also looked into the effect of α-viniferin on production of nitric oxide (NO), a key mediator in carrageenan-induced inflammation (*83*). Concurrent with α-viniferin treatment, NO content of the murine macrophage Raw 264.7 cells was augmented by LPS stimulation. Using nitrite concentration in the supernatant as an index, α-viniferin strongly inhibited NO production with IC_{50} value of 2.7 μM. However, such effect was not observed 12 hours after LPS stimulation, precluding its inhibition of inducible nitric oxide synthase (iNOS) (*77*). RT-PCR analysis revealed α-viniferin as a down-regulator of iNOS with IC_{50} value of 4.7 μM, apparently via similar mechanism on COX-2. Therefore, α-viniferin seems to display dual roles in anti-inflammation which are likely to enhance its overall efficacy. The inhibitory mechanism is different from that of resveratrol. Although resveratrol has been reported to execute its anti-inflammatory action by interfering with NF-kB signaling pathway to suppress NO production [84], and suppressing COX-2 synthesis (*85*), it preferentially inhibited activity of COX-1 (*86*). Whether oligostilbenes constituting of more than three monomeric units are better candidates as COX-2 selective inhibitors is still a vague proposition and there are definitely other factors affecting their selectivity between COX-2 and 1. For example, in another study looking at the COX inhibitory activity of resveratrol oligomers, resveratrol (E)-dehydrodimer and its glycoside, resveratrol (E)-dehydrodimer 11-*O*-β-D-glucopyranoside were demonstrated to inhibit COX-2 with IC_{50} values of 7.5 and 3.7 μM, respectively (*87*). This implied that β-glucoside at the *meta*-position, as well as the way of connection between the monomeric units significantly affect activity of this kind of dimmers. However, these two dimmers also inhibited COX-1 with comparable IC_{50} values, which would limit their potential in research of anti-inflammation drugs. Further investigations pertaining to the relative activities of different groups of oligostilbenes and the common structural features, which account for their selective inhibitory activity on COX-2 are required to give an appropriate ranking of these oligomers.

LOX, particularly 5-LOX, is another target in studies attempting to identify potent anti-inflammatory oligostilbenes. Several resveratrol dimers, *trans*-ε-viniferin and *cis*-ε-viniferin were also found to exhibit significant inhibitory activity against LOX (*88*). In another study, Huang and coworkers more specifically evaluated the effect of ten resveratrol tetramers, (+)-hopeaphenol, isohopeaphenol, vitisin A, (+)-vitisifuran A, heyneanol A and amurensins I-L, on the biosynthesis of leukotriene B_4 (LTB_4) (*25*). At concentrations of 10^{-5} mol L^{-1}, the former five tetramers elicited strong inhibition with inhibitory rates of 56, 60, 63, 72 and 76%, respectively. Several other resveratrol oligomers such as amurensins F, α-viniferin, vaticanol B, C and G were also reported to strongly inhibit biosynthesis or release of LTB_4 under different experimental conditions (*89, 79*). This suggested a suppressive effect of these oligomers on 5-LOX. As an approach to seek an understanding of the effect of oligostilbenes on the LOX pathway in eicosanoid production, Riaz and coworkers evaluated the potency of paeoninol relative to that of paeonin C, a monoterpene galactoside (*76*). Paeoninol displayed strong inhibitory potential against LOX with an IC_{50} value of 0.77 ± 0.004 μM and is about 130 times more potent than paeonin C. This led to the proposition that greater number of phenolic groups may be critical in amplifying inhibitory effect on LOX (*76*). It has been proposed that a greater number of phenolic groups could enhance the electron donating capacity of the molecule to reduce the Fe^{3+} active form of the enzyme to its Fe^{2+} inactive form (*76*). However, there still lacks consensus on how much this structural feature contributes to the final activity of oligostilbenes. As a contradictory example, amurensins I-M displayed very weak inhibitory effect on LOX despite possessing large number of phenolic groups (*25*). Therefore, other factors such as the pattern of substitution of the hydroxyl groups may be more important determinants in certain classes of oligostilbenes.

Inhibiting Release of TNF-α and Histamine

Apart from PGs and LTs, effects of oligostilbenes on the release or activity of other inflammatory mediators like TNF-α and histamine have also been examined. Liu and coworkers reported that vaticanol (B, C and G), α-viniferin, and hopeaphenol potently inhibited release of TNF-α and histamine from bone marrow mast cells without significant influence on the cell viability (*79*). Most of them were effective under both immunological (with IgE as stimulator) and nonimmunological (with calcium ionophore A23187 as sitmulator) conditions. Some oligostilbenes have also been reported to exhibit dose-dependent suppression of TNF-α production by murine peritoneal macrophages stimulated with LPS (*90*). On the other hand, a resveratrol trimer, gnetumontanin B has been shown to be potent inhibitor of TNF-α with an IC50 value of 1.49×10^{-6} mol L^{-1} (*91*) and at a concentration of 10^{-5} mol/L, gnetupendin C and gnetin D demonstrated 59.50% and 67.23% inhibition against this inflammation factor, respectively (*92*). Gnetuhainin M (trimer), N (trimer) and R (tetramer), displayed potent activity as a histamine receptor antagonist (*34,51*).

Inhibition of blood platelets

Platelets are involved in inflammatory processes that may contribute to the development of thrombosis and atherosclerosis (*93*). Several oligostilbenes such as vitisinol A, ampelopsin C, miyabenol A, (+)-vitisin A, and (+)-vitisin C have been shown to potently inhibit arachidonic acid- and 9,11-dideoxy-11α,9α-epoxy-methanoprostaglandin $F_{2\alpha}$-induced platelet aggregation (*78*). In particular, (+)-vitisin C was most effective against AA-induced platelet aggregation with an IC_{50} value of 5.7 ± 1.3 μM, whilst (-)-viniferal was most effective against U46619-induced aggregation with an IC_{50} value of 3.1 ± 2.5μM. Moreover, their antiplatelet activities were much stronger than aspirin (*78*).

Conclusion

The large number of *Gnetum* species and the diversity of oligostilbenes that have been discovered from them offer this genus an excellent botanical source of stilbene compounds for further *in vitro* and *in vivo* studies. Based on pharmacological studies, oligostilbenes have tremendous potential in the treatment and prevention of cancers and inflammation-related diseases, as they have demonstrated superior cytotoxic, apoptotic and antiproliferative activities against various cancer cell lines, were able to suppress inflammation via inhibition of LOX and COX, and inhibit release of various inflammatory mediators and platelet aggregation. However, the bioactivity varies widely with different structures (number of monomer, etc) and only a number of compounds have been extensively studied. More in depth structure-activity relationship studies are demanded to define the structures responsible for their bioactivity and bioavailability.

References

1. Gorham, J.; Tori, M.; Asakawa, Y. In *Biochemistry of the stilbenoids;* Chapman & Hall: London, 1995; pp 1-6.
2. Hain, R.; Bieseler, B.; Kindl, H.; Schroder, G.; Stocker, R. *Plant Mol. Biol.* **1990**, *15*, 325-335.
3. Cuendet, M.; Potterat, O.; Salvi, A.; Testa, B.; Hostettmann, K. *Phytochemistry* **2000**, *54*, 871-874.
4. Chen, C.Y.; Jang, J.H.; Li, M.H.; Surh, Y.J. *Biochem. Biophys. Res. Comm.* **2005**, *331*, 993-1000.
5. Park, E. J.; Min, H. Y.; Ahn, Y. H.; Bae, C. M.; Pyee, J. H.; Lee, S. K. *Bioorg. Med. Chem. Lett.* **2004**, *14*, 5895-5898.
6. Ko, S. K.; Lee, S. M.; Whang, W. K. Anti-platelet aggregation activity of stilbene derivatives from *Rheum undulatum. Pharmacol. Res.* **1999**, *22*, 401-403.
7. Cushman, M.; Nagarathnam, D.; Gopal, D.; Chakraborti, A. K.; Lin, C. M.; Hamel, E. *J. Med. Lett.* **1991**, *34*, 2579-2588.
8. Bradamante, S.; Barenghi, L.; Villa, A. *Cardiovasc. Drug Rev.* **2004**, *22 (3)*, 169-188.
9. Olas, B.; Wachowicz, B. *Platelets* **2005**, *6*, 251-260.
10. Delmas, D.; Jannin, B.; Latruffe, N. *Mol. Nutr. Food Res.* **2005**, *49*, 377-395.
11. Ulrich, S.; Wolter, F.; Stein, J. M. *Mol. Nutr. Food Res.* **2005**, *49*, 452-461.
12. de la Lastra, C. A.; Villegas, I. *Mol. Nutr. Food Res.* **2005**, *49*, 405-430.

13. Bhat, K. P. L.; Kosmeder, J. W.; Pezzuto, J. M. *Antioxid. Redox Signal.* **2001**, *3*, 1041-1064.
14. Fremont, L. *Life Sci.* **2000**, *66*, 663-673.
15. Bianchini, F.; Vainio, H. *Eur. J. Cancer Prev.* **2003**, *12*, 417-425.
16. Aggarwal, B. B.; Bhardwaj, A.; Aggarwal, R. S.; Seeram, N. P.; Shishodia, S.; Takada, Y. *Anticancer Res.* **2004**, *24*, 3-60.
17. Langcake, P.; Pryce, F. J. *Experientia* **1977**, *33*, 151-152.
18. Kurihara, H.; Kawabata, J.; Ichikawa, S.; Mizutani, J. *Agric. Biol. Chem.* **1990**, *54 (4)*, 1097-1099.
19. Ono, M.; Ito, Y.; Kinjo, J.; Yahara, S.; Nohara, T.; Niho, Y. *Chem. Pharm. Bull.* **1995**, *43 (5)*, 868-871.
20. Dai, J. R.; Hallock, Y. F.; Cardellina, J. H.; Boyd, M. R. *J. Nat. Prod.* **1998**, *61,* 351-353.
21. Sarker, S. D.; Whiting, P.; Dinan, L. *Tetrahedron.* **1999**, *55*, 513-524.
22. Gu, H. M.; Li, W. W.; Li, B. G. *Chin. J. Appl. Env. Biol.* **2000**, *6*, 83-85.
23. Lou, H. X.; Zhao, Y.; Ren, D. M.; Fan, P. H.; Ji, M. *Chin. J. Med. Chem.* **2004**, *14*, 202-208.
24. *Kim, H. J.; Chang, E. J.; Cho, S. H.; Chug, S. K.; Park, H. D.; Choi, S. W.* Biosci. Biotechnol. Biochem. **2002**, *66*, 1990-1993.
25. Huang, K. S.; Lin, M.; Cheng, G. F. *Phytochemistry* **2001**, *58*, 357-362.
26. Iinuma, M.; Tanaka, T.; Nakaya, K.; Maruyama, H.; Araki, Y.; Sakamoto, T.; Mishima, S. *Jpn. Kokai Tokkyo Koho.* **2002**, pp 11.
27. Mishima, S.; Matsumoto, K.; Futamura, Y.; Araki, Y.; Ito. T.; Tanaka, T.; Iinuma, M.; Nozawa, Y.; Akao, Y. *J. Exp. Ther. Oncol.* **2003**, *3*, 283-288.
28. Cassidy, A.; Hanley, B.; Lamuela-Raventos, R. M. *J. Sci. Food Agric.* **2000**, *80*, 1044-1062.
29. Mialoundamy, F.; In *Tropical forest, people and food: biocultural interactions and applications to development;* Hladika, Hladik, C. M.; Linares, O. F.; Pagezy, H.; Semple, A.; Hadley, M., Eds.; Partheron publishing group: Paries, 1993; pp 177-182.
30. Iliya, I.; Yanaka, T.; Iinuma, M.; Ali, Z.; Furasawa, M.; Nakaya, Ken-ichi.; Murata, J.; Darnaedi, D. *Helv. Chim. Acta* **2002**, *85*, 2538-2546.
31. Biswas, C.; Johri, B. M. In *The Gymnosperms*; Springer-Verlag: Berlin, 1997; pp 1-12.
32. Burkill, H. M. The useful plants of West Africa. *Royal botanical garden kew.* **1994**, *2*, 168.
33. Sotheeswaran, S.; Pasupathy, V. *Phytochemistry.* **1993**, *32*, 1083-1092.
34. Huang, K. S.; Li, R. L.; Wang, Y. H.; Lin, M. *Planta Med.* **2001**, *67*, 61-64.
35. Huang, K. S.; Wang, Y. H.; Li, R. L.; Lin, M. *Phytochemistry* **2000**, *54,* 875-881.
36. Lins, A. P.; Ribiero, M. N. DeS.; Gottlieb, O. R.; Gottlieb, H. E. *J. Nat. Prod.* **1982**, *45*, 754-761.

37. Boralle, N.; Gottlieb, H. E.; Gottlieb, O. R.; Kubitzki, K.; Lopes, L. M. X.; Yoshida, M.; Young, M. C. M. *Phytochemistry* **1993**, *34*, 1403-1407.
38. Xiang, W.; Jiang, B.; Li, X. M.; Zhang, H. J.; Zhao, Q. S.; Li, S. H.; Sun, H. D. *Fitoterapia* **2002**, *73*, 40-42.
39. Iliya, I.; Yanaka, T.; Iinuma, M.; Ali, Z.; Furasawa, M.; Nakaya, Ken-ichi.; Shirataki, Y.; Murata, J.; Darnaedi, D. *Chem. Pharmacol. Bull.* **2002**, *50*, 796-801.
40. Tanaka, T.; Iliya, I.; Ito, T.; Furusawa, M.; Nakaya, Ken-ichi.; Iinuma, M.; Shirataki, Y.; Matsuura, N.; Ubukata, M.; Murata, J.; Simozono, F.; Hirai, K. *Chem. Pharm. Bull.* **2001**, *49*, 858-862.
41. Ros, Barcelo, A.; Pomar, F. In *Studies in Natural Products Chemistry,* Vol. 27 (Part II); Atta-ur-Rahman., Eds.; Elsevier Science Publishers, 2002; pp 735-792.
42. Yao, C. S.; Lin, M. *Nat. Prod. Res.* **2005**, *19*, 443-448.
43. Iliya, I.; Ali, Z.; Yanaka, T.; Iinuma, M.; Furasawa, M.; Nakaya, Ken-ichi.; Murata, J.; Darnaedi, D. *Phytochemistry*. **2002**, *61*, 959-961.
44. Iliya, I.; Yanaka, T.; Iinuma, M.; Ali, Z.; Furasawa, M.; Nakaya, Ken-ichi.; Murata, J.; Darnaedi, D. *Helv. Chim. Acta* **2002**, *85*, 2394-2402.
45. Gorham, J. In *Methods in plant biochemistry.* Vol. 1; Dey, P. M.; Harborne, J. B., Eds.; Academic press limited: London, 1989; pp 159-196.
46. Chen, H.; Lin, M. *Chin. Chem. Lett.* **1998**, *9*, 1013-1015.
47. Chen, H.; Lin, M. *Chin. Chem. Lett.* **1999**, *10*, 579-582.
48. Lin, M.; Li, J. B.; Li, S. Z.; Yu, D. Q.; Liang, X. T. *Phytochemistry* **1992**, *31*, 633-638.
49. Iliya, I.; Ali, Z.; Yanaka, T.; Iinuma, M.; Furasawa, M.; Nakaya, Ken-ichi.; Murata, J.; Darnaedi, D.; Matsuura, N.; Makoto, U. *Phytochemistry* **2003**, *62*, 601-606.
50. Li, X. M.; Wang, Y. H.; Lin, M. *Chin. Chem. Lett.* **2001**, *12*, 611-612.
51. Huang, K. S.; Zhou, S.; Lin, M.; Wang, Y. H. Planta Med. **2002**, *68*, 916-920.
52. Ito, T.; Akao, Y.; Yi, H.; Ohguchi, K.; Matsumoto, K.; Toshiyuk, T.; Iinuma, M.; Nozawa, Y. *Carcinogenesis* **2003**, *24*, 1489-1497.
53. Kim, H. J.; Chang, E. J.; Bae, S. J.; Shim, S. M.; Park, H. D.; Rhee, C. H.; Park, J. H.; Choi, S. W. *Pharmacol. Res.* **2002**, *25*, 293-299.
54. Ohyama, M.; Tanaka, T.; Ito, T.; Iimuma, M.; Bastow, K. F.; Lee, K. H. *Bioorg. Med. Chem. Lett.* **1999**, 3057-3060.
55. Kim, H. J.; Lee, W. J.; Park, Y. H.; Cho, S. H.; Choi, S. W. *J. Food Sci. Nutr.* **2003**, *8*, 356-364.
56. Ito, T.; Akao, Y.; Tanaka, T.; Iinuma, M.; Nozawa, Y. *Biol. Pharm. Bull.* **2002**, *25*, 147-148.

57. Seo, E. K.; Chai, H.; Constant, H. L.; Santisuk, T.; Reutrakul, V.; Beecher, C. W. W.; Farnsworth, N. R.; Cordell, G. A.; Pezzuto, J. M.; Kinghorn, A. D. *J. Org. Chem.* **1999**, *64*, 6976-6983.
58. Kang, J. H.; Park, Y. H.; Choi, S. W.; Yang, E. K.; Lee, W. J. *Exp. Mol. Med.* **2003**, *35*, 467-474.
59. Chowdhury, S. A.; Kishino, K.; Satoh, R.; Hashimoto, K.; Kikuchi, H.; Nishikawa, H.; Shirataki, Y.; Sakagami, H. *Anticancer Res.* **2005**, *25*, 2055-2064.
60. Hiroshi, N.; Hiroshi, H.; Kenichiro, H.; Susumu, O.; Shogo, I.; Munekazu, I.; Tanaka, T.; Ohyama, M.; Ken, T.; Kimiko, T.; Daisuke, T.; Masaji, Y. *Tennen Yuki Kagobutsu Toronkai Koen Yoshishu.* **1997**, *39*, 571-576.
61. Kulanthaivel, P.; Janzen, W. P.; Ballas, L. M.; Jiang, J. B.; Hu, C. Q.; Darges, J. W.; Seldin, J. C.; Cofield, D. J.; Adams, L. M. *Planta Med.* **1995**, *61*, 41-44.
62. Sheng, Z.; Tian, C. Y.; Wang, Y. X.; Huang, H.; Guang, X. *Shanghai Yike Daxue Xuebao.* **1999**, *26*, 395-399.
63. Mackay, H. J.; Twelves, C. J. Protein kinase C: a target for anticancer drugs? *Endocr. Relat. Cancer* **2003**, *10*, 389-396.
64. Hofmann, J. *Curr. Cancer Drug Targets* **2004**, *4*, 125-146.
65. Jason, H.; Gill and Caroline, Dive. In *Apoptosis in Toxicology*; Roberts, R., Eds.; Taylor & Francis: UK, 2000; pp. 1-19.
66. McKenna, S. L.; McGowan, A. J.; Cotter, T. G. In *Advances in Biochemical Engineering Biotechnology*; Al-Rubeai, M., Eds.; Springer: Berlin, 1998; pp. 1-32.
67. Tewari, M.; Quan, L. T.; O'Rourke, K.; Desnoyers, S.; Zeng, Z.; Beidler, D. R.; Poirier, G. G. Salvesen, G. S.; Dixit, V. M. *Cell.* **1995**, *81,* 801-809
68. Nicholson, D. W.; Ali, A.; Thornberry, N. A.; Villancout, J. P.; Ding, C. K.; Gallant, M.; Gareau, Y.; Griffin, P. R.; Labelle, M.; Lazebnik, Y. A.; Munday, N. A.; Raju, S. A.; Smulson, M. E.; Yamin, T-T.; Yu, V. L.; Miller, D. K. *Nature* **1995**, *376*, 37.
69. Harvey, N. L.; Kumar, S. In *Advances in biochemical engineering biotechnology;* Al-Rubeai, M. Eds.; Springer: Berlin, 1998; pp. 107-128.
70. Grabley, S.; Thiericke, R. In *Drug Discovery from Nature*; Grabley, S.; Thiericke, R., Eds.; Springer: Berlin, 1999; pp. 3-37.
71. Gormally, E.; Hainaut, P. In *Mechanism in Carcinogenesis and Cancer Prevention*; Harri, U.; Vainio and Eino, K. Hietanen., Eds.; Springer: Berlin, 2003; pp. 57-82.
72. Ohguchi, K.; Akao, Y.; Matsumoto, K.; Tanaka, T.; Ito, T.; Iinuma, M.; Nozawa, Y. *BMC Biochem.* **2005**, *69*, 353-356.
73. Manson, M. M.; Howells, L. M.; Hudson, E. A. In *Mechanism in Carcinogenesis and Cancer Prevention*; Harri, U.; Vainio and Eino K. Hietanen., Eds.; Springer: Berlin, 2003; pp. 41-56.

74. Peter, G. F.; Chun, Y. J.; Kim, D.; Gillam, E, M. J.; Shimada, T.; *Mutation Res.* **2003**, *523-524*, 173-182.
75. McFadyen, M.; Murray, G. I.; *Future Oncol.* **2005**, *1*, 259-263.
76. Riaz, N.; Malik, A.; Rehman, A.; Ahmed, Z.; Muhammad, P.; Nawaz, S. A.; Siddiqui, J.; Choudhary, M. I. *Phytochemistry* **2004**, *65*, 1129-1135.
77. Chung, E. Y.; Kin, B. H.; Lee, M. K.; Yun, Y. P.; Lee, S. H.; Kyung, R. M.; Kim, Y. S. *Planta Med.* **2003**, *69*, 710-714.
78. Huang, Y. L.; Tsai, W. J.; Shen, C. C.; Chen, C. C. *J. Nat. Prod.* **2005**, *68*, 217-220.
79. Liu, B. L.; Inami, Y.; Tanaka, H.; Inagaki, N.; Iinuma, M.; Nagai, H. *Zhongguo Tianran Yaowu* **2004**, *2*, 176-183.
80. Kelloff, G. J.; Sigman, C. C. In *Mechanism in Carcinogenesis and Cancer Prevention*; Harri, U.; Vainio and Eino, K. H., Eds.; Springer: Berlin, 2003; pp. 187-210.
81. Murphy, R. C.; Bowers, R. C.; Dickinson, J.; Berry, K. Z. In *The Eicosanoids*; Peter C. P., Eds.; England: John Wiley & Sons, 2004; pp. 3-16.
82. Lee, S. H.; Shin, N. H.; Kang, S. H.; Park, J. S.; Chung, S. R.; Min, K. R.; Kim, Y. S. *Planta Med.* **1998**, *64,* 204-207.
83. Salvemini, D.; Wang, Z. Q.; Wyatt, P. S.; Bourdon, D. M.; Marino, M. H.; Manning, P. T.; Currie, M. G. *Br. J. Pharmacol.* **1996**, *118*, 829-838.
84. Tsai, S. H.; Lin-Shiau, S. Y.; Lin, J. K. *Br. J. Pharmacol.* **1999**, *126*, 673-680.
85. Subbaramaiah, K.; Chung, W. J.; Michaluart, P.; Telang, N.; Tanabe, T.; Inoue, H.; Jang, M.; Pezzuto, J. M.; Dannenberg, A. J. *J. Biol. Chem.* **1998**, *273*, 21875-21882.
86. Szewczuk, L. M.; Forti, L.; Stivala, L. A.; Penning, T. M. *J. Biochem. Chem.* **2004**, *279*, 22727-22737.
87. Waffo-Teguo, P.; Lee, D.; Cuendet, M.; Merillon, J. M.; Pezzuto, J. M.; Kinghorn, A. D. *J. Nat. Prod.* **2001**, *64*, 136-138.
88. Kim, H. J.; Ha, S. C.; Choi, S. W. *Nutraceuticals and Food* **2002**, *7*, 447-450.
89. Huang, K. S.; Lin, M.; Yu, L. N.; Kong, M. *Tetrahedron* **2000**, *56*, 1321-1329.
90. Li, J.; Cheng, G. F.; Zhu, X. Y. Yaoxue Xuebao **2000**, *35*, 335-338.
91. Li, X. M.; Lin, M.; Wang, Y.; Liu, X. *Planta Med.* **2004**, *70*, 160-165.
92. Li, X. M.; Lin, M.; Wang, Y. H. *J Asian Nat Prod Res.* **2003**, *5*, 113-119.
93. Peerschke, E. I. B.; Ghebrehiwet, B. *Immunol. Rev.* **2001**, *180*, 56-64.
94. Siddiqui, Z. A.; Rahman, M.; Khan, M. A.; Zaman, A.; Lavaud, C.; Massiot, G.; Nuzillard, J. M. *Tetrahedron,* **1993**, *49*, 10393-10396.
95. Huang, K. S.; Wang, Y. H.; Li, R. L.; Lin, M. *J. Nat. Prod.* **2000**, *63*, 86-89.

96. Wang, Y. H.; Huang, K. S.; Lin, M. *J. Asian Nat. Prod. Res.* **2001**, *3*, 169-176.
97. Iliya, I.; Tanaka, T.; Furasawa, M.; Ali, Z.; Nakaya, K.; Iinuma, M.; Shirataki, Y.; Murata, J.; Darnaedi, D. *Heterocycles* **2001**, *55*, 2123-2130.
98. Zhou, L. X.; Lin, M. *J. Asian Nat. Prod. Res.* **2000**, *2*, 169-175.
99. Iliya, I.; Tanaka, T.; Iinuma, M.; Ali, Z.; Furasawa, M.; Nakaya, K. *Heterocycles* **2002**, *57*, 1057-1062.
100. Iliya, I.; Tanaka, T.; Iinuma, M.; Ali, Z.; Furasawa, M.; Nakaya, K.; Matsuura, N.; Ubukata, M. *Heterocycles* **2002**, *57*, 1507-1512.
101. Oshima, Y.; Ueno, Y. *Phytochemistry* **1993**, *33*, 179-182.
102. Iliya, I.; Ali, Z.; Yanaka, T.; Iinuma, M.; Furasawa, M.; Nakaya, Ken-ichi.; Shirataki, Y.; Murata, J.; Darnaedi, D.; Matsuura, N. *Chem. Pharmacol. Bull.* **2003**, *51*, 85-88.
103. Ali, Z.; Tanaka, T.; Iliya, I.; Iinuma, M.; Furusawa, M.; Ito, T.; Nakaya, Ken-Ichi.; Murata, J.; Darnaedi, D. *J. Nat. Prod.* **2003**, *66*, 558-560.
104. Ito, T.; Tanaka, T.; Nakaya, Ken-ichi.; Iinuma, M.; Takahashi, Y.; Naganawa, H.; Ohyama, M.; Nakanishi, Y.; Bastow, K. F.; & Lee, K. H. *Tetrahedron* **2001**, *57*, 7309-7321.
105. Ito, T.; Tanaka, T.; Nakaya, Ken-ichi.; Iinuma, M.; Takahashi, Y.; Naganawa, H.; Ohyama, M.; Nakanishi, Y.; Bastowd, K. F. Lee, K. H. *Tetrahedron Lett.* **2001**, *42*, 5909-5912.
106. Ito, T.; Tanaka, T.; Ali, Z.; Akao, Y.; Nozawa, Y.; Takahashi, Y.; Sawa, R.; Nakaya, Ken-ichi.; Murata, J.; Darnaedi, D.; Iinuma, M. *Heterocycles* **2004**, *63*, 129-136.

Chapter 5

Phytochemistry and Quality Control of Black Cohosh (*Actaea racemosa*), Medicinal *Cimicifuga* Species and Their Derived Products

Liang Zhou[1], Qing-Li Wu[2], James E. Simon[2], Chun-Yu Liu[1], Jun-Shan Yang[1], and Yong-Hong Liao[1]

[1]Institute of Medicinal Plant Development (IMPLAD), Chinese Academy of Medical Sciences and Peking Union Medical College, Xi Bei Wang, Hai Dian District, Beijing 100094, People's Republic of China
[2]New Use Agriculture and Natural Plant Products Program, Department of Plant Biology and Plant Pathology, Rutgers, The State University of New Jersey, 59 Dudley Road, New Brunswick, NJ 08901

Black Cohosh, *Actaea racemes* L. formerly known as *Cimicifuga racemosa* L., and other medicinal *Cimicifuga* species and their derived products have been used for the treatment of menopausal disorders and other conditions. Phytochemical studies on medicinal *Cimicifuga* species have led to the isolation and structural elucidation of over 180 compounds, mainly triterpene glycosides and phenolics. Numerous analytical methods have been developed for the qualitative and quantitative analysis of main bio-active compounds in black cohosh and *Cimicifuga* plants and derived products yet their analysis for quality control and quantitation remains challenging. This article reviews the advances in the phytochemistry and analytical methods concerning *Cimicifuga* species and derived products, with a particular view in summarizing the feasibility of using chemical assays for the purpose of quality control, such as the tests of identity, quantity and stability, for *Cimicifuga* herbs and derived products.

Black cohosh (*Actaea racemosa* L., formerly classified as *Cimicifuga racemosa* L.,) is a perennial herb in the Ranunculaceae family and native to North America. The roots and rhizomes have traditionally been used for gynecological complaints and rattlesnake bites by Native Americans. Currently, many black cohosh derived preparations are commercially available as either herbal medicinal products or dietary supplements for the treatment of menopausal disorders in Europe and America. In Asia, the rhizomes of other *Cimicifuga* species herbs, such as *C. dahurica, C. foetida, C. heracleifolia, C. japonica, C. acerina* and *C. simplex*, have also been used in traditional or folk medicine systems. The former three species are used in traditional Chinese medicine as "Sheng-ma" and recorded in Pharmacopoeia of the People's Republic of China (*1*) for inducing the eruption of measles at the onset of symptoms, and the treatment of headache due to pathogenic "windheat", sore throat, and uterine and rectal prolapses.

Phytochemical studies on medicinal *Cimicifuga* species have led to the isolation and structural elucidation of over 180 compounds, many being triterpene glycosides and phenolics. However, despite the extensive number of pharmacological and clinical studies which have been conducted on these species, neither the active principles nor the mode of action of black cohosh and other related species are well understood. As a result, the commercial black cohosh products are generally standardized based on the total content of triterpene glycosides, calculated as 23-epi-26-deoxyactein (**117**), whereas ferulic acid (**168**) and isoferulic acid (**169**) are utilized as markers for qualitative and quantitative analysis of "Shengma" as described in Pharmacopoeia of the People's Republic of China (*1*). The present paper focuses on the advances in the phytochemistry and analytical methods concerning black cohosh and *Cimicifuga* species and derived products, with a particular view in providing an assessment on the various methods used in the determination of quality and natural product content. In this review, we use the former name of *Cimicifuga* rather than *Actaea* to avoid confusion with the published literature.

Chemical Components from Medicinal *Cimicifuga* Species

Many types of phytochemical compounds including cyclolanstane triterpenes and their glycosides, aromatic acids and related derivatives, chromones and other compounds, have been isolated and elucidated from *Cimicifuga* plants.

Cyclolanstane Triterpenes and Their Glycosides

Triterpenes and their glycosides have been regarded as basic and characteristic components in *Cimicifuga* plants and have drawn considerable attention. Approximately 150 compounds (**1-150**) of this type (Table I, for structures of compounds, see Scheme 1) have been isolated from *Cimicifuga* species, mainly coming from the rhizomes of *C. racemosa*, *C. simplex*, *C. foetida* and *C. dahurica*. All of these compounds belong to 9,19-cyclolanostane triterpenes, with a 9,19-cyclopropane in ring B and a side-chain being generally highly oxygenated. In addition, the sugar moieties of the glycosides are mostly xylose, together with glucose, arabinose and galactose. The largest group of cyclolanstane triterpenes reported contain cimigenol (**1**) as an aglycone. Of these type of compounds, cimigenol xyloside (**2**), detected in many *Cimicifuga* species, has often been referred to in many published papers as cimicifugoside. However, in our review, cimicifugoside is referred to compound **126**.

The triterpene glycoside containing extracts of *C. foetide* and *C. racemosa* have been extensively studied *in vitro*, *in vivo* and some clinical trials, in particular on their role in the relief of menopausal symptoms (see review *50*). In addition, purified triterpene glycosides from *Cimicifuga* plants have also been reported to show various biological activities. For example, in an early study on the endocrine effects of the ingredients of *C. racemosa*, actein (**125**) and cimigenol xyloside (**2**) were found to reduce the serum luteinizing hormone levels *in vivo* (*51*). In another study, cimigenol (**1**) and some related cycloartane triterpenoids appeared to inhibit Epstein–Barr virus early antigen (EBV-EA) activation in Raji cells (*52*). Purified triterpene glycosides were also found to display cytotoxicities (*12,53,54*), anti-HIV activity (*55*), antimalarial activity and nucleoside transport inhibitory activity (*56*).

Aromatic Acids and Related Derivatives

Six aromatic acids (**165-170**) and 15 derived esters (**151-164, 185**) have presently been documented from *Cimicifuga* plants (Table II). These compounds have been demonstrated to show various biological activities including antioxidant (*60,61*), anti-inflammatory (*62*), vasoactivity (*63*), estrogenic activity (*64*), antihyperglycemic activity (*65*), and inhibiting neutrophil elastase activity and collagenolysis (*66,67*). The phenolic acids, in particular, ferulic acid (**168**) and isofurulic acid (**169**), are generally regarded as active principles of *Cimicifuga* herbs used as anti-inflammatory medicines.

Chromones and Other Compounds

Ten chromones (**171-180**) (Table III) and several alkaloids (**181-184**) (*73,74,75*) have also been isolated from *Cimicifuga* species. A new cyclic guanidine alkaloid, cimipronidine (**181**), isolated from the fraction of *C.*

Table I. Cyclolanostane Triterpenes and Their Glycosides from *Cimicifuga* Species

No.	*Compound*	*Formula (MW)*	*Source*	*Ref.*
1	cimigenol	$C_{30}H_{48}O_5$ (488)	a,b	*2,3,4*
2	cimigenol xyloside	$C_{35}H_{56}O_9$ (620)	b,c,g	*3,5,6,7*
3	25-*O*-acetylcimigenoside	$C_{37}H_{58}O_{10}$ (662)	a,b,c,e,g	*5,6,7,8,9*
4	25-*O*-acetylcimigenol-3-*O*-*β*-D-glucopyranosyl-(1→3)-*β*-D-xylopyranoside	$C_{43}H_{68}O_{15}$ (824)	e	*9*
5	12*β*-hydroxycimigenol-3-*O*-*β*-D-xylopyranoside	$C_{35}H_{56}O_{10}$ (636)	b,c,e	*6,10,11*
6	25-*O*-acetyl-12*β*-hydroxycimigenol-3-*O*-*α*-L-arabinopyranoside	$C_{37}H_{58}O_{11}$ (678)	g	*12*
7	12*β*,21-dihydroxycimigenol-3-*O*-*α*-L-arabinopyranoside	$C_{35}H_{56}O_{11}$ (652)	g	*12*
8	16,23*R*:16,24*S*-diepoxy-3*β*,12*β*,15*α*,25-tetrahydroxy-cycloart-7-ene-3-*O*-*α*-L-arabinopyranoside	$C_{35}H_{54}O_{10}$ (634)	e	*13*
9	3-*O*-*α*-L-arabinopyran-cimigenol-15-*O*-*β*-D-glupyranoside	$C_{41}H_{60}O_{14}$ (776)	b	*4*
10	12*β*-hydroxycimigenol-3-*O*-*α*-L-arabinopyranoside	$C_{35}H_{56}O_{10}$ (636)	c,e,g	*10,11,12*
11	7*β*-hydroxycimigenol-3-*O*-*β*-D-xylopyranoside	$C_{35}H_{56}O_{10}$ (636)	b, e	*4,11*
12	1*α*-hydroxycimigenol-3-*O*-*β*-D-xylopyranoside	$C_{35}H_{56}O_{10}$ (636)	e	*14*
13	1*α*-hydroxycimigenol-3-*O*-*β*-D-galbinopyranoside	$C_{36}H_{58}O_{11}$ (656)	e	*14*
14	25-*O*-acetyl-7*β*-hydroxycimigenol-3-*O*-*β*-D-xylopyranoside	$C_{37}H_{58}O_{11}$ (678)	e	*11*
15	25-*O*-acetylcimigeol	$C_{32}H_{50}O_6$ (530)	f	*15*
16	7,8-didehydrocimigenol-3-*O*-*β*-D-galbinopyranoside	$C_{36}H_{56}O_{10}$ (648)	e	*14*
17	cimigenol-3-*O*-*β*-D-galbinopyranoside	$C_{36}H_{58}O_{10}$ (650)	c, e	*10,14*
18	25-*O*-methylcimigenol-3-*O*-*β*-D-galbinopyranoside	$C_{37}H_{60}O_{10}$ (664)	e	*16*
19	25-*O*-acethylcimigenol-3-*O*-*β*-D-galbinopyranoside	$C_{38}H_{60}O_{11}$ (692)	e	*16*
20	25-*O*-acethylcimigenol-3-*O*-*β*-D-glubinopyranoside	$C_{38}H_{60}O_{11}$ (692)	e	*16*

Table I. *Continued.*

No.	*Compound*	*Formula (MW)*	*Source*	*Ref.*
21	cimiside B	$C_{40}H_{64}O_{13}$ (752)	b	*6*
22	25-*O*-acetyl-1α-hydroxycimigenol-3-*O*-β-D-xylopyranoside	$C_{37}H_{58}O_{11}$ (678)	e	*17*
23	1α-hydroxycimigenol-3-*O*-β-D-xylopyranoside	$C_{35}H_{56}O_{10}$ (636)	e	*17*
24	22-hydroxycimigenol-3-*O*-β-D-xylopyranoside	$C_{35}H_{56}O_{10}$ (636)	d	*18*
25	25-*O*-acetylcimigenol-3-*O*-β-D-xylopyranoside	$C_{36}H_{58}O_9$ (634)	a, b, c,g	*8,19*
26	25-*O*-methylcimigenol-3-*O*-α-L-arabinopyranoside	$C_{36}H_{58}O_9$ (634)	g	*12,20*
27	cimigenol-3-*O*-β-D-glubinopyranoside	$C_{36}H_{58}O_{10}$ (650)	c	*5*
28	cimiracemoside A	$C_{35}H_{56}O_{10}$ (636)	g	*21*
29	cimiracemoside B	$C_{35}H_{56}O_{10}$ (636)	g	*21*
30	cimiracemoside C	$C_{35}H_{56}O_9$ (620)	g	*7,21*
31	cimiracemoside D	$C_{37}H_{58}O_{11}$ (678)	g	*21*
32	25-*O*-acetylcimigenol-3-*O*-β-D-glucopyranosyl-(1→2)-β-D-xylopyranoside	$C_{43}H_{68}O_{15}$ (824)	c	*22*
33	cimigenol-3-*O*-β-D-glucopyranosyl-(1→3)-β-D-xylopyranoside	$C_{41}H_{66}O_{14}$ (782)	c	*22*
34	cimigenol-3-*O*-β-D-glucopyranosyl-(1→2)-β-D-xylopyranoside	$C_{41}H_{66}O_{14}$ (782)	c	*23*
35	cimigenol-3-*O*-β-D-glucopyranosyl-(1→2) -β-D-glucopyranosyl-(1→2)-β-D-xylopyranoside	$C_{47}H_{76}O_{19}$ (944)	c	*22*
36	12β-hydroxycimigenol-3-*O*-β-D-galactopyranoside	$C_{36}H_{58}O_{11}$ (666)	c	*10*
37	Cimigol	$C_{35}H_{48}O_5$ (488)	a,b,c	*24*
38	24-*epi*-7,8-didehydrocimigenol	$C_{30}H_{46}O_5$ (486)	f	*15*
39	3-keto-24-*epi*-7,8-didehydrocimigenol	$C_{30}H_{44}O_5$ (484)	f	*15*
40	2′,4′-*O*-diacetyl-24-*epi*-7,8-didehydrocimignol-3-*O*-β-D-xylopyranoside	$C_{39}H_{58}O_{11}$ (702)	f	*15*

Continued on next page.

Table I. *Continued.*

No.	*Compound*	*Formula (MW)*	*Source*	*Ref.*
41	3′-*O*-acetyl-24-*epi*-7,8-didehydrocimigenol-3-*O*-*β*-D-xylopyranoside	$C_{37}H_{56}O_{10}$ (660)	f	*15*
42	24-*epi*-7,8-didehydrocimigenol-3- *O*-*β*-D-xylopyranoside	$C_{35}H_{54}O_9$ (618)	f	*15*
43	7,8-didehydrocimigenol-3-*O*-*α*-L-arabinopyranoside	$C_{35}H_{54}O_9$ (618)	e	*14*
44	25-*O*-acetyl-7,8-didehydrocimigenol-3-*O*-*β*-D-xylopyranoside	$C_{37}H_{56}O_{10}$ (660)	b,e	*4,14*
45	25-*O*-acetyl-7,8-didehydrocimigenol-3-*O*-*α*-L-arabinopyranoside	$C_{37}H_{56}O_{10}$ (660)	e	*14*
46	25-anhydrocimigenol-3-*O*-*β*-D-xylopyranoside	$C_{35}H_{54}O_8$ (602)	c,g	*5,7,25*
47	7,8-didehydrocimigenol	$C_{30}H_{46}O_5$ (486)	b, f	*4,15*
48	7,8-didehydrocimigenol-3-*O*-*β*-D-xylopyranoside	$C_{35}H_{54}O_9$ (618)	b, e	*4,14*
49	25-*O*-acetyl-7,8-didehydrocimigeol	$C_{32}H_{48}O_6$ (528)	b, f	*4,15*
50	25-anhydrocimigenol-3-*O*-*β*-D-galactopyranoside	$C_{36}H_{56}O_9$ (632)	c	*22*
51	25-anhydrocimigenol-3-*O*-*α*-L-arabinopyranoside	$C_{35}H_{54}O_8$ (602)	g	*7*
52	cimiracemoside J	$C_{37}H_{56}O_{10}$ (660)	g	*7*
53	cimiracemoside K	$C_{37}H_{56}O_{10}$ (660)	g	*7*
54	25-*O*-methyl-24-*O*-acetylhydroshengmanol-3-*O*-*β*-D-xylopyranoside	$C_{38}H_{62}O_{11}$ (694)	e	*26,27*
55	25-*O*-methyl-7*β*-hydroxy-24-*O*-acetylhydroshengmanol-3-*O*-*β*-D-xylopyranoside	$C_{38}H_{62}O_{12}$ (710)	e	*26*
56	25-*O*-methyl-1*α*-hydroxy-24-*O*-acetylhydroshengmanol-3-*O*-*β*-D-xylopyranoside	$C_{38}H_{62}O_{12}$ (710)	e	*26*
57	7*β*-hydroxy-24-*O*-acetylhydroshengmanol-3-*O*-*β*-D-xylopyranoside	$C_{37}H_{60}O_{12}$ (696)	e	*16*
58	7,8-didehydro-24-*O*-acetylhydroshengmanol-3-*O*-*β*-D-xylopyranoside	$C_{37}H_{58}O_{11}$ (678)	f	*15*

Table I. *Continued.*

No.	*Compound*	*Formula (MW)*	*Source*	*Ref.*
59	24-*O*-acetyl-7,8-didehydroshengmanol-3-*O*-α-L-arabinopyranoside	$C_{37}H_{58}O_{11}$ (678)	e	*14*
60	24-*O*-acetyl-25-*O*-methyl-7,8-didehydroshengmanol-3-*O*-β-D-xylopyranoside	$C_{38}H_{60}O_{11}$ (692)	e	*14*
61	3-arabinosyl-24-*O*-acetylhydroshengmanol-15-*O*-β-D-glucopyranoside	$C_{43}H_{70}O_{16}$ (842)	b	*28*
62	3-xylosyl-24-*O*-acetylhydroshengmanol-15-*O*-β-D-glucopyranoside	$C_{43}H_{70}O_{16}$ (842)	b	*28*
63	24-*O*-acetylhydroshengmanol-3-*O*-β-D-xylopyranoside	$C_{37}H_{60}O_{11}$ (680)	g	*7*
64	24-*epi*-7β-hydroxy-24-*O*-acetylhydroshengmanol-3-*O*-β-D-xylopyranoside	$C_{37}H_{60}O_{12}$ (696)	e	*26*
65	24-*epi*-24-*O*-acetyl-7,8-didehydroshengmanol-3-*O*-β-D-galopyranoside	$C_{38}H_{60}O_{12}$ (708)	e	*16*
66	24-*epi*-24-*O*-acetyl-7,8-didehydroshengmanol-3-*O*-β-D-xylopyranoside	$C_{37}H_{58}O_{11}$ (678)	e	*14*
67	24-*epi*-24-*O*-acetyl-7,8-didehydroshengmanol-3-*O*-α-L-arabinopyranoside	$C_{37}H_{58}O_{11}$ (678)	e	*26*
68	24-*epi*-24-*O*-acetyl-7,8-didehydroshengmanol-3-*O*-(2′-*O*-malonyl)-β-D-xylopyranoside	$C_{40}H_{60}O_{14}$ (764)	e	*27*
69	24-*epi*-24-*O*-acetylhydroshengmanol-3-*O*-β-D-galopyranoside	$C_{38}H_{62}O_{12}$ (710)	e	*16*
70	cimiside C	$C_{43}H_{70}O_{16}$ (842)	b	*29*
71	cimiside D	$C_{43}H_{70}O_{16}$ (842)	b	*29*
72	24-*O*-acetylshengmanol-3-*O*-β-D-glucopyranosyl-(1$\rightarrow$2)-β-D-xylopyranoside	$C_{43}H_{70}O_{16}$ (842)	c	*30*

Continued on next page.

Table I. *Continued.*

No.	*Compound*	*Formula (MW)*	*Source*	*Ref.*
73	23-*O*-acetylshengmanol-3-*O*-*β*-D-xylopyranoside	$C_{37}H_{58}O_{10}$ (662)	d,e,g	*9,17;18, 19*
74	23-*O*-acetyl-7,8-didehydroshengmanol-3-*O*-*β*-D-galopyranoside	$C_{38}H_{58}O_{11}$ (690)	e	*14*
75	23-*O*-acetyl-7,8-didehydroshengmanol-3-*O*-(2′-*O*-malonyl)-*β*-D-xylopyranoside	$C_{40}H_{58}O_{13}$ (746)	e	*27*
76	23-*O*-acetyl-shengmanol-3-*O*-(2′-*O*-malonyl)-*β*-D- xylopyranoside	$C_{40}H_{60}O_{13}$ (748)	e	*27*
77	23-*O*-acetyl-7,8-didehydroshengmanol-3-*O*-*β*-D-xylopyranoside	$C_{37}H_{56}O_{10}$ (660)	e	*27*
78	23-*O*-acetylshengmanol-3-*O*-*β*-D-glucopyranosyl-(1→3)-*β*-D-xylopyranoside	$C_{43}H_{68}O_{15}$ (824)	e	*9*
79	23-*O*-acetyl-7,8-didehydroshengmanol-3-*O*-*α*-L-arabinopyranoside	$C_{37}H_{56}O_{10}$ (660)	e	*16*
80	23-*O*-acetyl-shengmanol-3-*O*-*α*-L-arabinopyranoside	$C_{37}H_{58}O_{10}$ (662)	g	*12;7*
81	23-*O*-acetyl-1*α*-hydroxyshengmanol-3-*O*-*β*-D-xylopyranoside	$C_{37}H_{58}O_{11}$ (678)	e	*17*
82	7*β*-hydroxy-23-*O*-acetylshengmanol-3-*O*-*β*-D-xylopyranoside	$C_{37}H_{58}O_{11}$ (678)	e	*31*
83	cimiracemoside L	$C_{39}H_{60}O_{11}$ (704)	g	*7*
84	cimiracemoside M	$C_{39}H_{60}O_{11}$ (704)	g	*7*
85	cimicifugoside H-1	$C_{35}H_{52}O_{9}$ (616)	g, h	*7,32*
86	bugbanoside D	$C_{37}H_{54}O_{11}$ (674)	e	*13*
87	bugbanoside E	$C_{37}H_{54}O_{10}$ (658)	e	*13*
88	cimifugoside H-5	$C_{35}H_{52}O_{10}$ (632)	h	*32*
89	cimicidanol-3-*O*-arabinoside	$C_{35}H_{52}O_{9}$ (616)	c	*32*
90	cimicidanol	$C_{30}H_{44}O_{5}$ (484)	c	*25*
91	cimicifol	$C_{37}H_{54}O_{10}$ (658)	c	*25*
92	(20R,24R)-24,25-epoxy-3*β*-(*β*-D-xylopyranosyloxy)-9,19-cyclolanost-7-ene-16,23-dione	$C_{35}H_{52}O_{8}$ (600)	b	*33*

Table I. *Continued.*

No.	*Compound*	*Formula (MW)*	*Source*	*Ref.*
93	(20R,24R)-24,25-epoxy-15α-hydroxy-3β-(β-D-xylopyranosyloxy)-9,19-cyclolanost-7-ene-16,23-dione	$C_{35}H_{52}O_8$ (616)	b	*33*
94	cimicifugoside H-2	$C_{35}H_{54}O_{10}$ (634)	c, g, h	*7,25,32*
95	bugbanoside C	$C_{37}H_{56}O_{12}$ (692)	e	*13*
96	15α-hydroxycimicidol-3-*O*-β-D-xylbinopyranoside	$C_{35}H_{54}O_{11}$ (650)	c, h	*25,32*
97	24-hydroxy-12β-acetoxy-25,26,27-trinorcycloart-16,23-dione-3-*O*-α-L-arabinopyranoside	$C_{34}H_{52}O_{10}$ (620)	b	*4*
98	cimifugoside H-3	$C_{32}H_{48}O_9$ (576)	h	*34*
99	dahurinol	$C_{30}H_{48}O_5$ (488)	b, c	*35,36*
100	cimiracemoside E	$C_{35}H_{58}O_{10}$ (662)	g	*21*
101	isodahurinol	$C_{30}H_{48}O_5$ (488)	b, c	*35,36*
102	25-*O*-methylisodahurinol	$C_{31}H_{50}O_5$ (502)	b, c	*35,36*
103	dehydroxydahurinol	$C_{30}H_{46}O_4$ (470)	b, c	*35,36*
104	24-*O*-acetylisodahurinol-3-*O*-β-D-galactopyranoside	$C_{38}H_{60}O_{11}$ (692)	c	*22*
105	shengmanol xyloside	$C_{35}H_{58}O_{10}$ (638)	a, b, c	*37*
106	cimiaceroside A	$C_{35}H_{54}O_9$ (618)	a, g	*19,38*
107	cimiaceroside B	$C_{35}H_{56}O_9$ (620)	a	*38*
108	2′-*O*-malonylcimiaceroside B	$C_{38}H_{58}O_{12}$ (706)	e	*27*
109	cimiracemoside F	$C_{37}H_{56}O_{11}$ (676)	g	*19,21*
110	cimiracemoside G	$C_{37}H_{56}O_{11}$ (676)	g	*21*
111	cimiracemoside H	$C_{37}H_{58}O_{11}$ (678)	g	*21*
112	20(*S*),22(*R*),23(*S*),24(*R*)-16β:23;22:25-diepoxy-12β-acetoxy-3β,23,24-trihydroxy-9,19-cycloanostane-3-*O*-α-L-arabinopyranoside	$C_{37}H_{58}O_{11}$ (678)	g	*12*
113	20(*S*),22(*R*),23(*S*),24(*R*)-16β:23;22:25-diepoxy-3β,23,24-triol-9,19-cycloanostane-3-*O*-β-D-glucopyranosyl-(1→2)-β-D-glucopyranosyl-(1→2)-β-D-xylopyranoside	$C_{47}H_{76}O_{19}$ (944)	h	*39*

Continued on next page.

Table I. *Continued.*

No.	*Compound*	*Formula (MW)*	*Source*	*Ref.*
114	20(*S*),22(*R*),23(*S*),24(*R*)-16*β*:23;22:25-diepoxy-3*β*,23,24-triol-9,19-cycloanostane-3-*O*-(6-*O*-trans-isoferuloyl-*β*-D-glucopyranosyl)-(1→2)-*β*-D-glucopyranosyl-(1→2)-*β*-D-xylopyranoside	$C_{57}H_{84}O_{21}$ (1120)	h	*39*
115	27-deoxyacteol	$C_{32}H_{48}O_6$ (528)	c	*25*
116	acteol-3-*O*-arabinopyranoside	$C_{37}H_{56}O_{11}$ (676)	c	*25*
117	23-*epi*-26-deoxyactein	$C_{37}H_{56}O_{10}$ (660)	c, g	*29,32,40*
118	cimiracemoside I	$C_{35}H_{52}O_8$ (600)	g	*7*
119	cimiracemoside N	$C_{37}H_{56}O_{10}$ (660)	g	*7*
120	26-deoxycimicifugoside	$C_{37}H_{54}O_{10}$ (658)	e, g	*7,14*
121	2′-*O*-malonylcimicifugoside	$C_{40}H_{60}O_{14}$ (764)	e	*14*
122	2′-*O*-acetylactein	$C_{39}H_{58}O_{11}$ (702)	c, g	*7,41*
123	2′-*O*-acetyl-26-deoxyactein	$C_{39}H_{58}O_{10}$ (686)	c	*41*
124	26- deoxyactein	$C_{37}H_{56}O_{10}$ (660)	g, h	*19,32,40*
125	actein	$C_{37}H_{56}O_{11}$ (676)	e, g, h, c	*5,32,40, 42*
126	cimicifugoside	$C_{37}H_{54}O_{11}$ (674)	e, g	*42*
127	bugbanoside A	$C_{35}H_{52}O_{10}$ (632)	e	*43*
128	bugbanoside B	$C_{35}H_{54}O_{10}$ (634)	e	*43*
129	cimiracemoside O	$C_{39}H_{58}O_{12}$ (718)	g	*7*
130	cimiracemoside P	$C_{37}H_{54}O_{11}$ (674)	g	*7*
131	acerionol	$C_{30}H_{46}O_5$ (486)	a, b, c	*44*
132	24-*O*-acetylacerionol	$C_{32}H_{48}O_6$ (528)	a, b, c	*44*
133	heracleifolinol	$C_{32}H_{50}O_7$ (546)	f	*15*
134	cimicinol	$C_{35}H_{52}O_8$ (600)	c	*25*
135	24-*epi*-acerinol	$C_{30}H_{46}O_5$ (486)	f	*15*
136	acerinol	$C_{30}H_{46}O_5$ (486)	a	*37*
137	25-*O*-methylacerinol	$C_{31}H_{48}O_5$ (500)	a	*37*
138	foetidinol	$C_{27}H_{40}O_5$ (444)	c	*25,45*
139	16*α*,24*α*-dihydroxy-12*β*-acetoxy-25,26,27-trinor-16,24-cycloartan-23-one-3-*O*-*α*-L-arabinopyranoside	$C_{34}H_{52}O_{10}$ (620)	b	*4*
140	cimicifugoside H-4	$C_{32}H_{48}O_9$ (576)	c, h	*25,34,46*
141	cimicifugoside H-6	$C_{32}H_{48}O_{10}$ (592)	c, h	*25,34*
142	cimilactone A	$C_{33}H_{50}O_9$ (590)	b	*47*
143	cimilactone B	$C_{33}H_{48}O_9$ (588)	b	*47*

Table I. *Continued.*

No.	*Compound*	*Formula (MW)*	*Source*	*Ref.*
144	12β-acetoxy-15-oxo-shengmanol-3-*O*-*β*-D-xylopyranoside	$C_{37}H_{58}O_{12}$ (694)	b	*33*
145	12β-acetoxy-15-oxo-7,8-didehydroshengmanol-3-*O*-*β*-D-xylopyranoside	$C_{37}H_{56}O_{12}$ (692)	b	*33*
146	(23*R*,24*S*)-15-oxo-16-enol-9,19-cyclolanstane-3-*O*-*β*-D-xylopyranoside	$C_{35}H_{56}O_{10}$ (636)	b	*48*
147	(23*R*,24*S*)-15-oxo-16-enol-9,19-cyclolanstane-7-ene-3-*O*-*β*-D-xylopyranoside	$C_{35}H_{54}O_{10}$ (634)	b	*48*
148	(23*R*,24*S*)-12β-acetoxy-15-oxo-16-enol-9,19-cyclolanstane-3-*O*-*β*-D-xylopyranoside	$C_{37}H_{58}O_{12}$ (694)	b	*48*
149	(23*R*,24*S*)-12β-acetoxy-15-oxo-16-enol-9,19-cyclolanstane-7-ene-3-*O*-*β*-D-xylopyranoside	$C_{37}H_{56}O_{12}$ (692)	b	*48*
150	actaeaepoxide-3-*O*-*β*-D-xylopyranoside	$C_{35}H_{58}O_{10}$ (638)	g	*49*

a: *Cimicifuga acerina* b: *Cimicifuga dahurica* c: *Cimicifuga foetida*
d: *Cimicifuga japonica* e: *Cimicifuga simplex* f: *Cimicifuga heracleifolia*
g: *Cimicifuga racemosa* h: Cimicifuga Rhizome species unidentified.

racemosa roots showing 5-HT7 receptor binding activity, was proposed as a potentially significant marker for black cohosh (*28*).

Quality Control of *Cimicifuga* Species and Derived Products

The modern definition of drug and drug product quality requires that the product meet strict standards of identity, potency, purity, stability and bioavailability (*76*). Toward this, recent guidelines on the quality of herbal medicinal products (Evidence for quality of finished natural health products, Health Canada, http://www.hc-sc.gc.ca/hpfbdgspa/nhpd-dpsn/regs_cg2_sp_e.html; Guidance on quality of herbal medicinal Products, EMEA, http://www.emea.eu.int/pdfs/human/qwp/281900en.pdf) also require that all finished herbal medicinal products should ensure the identity, quantity/potency, purity and stability by proper testing methods. Numerous analytical

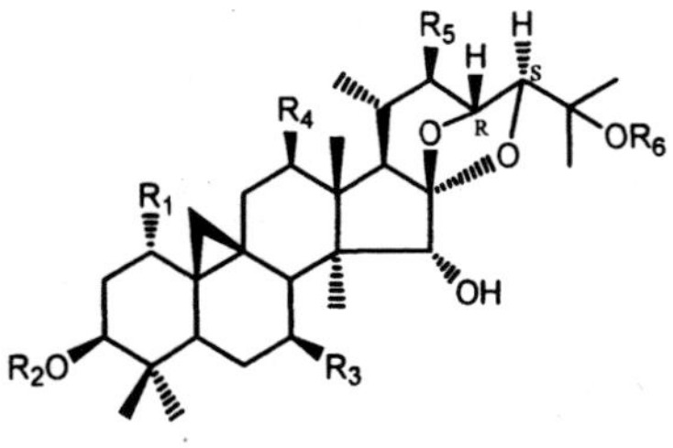

1-36

	R1	R2	R3	R4	R5	R6
1	H	H	H	H	H	H
2	H	β-D-xyl(p)	H	H	H	H
3	H	β-D-xyl(p)	H	H	H	Ac
4	H	β-D-glu(p)-(1→3)-β-D-xyl(p)	H	H	H	Ac
5	H	β-D-xyl(p)	H	OH	H	H
6	H	α-L-ara(p)	H	OH	H	Ac
7	H	α-L-ara(p) (21-CH_2OH)	H	OH	H	H
8	H	α-L-ara(p)	$\Delta^{7,8}$	OAc	H	H
9	H	α-L-ara(p) (15-*O*-β-D-glu)	H	H	H	H
10	H	α-L-ara(p)	H	OH	H	H
11	H	β-D-xyl(p)	OH	H	H	H
12	OH	α-L-ara(p)	H	H	H	H
13	OH	β-D-gal(p)	H	H	H	H
14	H	β-D-xyl(p)	OH	H	H	Ac
15	H	H	H	H	H	Ac
16	H	β-D-gal(p)	$\Delta^{7,8}$	H	H	H
17	H	β-D-gal(p)	H	H	H	H
18	H	β-D-gal(p)	H	H	H	Me
19	H	β-D-gal(p)	H	H	H	Ac
20	H	β-D-xyl(p)	H	H	H	Ac
21	H	β-D-xyl(p)-(1→3)-β-D-xyl(p)	H	H	H	H
22	OH	β-D-xyl(p)	H	H	H	Ac
23	OH	β-D-xyl(p)	H	H	H	H
24	H	β-D-xyl(p)	H	H	OH	H
25	H	α-L-ara(p)	H	H	H	Ac
26	H	α-L-ara(p)	H	H	H	Me
27	H	β-D-glu(p)	H	H	H	H
28	H	α-L-ara(p) (21-CH_2OH)	H	H	H	H
29	H	β-D-xyl(p) (21-CH_2OH)	H	H	H	H
30	H	α-L-ara(p)	H	H	H	H
31	H	α-L-ara(p)	H	OAc	H	H
32	H	β-D-glu(p)-(1→2)-β-D-xyl(p)	H	H	H	Ac
33	H	β-D-glu(p)-(1→3)-β-D-xyl(p)	H	H	H	H
34	H	β-D-glu(p)-(1→2)-β-D-xyl(p)	H	H	H	H
35	H	β-D-glu(p)-(1→2)-β-D-glu(p)-(1→2)-β-D-xyl(p)	H	H	H	H
36	H	β-D-gal(p)	H	OH	H	H

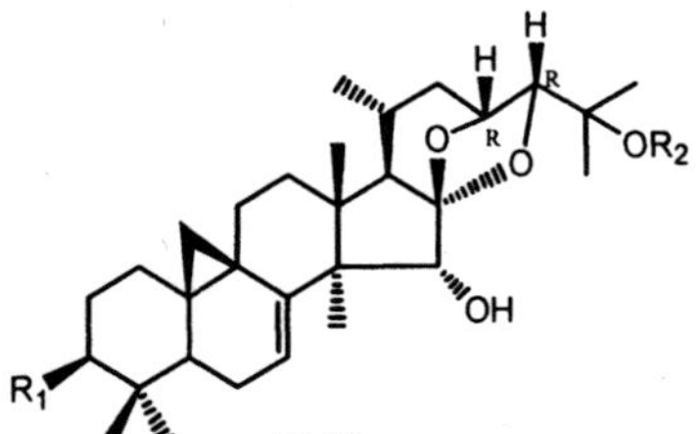

37-45

	R1	R2
37	OH 7,8-dihydro	H
38	OH	H
39	=O	H
40	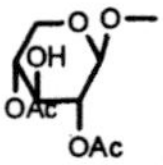	H
41	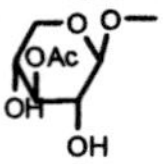	H
42	β-D-xyl(p)	H
43	α-L-ara(p) (24*S*)	H
44	β-D-xyl(p) (24*S*)	OAc
45	α-L-ara(p) (24*S*)	OAc

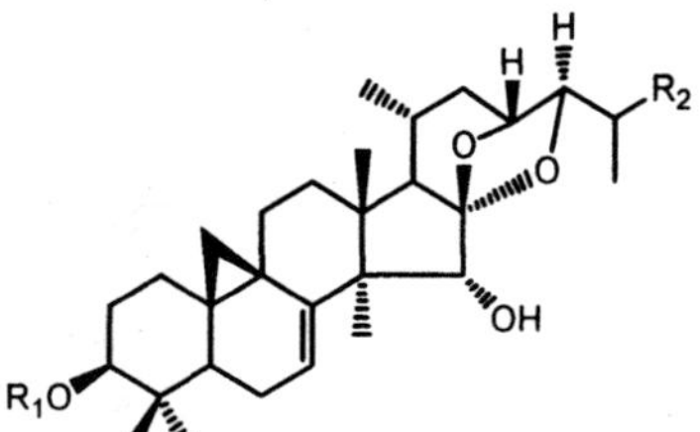

46-53

	R1	R2
46	β-D-xyl(p) 7,8-2H	$=CH_2$
47	H	
48	β-D-xyl(p)	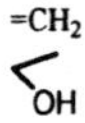
49	H	
50	β-D-gal(p)	$=CH_2$
51	α-L-ara(p) 7,8-2H	$=CH_2$
52	α-L-ara(p) 12β-OAc	$=CH_2$
53	β-D-xyl(p) 12β-OAc	$=CH_2$

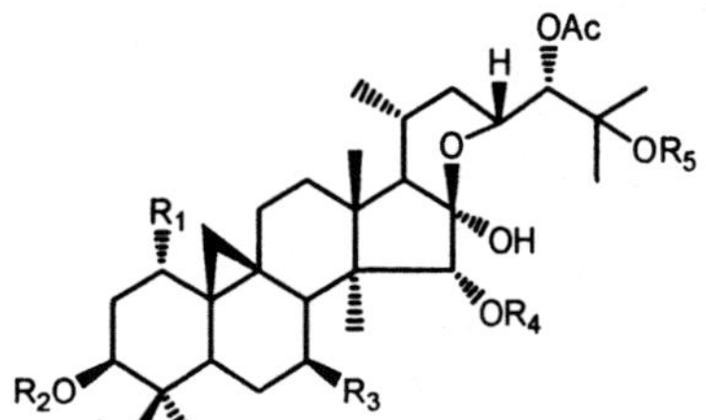

54-63

	R1	R2	R3	R4	R5
54	H	β-D-xyl(p)	H	H	Me
55	H	β-D-xyl(p)	OH	H	Me
56	OH	β-D-xyl(p)	H	H	Me
57	H	β-D-xyl(p)	OH	H	H
58	H	β-D-xyl(p)	$\Delta^{7,8}$	H	H
59	H	α-L-ara(p)	$\Delta^{7,8}$	H	H
60	H	β-D-xyl(p)	$\Delta^{7,8}$	H	Me
61	H	α-L-ara(p)	H	β-D-glu(p)	H
62	H	β-D-xyl(p)	H	β-D-glu(p)	H
63	H	β-D-xyl(p)	OH	H	H

Scheme 1. Structures of compounds identified in black cohosh.

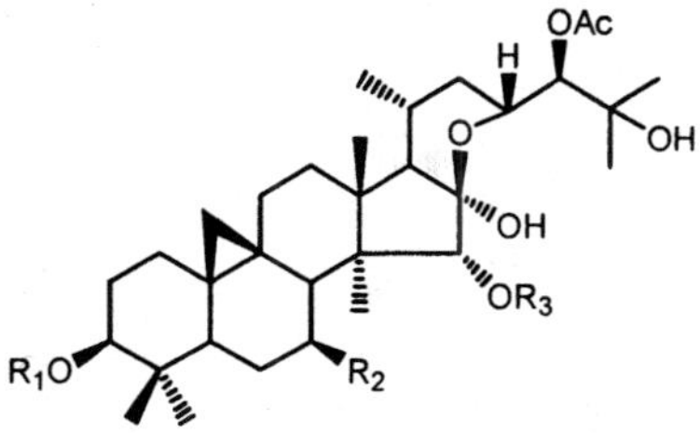

64-72

	R$_1$	R$_2$	R$_3$
64	β-D-xyl(p)	OH	H
65	β-D-gal(p)	$\Delta^{7,8}$	H
66	β-D-xyl(p)	$\Delta^{7,8}$	H
67	α-L-ara(p)	$\Delta^{7,8}$	H
68	(2-*O*-malonyl)β-D-xyl(p)	$\Delta^{7,8}$	H
69	α-L-ara(p)	H	H
70	β-D-xyl(p)	H	β-D-glu(p)
71	β-D-lyl(p)	H	β-D-glu(p)
72	β-D-glu(p)-(1→2)-β-D-xyl(p)	H	H

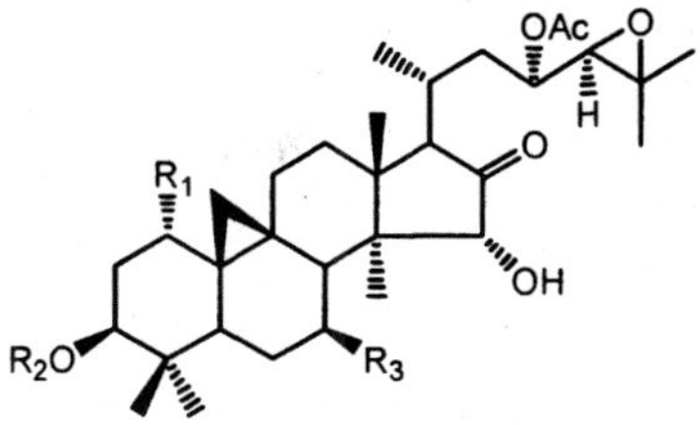

73-84

	R$_1$	R$_2$	R$_3$
73	H	β-D-xyl(p)	H
74	H	β-D-gal(p)	$\Delta^{7,8}$
75	H	(2-*O*-malonyl)β-D-xyl(p)	$\Delta^{7,8}$
76	H	(2-*O*-malonyl)β-D-xyl(p)	H
77	H	β-D-xyl(p)	$\Delta^{7,8}$
78	H	β-D-glu(p)-(1→3)-β-D-xyl(p)	H
79	H	α-L-ara(p)	$\Delta^{7,8}$
80	H	α-L-ara(p)	H
81	OH	β-D-xyl(p)	H
82	H	β-D-xyl(p)	OH
83	H	(4-acetoxy)α-L-ara(p)	H
84	H	(4-acetoxy)β-D-xyl(p)	H

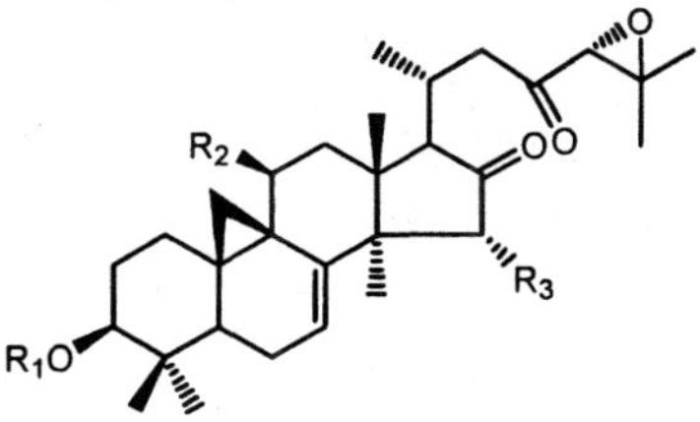

85-93

	R$_1$	R$_2$	R$_3$
85	β-D-xyl(p)	OH	H
86	α-L-ara(p) 12β-OAc	H	OH
87	α-L-ara(p) 12β-OAc	H	H
88	β-D-xyl(p)	OH	OH
89	α-L-ara(p)	OH	H
90	H	OH	H
91	β-D-xyl(p) 12β-OAc	H	H
92	β-D-xyl(p)	H	H
93	β-D-xyl(p)	H	OH

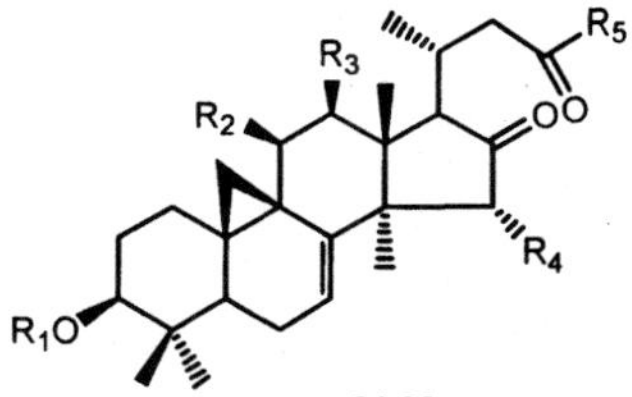

94-98

	R$_1$	R$_2$	R$_3$	R$_4$	R$_5$
94	β-D-xyl(p)	OH	H	H	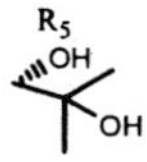
95	α-L-ara(p)	H	OAc	OH	
96	β-D-xyl(p)	OH	H	OH	
97	α-L-ara(p)	H, 7,8-dihydro	OAc	H	CH_2OH
98	β-D-xyl(p)	OH	H	H	CH_2OH

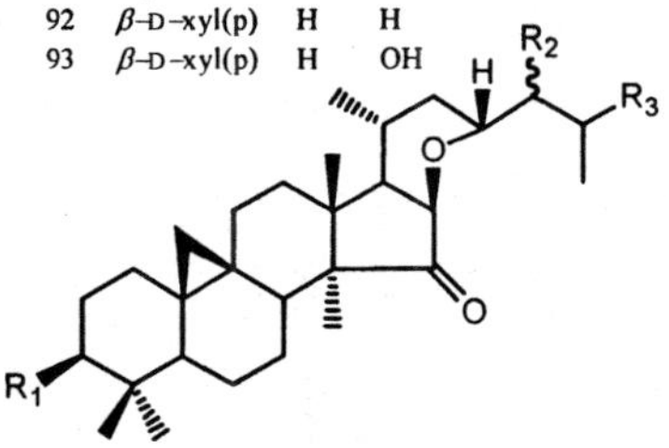

99-104

	R$_1$	R$_2$	R$_3$
99	H	OH (24*R*)	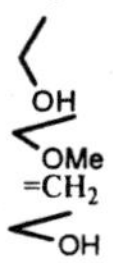
100	β-D-xyl(p)	OAc(24*S*)	
101	OH	OH (24*S*)	
102	OH	OH	
103	OH	OH	$=CH_2$
104	β-D-gal(p)	OAc(24*S*)	

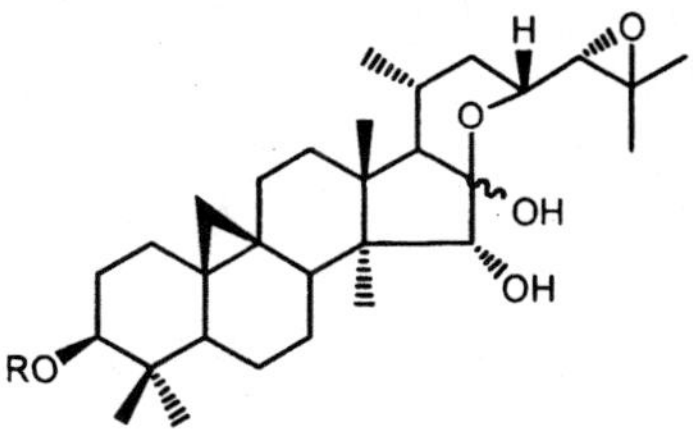

105 R=β-D-xyl(p)

Scheme 1. Continued. Continued on next page.

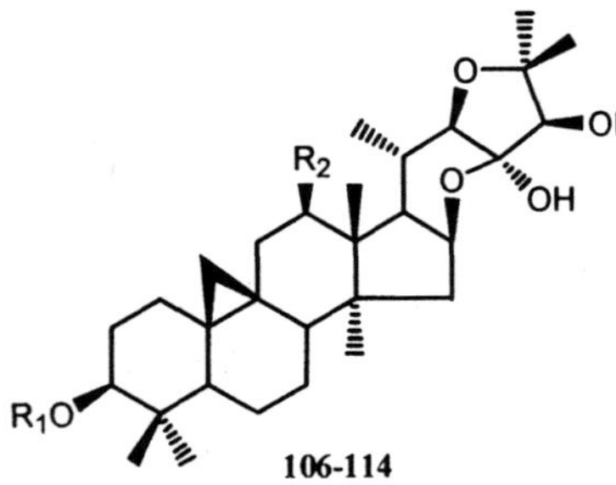

106-114

	R_1	R_2
106	β-D-xyl(p) $\Delta^{7,8}$	H
107	β-D-xyl(p)	H
108	(2-*O*-malonyl)β-D-xyl(p)	H
109	β-D-xyl(p) $\Delta^{7,8}$	OAc
110	α-L-ara(p) $\Delta^{7,8}$	OAc
111	β-D-xyl(p)	OAc
112	α-L-ara(p)	OAc
113	β-D-glu(p)-(1→2)-β-D-glu(p)-(1→2)-β-D-xyl(p)	H
114	β-D-(6-*O*-trans-isoferuloyl-glu)-(1→2)-β-D-glu(p)-(1→2)-β-D-xyl(p)	H

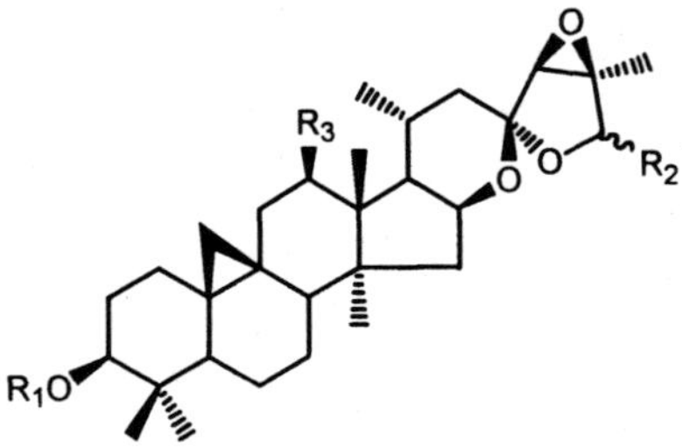

115-119

	R_1	R_2	R_3
115	H	H	OAc
116	α-L-ara(p)	OH	OAc
117	β-D-xyl(p)	H	OAc
118	β-D-xyl(p)$\Delta^{7,8}$	H	H
119	α-L-ara(p)	H	OAc

$\Delta^{7,8}$

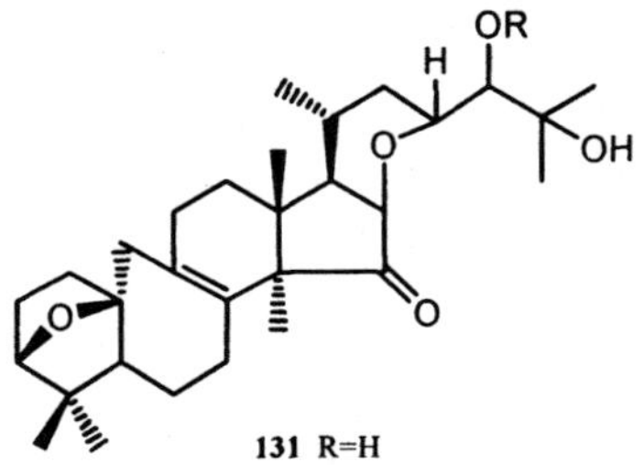

131 R=H
132 R=OAc

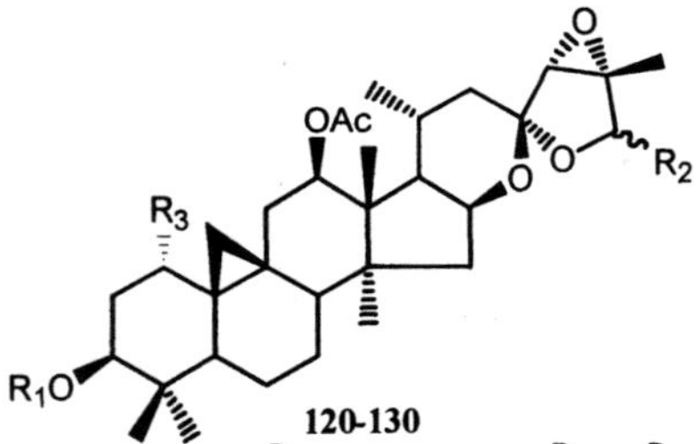

120-130

	R_1	R_2	R_3
120	β-D-xyl(p) $\Delta^{7,8}$	H	H
121	(2-*O*-malonyl)β-D-xyl(p)$\Delta^{7,8}$	OH	H
122	(2-*O*-acetyl)β-D-xyl(p)	OH	H
123	(2-*O*-acetyl)β-D-xyl(p)	H	H
124	β-D-xyl(p)	H	H
125	β-D-xyl(p)	OH	H
126	β-D-xyl(p)$\Delta^{7,8}$	H	H
127	β-D-xyl(p)$\Delta^{7,8}$	OH	OH
128	β-D-xyl(p)	OH	OH
129	(4-*O*-acetyl)β-D-xyl(p)	OH	H
130	β-D-xyl(p)	=O	H

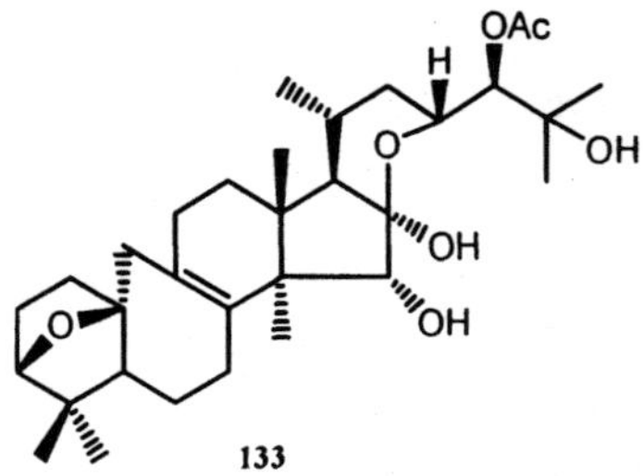

133

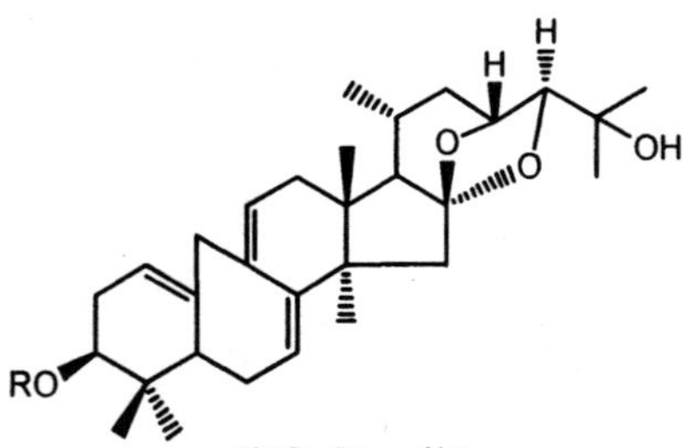

134 R=β-D-xyl(p)

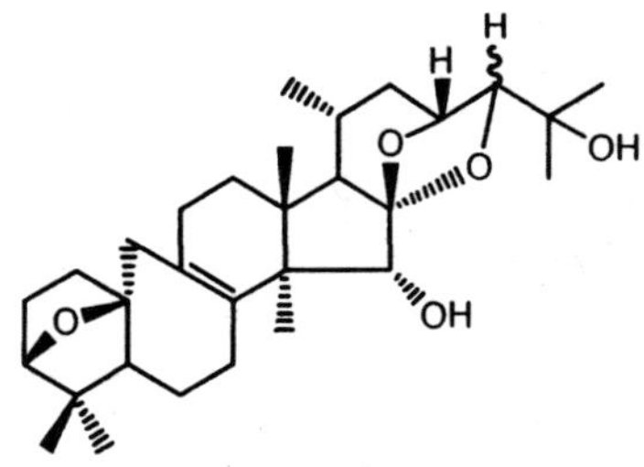

135 R=H (24*R*)
136 R=H (24*S*)
137 R=CH_3 (24*S*)

Scheme 1. Continued.

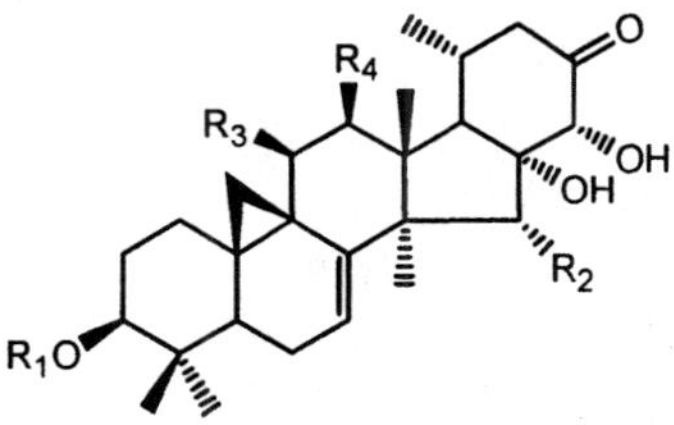

	R1	R2	R3	R4
138	H	H	OH	H
139	α-L-ara(p)	H	H	OAc
140	β-D-xyl(p)	H	OH	H
141	β-D-xyl(p)	OH	OH	H

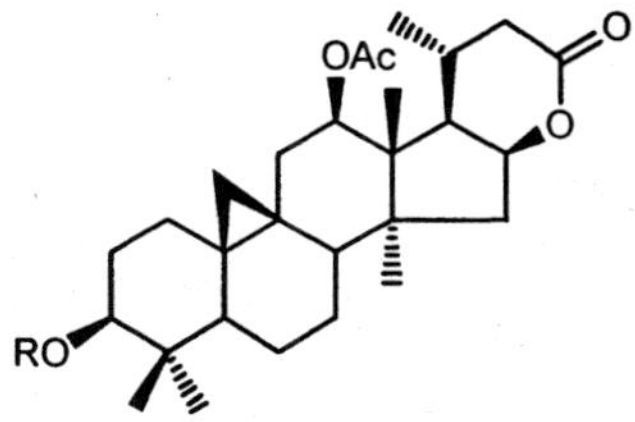

142 R=β-D-xyl(p)
143 R=β-D-xyl(p) $\Delta^{7,8}$

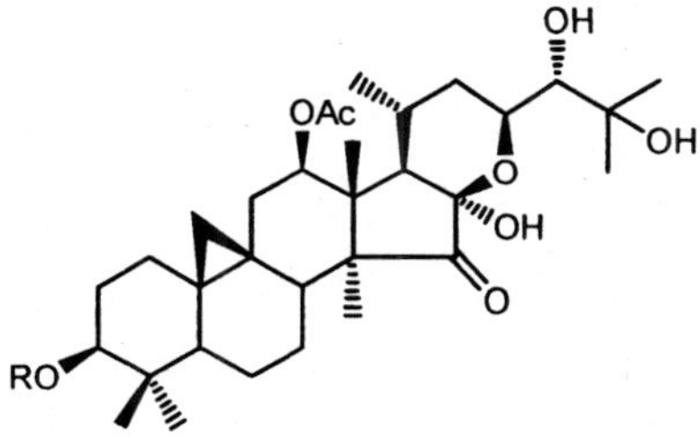

144 R=β-D-xyl(p)
145 R=β-D-xyl(p) $\Delta^{7,8}$

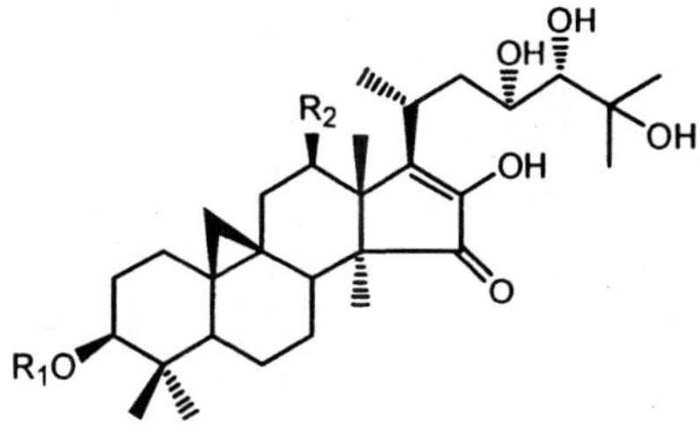

	R1	R2
146	β-D-xyl(p)	H
147	β-D-xyl(p)$\Delta^{7,8}$	H
148	β-D-xyl(p)	OAc
149	β-D-xyl(p)$\Delta^{7,8}$	OAc

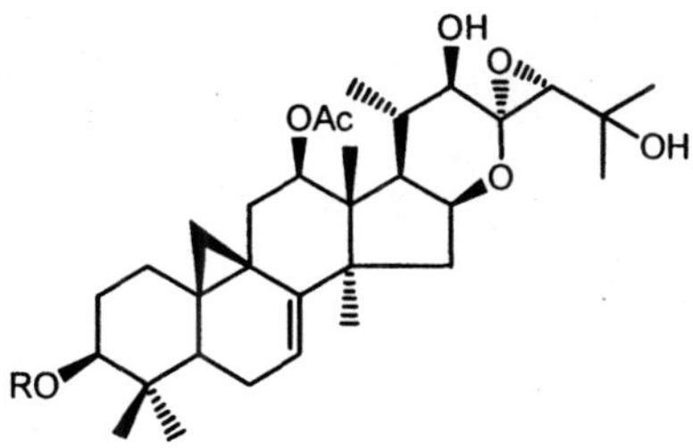

150 R=β-D-xyl(p)

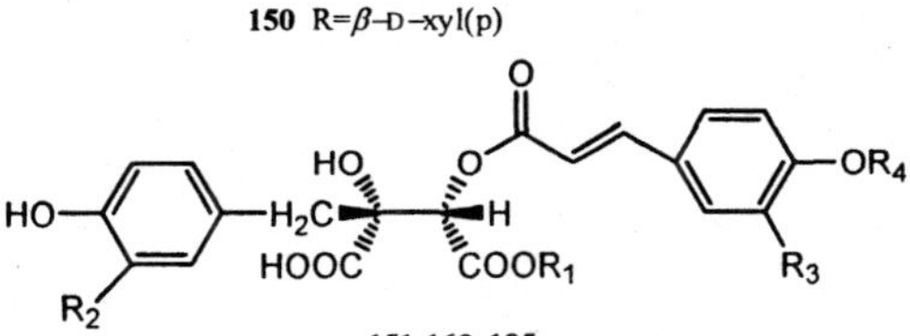

151-160, 185

	R1	R2	R3	R4
151	H	OH	OCH_3	H
152	H	OH	OH	CH_3
153	H	OH	H	H
154	H	H	OH	H
155	H	H	OCH_3	H
156	H	H	OH	CH_3
157	CH_3	OH	OCH_3	H
158	CH_3	OH	OH	CH_3
159	CH_3	H	OCH_3	H
160	CH_3	H	OH	CH_3
185	H	OH	OH	H

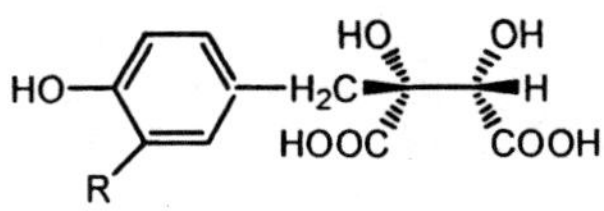

165 R=H
166 R=OH

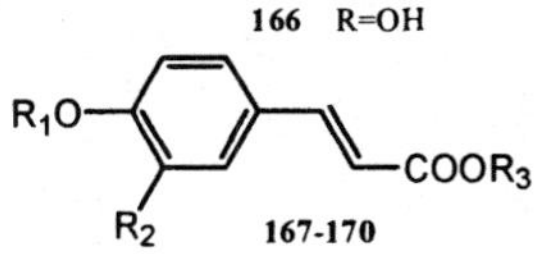

167-170

	R1	R2	R3
167	H	OH	H
168	H	OMe	H
169	Me	OH	H
170	H	H	H

Scheme 1. Continued. *Continued on next page.*

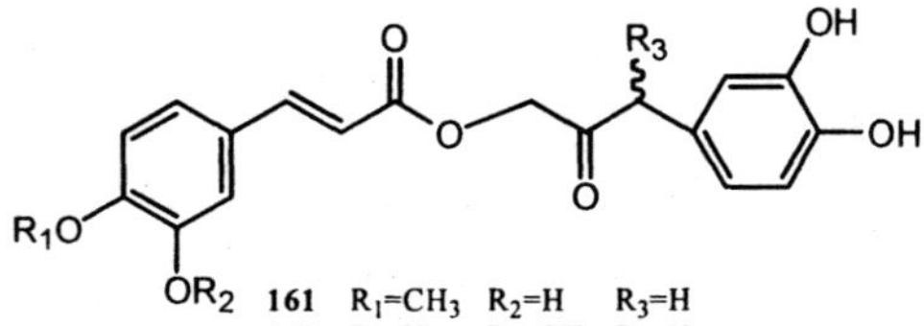

161 R_1=CH_3 R_2=H R_3=H
162 R_1=H R_2=CH_3 R_3=H
163 R_1=CH_3 R_2=H R_3=OCH_3
164 R_1=H R_2=CH_3 R_3=OCH_3

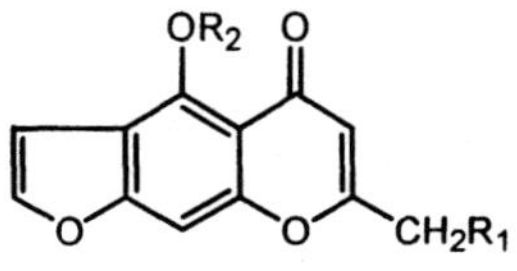

171 R_1=OH R_2=CH_3
172 R_1=H R_2=H
173 R_1=H R_2=CH_3

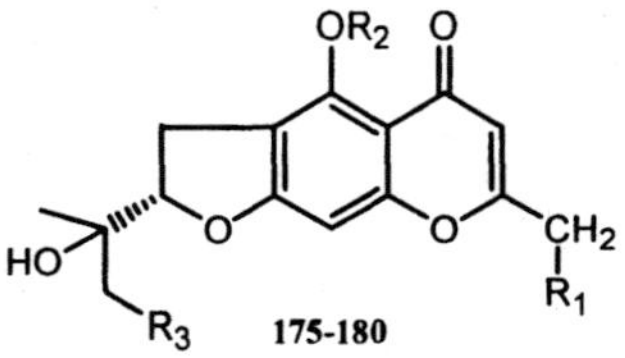

175-180

	R_1	R_2	R_3
175	OH	Me	H
176	*O*-*β*-D-glu(p)	Me	H
177	OH	H	H
178	H	H	H
179	OH	H	OH
180	*O*-*β*-D-glu(p)	H	H

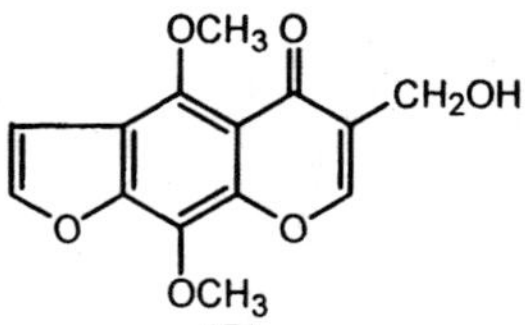

174

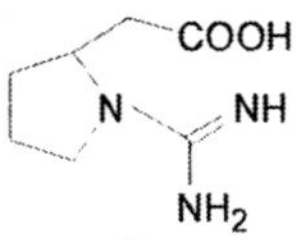

181

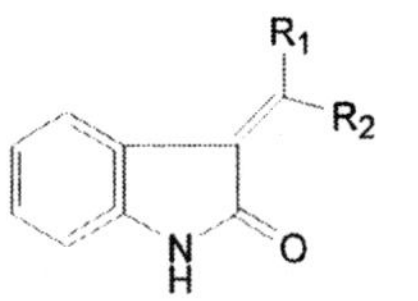

182 R_1= CH=C(CH_3)$_2$, R_2=H
183 R_1= H, R_2= CH=C(CH_3)$_2$

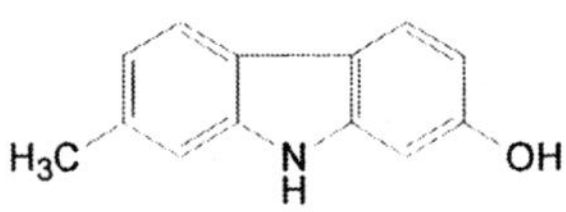

184

Scheme 1. Continued.

Table II. Aromatic acids and derivatives from *Cimicifuga* Species

No.	Compound	Formula (MW)	Source	Ref
151	cimicifugic acid A	$C_{21}H_{20}O_{11}$ (448)	e	*57*
152	cimicifugic acid B	$C_{21}H_{20}O_{11}$ (448)	e	*57*
153	cimicifugic acid C	$C_{20}H_{18}O_{10}$ (418)	e	*57*
154	cimicifugic acid D	$C_{20}H_{18}O_{10}$ (418)	e	*58*
155	cimicifugic acid E	$C_{21}H_{20}O_{10}$ (432)	e	*58*
156	cimicifugic acid F	$C_{21}H_{20}O_{10}$ (432)	e	*58*
157	cimicifugic acid A-1-methyl ester	$C_{22}H_{22}O_{11}$ (462)	e	*58*
158	cimicifugic acid B-1-methyl ester	$C_{22}H_{22}O_{11}$ (462)	e	*58*
159	cimicifugic acid E-1-methyl ester	$C_{22}H_{22}O_{10}$ (446)	e	*58*
160	cimicifugic acid F-1-methyl ester	$C_{22}H_{22}O_{10}$ (446)	e	*58*
161	cimiracemate A	$C_{19}H_{18}O_{7}$ (358)	g	*59*
162	cimiracemate B	$C_{19}H_{18}O_{7}$ (358)	g	*59*
163	cimiracemate C	$C_{20}H_{20}O_{8}$ (388)	g	*59*
164	cimiracemate D	$C_{20}H_{20}O_{8}$ (388)	g	*59*
165	piscidic acid	$C_{11}H_{12}O_{7}$ (256)	e	*57*
166	fukiic acid	$C_{11}H_{12}O_{8}$ (272)	e	*57*
167	caffeic acid	$C_{9}H_{8}O_{4}$ (180)	e	*57*
168	ferulic acid	$C_{10}H_{10}O_{4}$ (194)	e, g	*57,59*
169	isoferulic acid	$C_{10}H_{10}O_{4}$ (194)	e, g	*57,59*
170	*p*-coumaric acid	$C_{9}H_{8}O_{4}$ (180)	e	*57*
185	fukinolic acid	$C_{20}H_{18}O_{11}$ (434)	e	*58*

a: *Cimicifuga acerina* b: *Cimicifuga dahurica* c: *Cimicifuga foetida* d: *Cimicifuga japonica* e: *Cimicifuga simplex* f: *Cimicifuga heracleifolia* g: *Cimicifuga racemosa* h: Cimicifuga Rhizome species unidentified.

Table III. Chromones from Cimicifuga Species

No.	Compound	Formula (MW)	Source	Ref
171	Khellol	$C_{13}H_{10}O_{5}$ (246)	e	*68*
172	Norvisnagin	$C_{12}H_{8}O_{4}$ (216)	b, c	*25,69*
173	Visnagin	$C_{13}H_{10}O_{4}$ (230)	b	*69*
174	Ammiol	$C_{14}H_{12}O_{6}$ (276)	e	*68*
175	Cimifugin	$C_{16}H_{18}O_{6}$ (306)	c, e	*68,70,71*
176	cimifugin glucoside	$C_{22}H_{28}O_{11}$ (468)	c	*70*
177	Norcimifugin	$C_{15}H_{16}O_{6}$ (292)	c	*71,72*
178	Visamminol	$C_{14}H_{12}O_{6}$ (276)	b	*68*
179	6′-hydroxylangelicain	$C_{15}H_{16}O_{7}$ (308)	c	*71*
180	prim-*O*-glucosylangelicain	$C_{21}H_{26}O_{11}$ (454)	c	*71*

methods have been developed for the qualitative and /or quantitative analysis of the different ingredients including triterpene glycosides and phenolics in *Cimicifuga* plants and derived products. Since tests for purity, such as assays for microbiological contamination and the determination of heavy metals and pesticides, are generally similar for all plant derived products, such methods will not be addressed here. Rather, this section will mainly focus on the applicability of analytical methods, in particular chromatographic assays, to the quality control of *Cimicifuga* herbs and derived products. This would include tests for botanical and product identity, and natural product quantity and stability.

Confirmation of Identity

The botanical identity of the plant raw materials of *Cimicifuga* species may be confirmed using traditional pharmacognostic techniques such as taxonomical and organoleptic identification (*77*). In addition, molecular methods including RAPD-PCR analysis (*78*), DNA fingerprinting (*79*), and FT-IR spectroscopy (*80*) may also play a role in the identification of the species. This latter technique appears to of particular interest to industry as it could allow a rapid differentiation of *Cimicifuga* species without involving the pretreatment of plant samples including extraction and separation (*80*).

However, chromatographic methods remain the most robust method in the confirmation of botanical identity of the processed product, mostly because such techniques can provide both qualitative and quantitative analyses, and also because of the detailed chemical information gained by coupling with UV detector and/or MS spectrometry. For example, recently, Wang and colleagues (*81*) developed a LC/TIS-MS method to produce chemical fingerprinting chromatograms of seven *Cimicifuga species,* including *C. racemosa, C. dahurica, C. foetida, C. heracleifolia, C. japonica, C. acerina* and *C. simplex* (Figure 1a-g). In each chromatogram, 10 of the most characteristic peaks were specified based on the corresponding retention time and mass spectroscopy. In addition, not least characteristic peaks of *C. racemosa* and about half of the Chinese species were identified on the basis of their MS spectra and/or comparing with standard samples. Under the extraction and separation conditions utilized by the authors, the most characteristic peaks of *C. racemosa, C. dahurica, C. foetida* and *C. heracleifolia* exhibited retention times of more than 30 min, whereas those of *C. japonica, C. acerina* and *C. simplex* were observed in the range of 10 to 30 min. These results demonstrated that the fingerprint profiles of the seven herbs were markedly different from each other such that the species can be well differentiated. In another study, the identity of an 85-year-old sample was confirmed to be black cohosh on the basis that sample displayed a similar chemical fingerprint profile to a newly collected *C. racemosa* herb using either HPLC-PDA or LC-MS methods (*82*).

The confirmation of *Cimicifuga* species identity in commercial extracts and preparations is more challenging, and should no longer only use traditional pharmacognostic and molecular methods. In addition, the applicability of FT-IR spectroscopy may also be limited since it can be greatly affected by the presence of excipient matrix. As a result, chromatographic methods remain a most viable choice. Jiang et al. (*82*) utilized fingerprinting chromatograms produced by a LC/TIS-MS method for the examination of commercial products of black cohosh. In this study, six products which claimed to originate from black cohosh were tested, and 10 large peaks in the spectra were utilized as markers for comparison. Without similarity metrics being determined, five products qualitatively showed some similarities in terms of the characteristic peaks in the chromatograms, and they likely originated from *C. racemosa.* However, the utilization of similarity to determining the botanical identity in commercial products has limitations. For example, one product claimed to be from wild harvested (or wildcrafted) *C. racemosa* (*82*) displayed an apparent difference in characteristic peaks from those of another *C. racemosa* plant (Figure 1h), and yet the method did not exclude the product from being of black cohosh origin since such a difference might result from manufacturing process, rather than different species. Adulteration of another plant into a black cohosh product matrix may also go undetected.

In another study, He and others (*83*) utilized species-specific markers to develop a LC-MS method for identifying the botanical identity in commercial products. The markers utilized included four triterpene glycosides, namely, 23-epi-26-deoxyactein (**117**), actein (**125**), cimiracemoside C (**30**, also called cimicifugoside M), and cimigenol xyloside (**2**), and a chromone, cimifugin (**175**). Of the five compounds, cimigenol xyloside is found widely in *Cimicifuga* species, 23-epi-26-deoxyactein and actein only in *C. racemosa* and *C. foetida* (*81*), while cimiracemoside C and cimifugin were *C. racemosa*-specific and *C. foetida*-specific, respectively (Figure 2). Upon utilization of this method, commercial products containing 23-epi-26-deoxyactein (**117**) and actein (**125**) could be linked as coming from either *C. racemosa* or *C. foetida* origin. In addition, the presence of cimiracemoside C (**30**) could confirm the identity of *C. racemosa*, while the presence of cimifugin (**175**) could suggest the products originating from *C. foetida.* As a result, the authors (*83*) had successfully determined the botanical identity of over 30 commercial black cohosh products, and found that the botanical identity of those products could be attributable to three types of plant origin, *C. racemosa*, *C. foetida* and the mixture of the two species. With further validation, such a method could be of value as a rapid and accurate quality control technique for confirmation of botanical identity in commercial products.

Quantification of Chemical Markers

HPLC methods (as summarized in Table IV) have been widely utilized to analyze the content of triterpene glycosides and phenolics in *Cimicifuga* herbs,

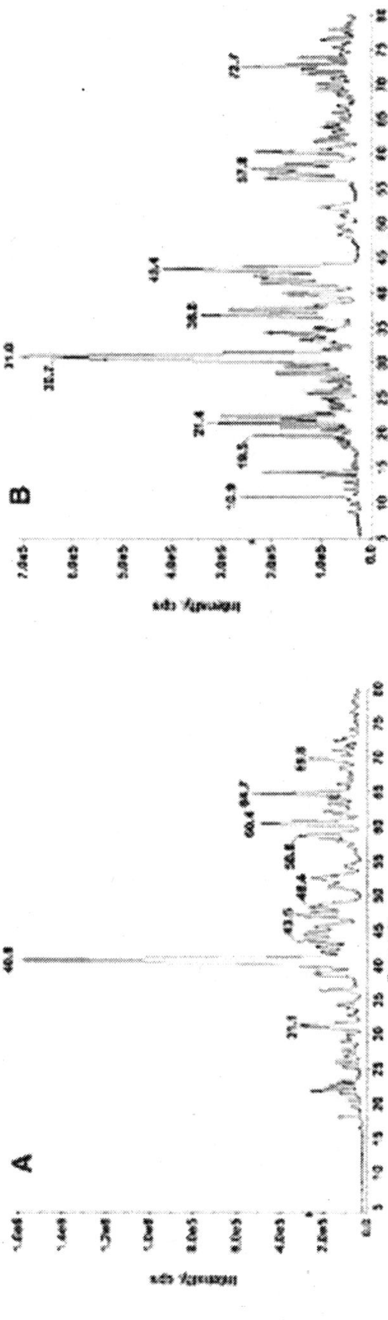
A
B
Intensity, cps
Time, min

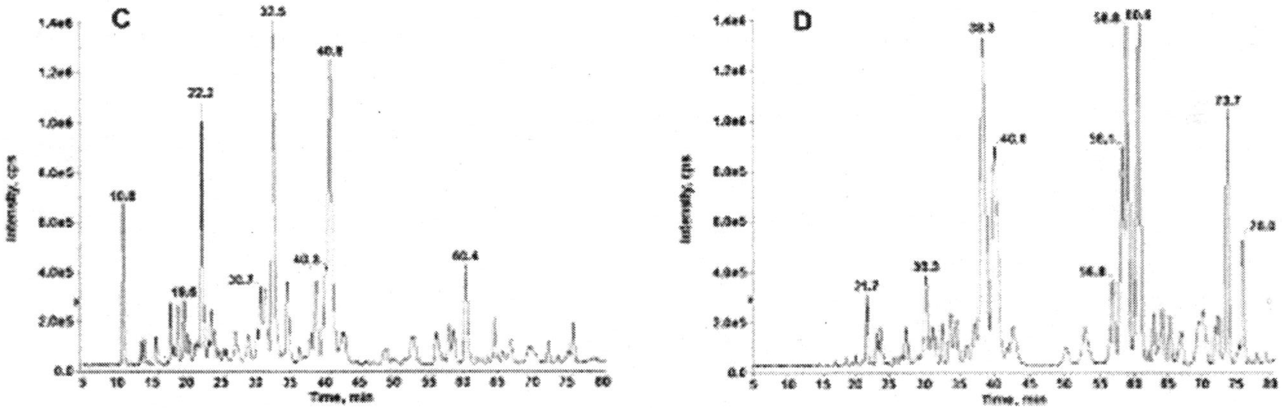

*Figure 1. Base peak ion chromatograms of Cimicifuga plants and a commercial product using mass spectrometry with TIS ion source. (**A**) C. racemosa, (**B**) C. dahurica, (**C**) C. foetida, (**D**) C. heracleifolia, (**E**) C. japonica, (**F**) C. acerina, and (**G**) C. simplex, (**H**) a commercial products claimed from wild crafted black cohosh. Adopted and modified from Wang et al., 2005.* *Continued on next page.*

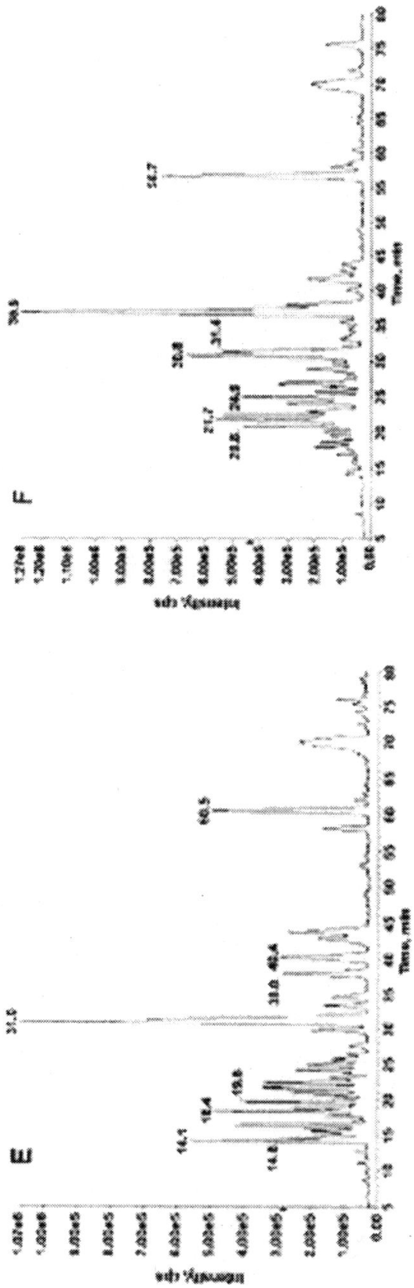
F
E
Intensity, cps
Time, min

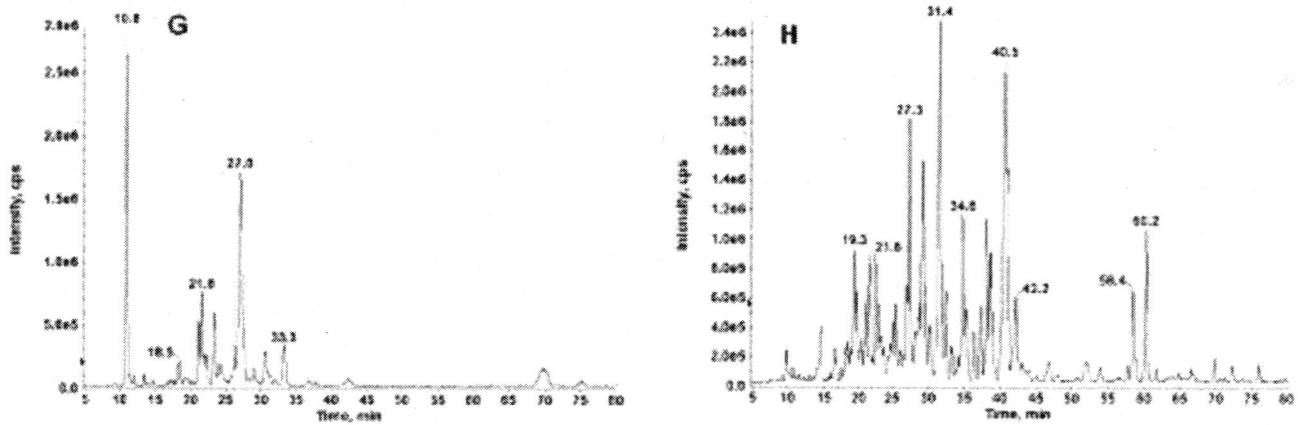

Figure 1. Continued.

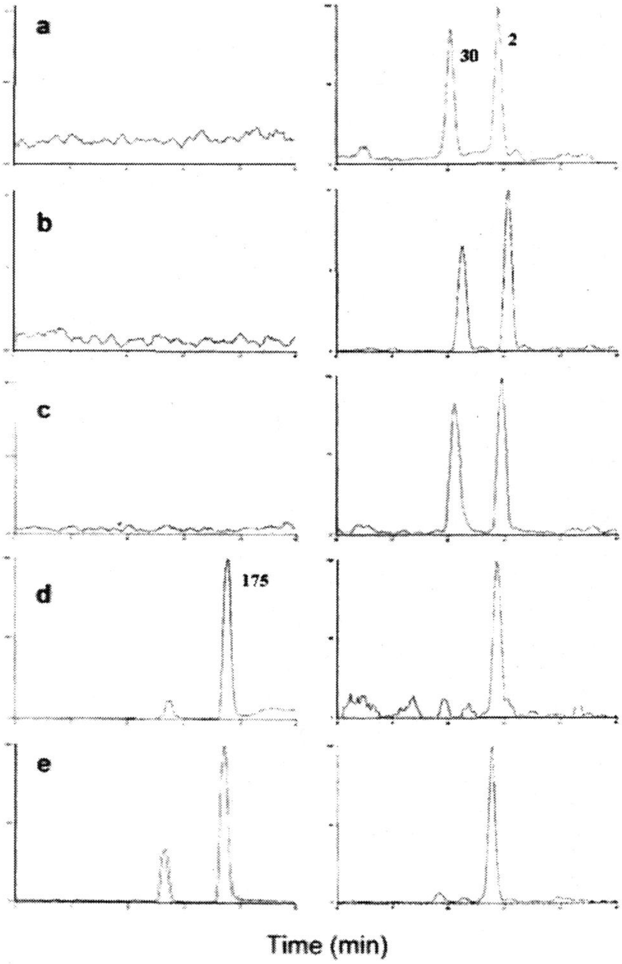

*Figure 2. MS chromatograms of selected ion mode at m/z 307 for cimifugin (**175**) and m/z 621 for cimiracemoside C (**30**) and cimigenol xyloside (**2**), a) C. racemosa, b) commercial product Remmifemin™, c) commercial product CimiPure®, d) C. foetida, e) A commercial product not named. Adopted and modified from He et al., 2000.*

their extracts and preparations. In these methods, the analytical separation of those markers is generally carried out using reversed-phase columns with gradient mobile phase systems consisting of methanol/ acetonitrile and water with or without small amount of an acid as a modifier (Table IV).

Due to lack of chromophores in triterpene glycosides, the use of ELSD has advantages in detection compared to UV detectors. Several research groups have developed and validated HPLC-ELSD methods to quantify triterpene glycosidic compounds (*85-87*). For example, Li *et al.* (*87*) developed such a method to quantitatively determine the content of 13 triterpene glycosides in a single run. In addition, they had also demonstrated the applicability of the method to the quantification of triterpene glycosides in commercial black cohosh products. Separately, Kong and others (*86*) reported that an HPLC-ELSD assay could effectively determine the content of triterpenes in *C. foetida.* The Institute for Nutraceutical Advancement reported a validated ELSD assay for the determination of triterpene glycosides in black cohosh by HPLC (Figure 3, http://www.nsf.org/ business/ina/blackcohosh.asp?program=INA). In this method, 10 triterpene glycosides, including 23-epi-26-deoxyactein (**117**), actein (**125**), 26-deoxycimicifugoside (**120**), cimiracemoside A (**28**), C (**30**), E (**100**) and F (**109**), acetyl shengmanol xyloside (**73**) and cimigenol xyloside (**2**), in *C. racemosa* root materials and powdered extracts were determined based on the reference standard response of 23-epi-26-deoxyactein (**117**). Nonetheless, it should be noted that ELSD does not generate any structural information and therefore peak identification usually requires either standard samples or mass spectroscopy.

Mass spectrometry with different ion sources has also been used for the measurement of triterpene glycosides in *Cimicifuga* samples (81,83,88). Recently, a highly sensitive LC-TIS-MS method for the quantification of 23-epi-26-deoxyactein (**117**) was developed by Wang *et al.* (*81*) with a limit of detection being as low as 2.5 ng/ml, which was lower than that provided by APCI-LC-MS methods (*88,89*). In addition, the LC-TIS-MS assay appeared feasible to the determination of triterpene glycosides in both *Cimicifuga* herbs and their commercial products.

Determination of phenolics in *Cimicifuga* herbs and their derived products is often by HPLC with UV or DAD detector due to the presence of chromophores with maximum UV absorption (λ_{max}) ranging from 230 to 360 nm. For example, Pan *et al.* (*90*) had developed and validated a simple HPLC/UV method for quantitation of ferulic acid (**168**) and isoferulic acid (**169**) in rhizoma Cimicifugae with a recovery being higher than 98% for both chemicals. Other studies (*84,87,88,91*) have also utilized HPLC–UV methods to determine phenolics, including aromatic acids and flavonoids, in *Cimicifuga* herbs and their derived products, although full validation were lacking. Apart from HPLC analysis of the phenolic components, Panossian *et al.* (*92*) developed and

Table IV. Chromatographic Analytical Methods Applied to Determine *Cimicifuga* Species

Column	*Mobile phase*	*Detector*	*Plant*	*Analytes*	*Ref*
Hypersil ODS (250×4 mm, 5 μm)	Water (A): acetonitrile (B):phosphoric acid (C) 0-2 min, 79.5% A:19.5% B:1% C; 2-18 min, 19.5% A:79.5% B:1% C to 99% B:1% C; 18-21 min, 99% B:1% C, at a flow rate of 0.8 ml/min	PDA at 299 and 360 nm	*C. racemosa*	Formononetine, kaempferol	*84*
Supelco Discovery C-18 (150×4.6 mm, 5 μm)	Gradient elution from 58% water (A): acetonitrile (B) 21%:reagent alcohol (C) 21% to 52% A:14% B:34% C within 35 min at a flow rate of 1.0 ml/min	ELSD	*C. racemosa*	Cimiracemoside A (**28**), 23-epi-26-deoxyactein (**117**), actein (**125**)	*85*
Hypersil ODS2 (200×4 mm, 5 μm)	Methanol in water: 0-8 min, 67.5%; 8-20 min 80%, at a flow rate of 0.8 ml/min.	ELSD	*C. foetida*	23-epi-26-deoxyactein (**117**), actein (**125**), cimigenol xyloside (**2**)	*86*
Phenomenex Synergi Hydro RP 80A C-18 (150×3.0 mm, 4 μm)	Acetonitrile in 10 mM ammonium acetate aqueous solution: 0-18 min, 5-28%; 18-35 min, 28-35%; 30-50 min, 35%; 50-70 min, 35-55%; 70-80 min, 55-75%, at a flow rate of 1.0 ml/min .	Mass spectrometry with TIS ion source	*C. racemosa, C. dahurica, C. foetida, C. heracleifolia, C. japonica, C. acerina* and *C. simplex*	23-epi-26-deoxyactein (**117**), cimicifugoside H-1 (**85**), acetylshengmanol xyloside (**73**), cimigenol xyloside (**2**), 25-acetylcimigenol xyloside (**3**)	*81*

Table IV. Continued

Column	*Mobile phase*	*Detector*	*Plant*	*Analytes*	*Ref*
Waters C-18 (150×3.9 mm, 5 µm)	Acetonitrile in water: 0-18 min, 5-18%; 18-36 min, 28-35%; 36-45 min, 35-55%; 45-55 min, 55-75%, at a flow rate of 1 ml/min.	PDA at 203 nm and Mass spectrometry with APCI source	*C. racemosa*	23-epi-26-deoxyactein (**117**), actein (**125**), cimiracemoside C (**30**), cimigenoside (**2**)	*88*
Phenomenex Aqua C-18 (250×4.6 mm, 5 µm)	Acetonitrile in 10% formuc acid aqueous solution: 0-15 min, 5-15%; 15-20 min, 15%; 20-50 min, 15-50%; 50-55 min, 50-100%, at a flow rate of 1 ml/min.	PDA at 320 nm	*C. racemosa*	Caffeic acid (**167**), ferulic acid (**168**), isoferulic acid (**169**), fukinolic acid (**185**), cimicifugic acid A (**151**) and B (**152**)	*88*
HP Hypersil ODS (125×4 mm, 5 µm)	Acetonitrile in water: 0-18 min, 5-18%; 18-36 min, 28-35%; 36-45 min, 35-55%; 45-55 min, 55-75%; 55-56 min, 75-100%; 56-65 min, 100%, at a flow rate of 1.0 ml/min.	Mass spectrometry with APCI source	*C. racemosa, C. foetida*	23-epi-26-deoxyactein (**117**), actein (**125**), cimiracemoside C (**30**)*, cimigenol xyloside (**2**) and cimifugin (**175**)	*83*
XTerra MS C18 (100×3.9 mm, 3.5 µm)	Linear gradient elution from 20 to 50% methanol in water within 12 min at a flow rate of 1.0 ml/min.	PDA at 325 nm	*C. racemosa*	Caffeic acid (**167**), ferulic acid (**168**), isoferulic acid (**169**), fukinolic acid (**185**), cimicifugic acid A (**151**) and B (**152**)	*91*

Continued on next page

Table IV. Continued

Column	*Mobile phase*	*Detector*	*Plant*	*Analytes*	*Ref*
Hypersil ODS (250 mm×4.6 mm, 5 μm)	methanol-water-phosphoric acid (50: 150: 0.1) at a flow rate of 1.0 ml/min	UV at 320 nm	*C. foetida*	Caffeic acid (**167**), ferulic acid (**168**)	90
Waters YMC ODS-AQ RP-18 (250 mm×4.6 mm, 5 μm)	0.05% TFA (A), acetonitrile (B) and water (C): 0–8 min, 80% A, 20% B; 8–15 min, 32% B, 68% C; 15–55 min, 64% B, 36% C; 55–65 min, 95% B, 5% C; 65-70 min, 95% B, 5% C, at a flow rate of 1.6 ml/min.	PDA and ELSD	*C. racemosa*	caffeic acid (**167**), ferulic acid (**168**), isoferulic acid (**169**), cimicifugoside H-1(**85**), cimiracemoside A (**28**), cimicifugoside H-2 (**94**), (26*R*)-actein (**125**), 26-deoxycimicifugoside (**120**), (26*S*)-actein (**125**), 23-*epi*-26-deoxyactein (**117**), 23-acetyl-shengmanol-xyloside (**73**), 26-deoxyactein (**124**), 25-acetyl-cimigenol-arabinoside (**25**), 25-acetyl-cimigenol- xyloside (**3**), cimigenol- arabinoside (**30**), cimigenol- xyloside (**2**)	*87*

Table IV. Continued

Column	*Mobile phase*	*Detector*	*Plant*	*Analytes*	*Ref*
Phenomenex Prodigy ODS (250 mm×4.6 mm, 5 μm)	Acetonitrile in 0.1% formic acid aqueous solution: 0-30 min, 30-40%; 30-60 min, 40-60%, at a flow rate of 1.0 ml/min.	ELSD	*C. racemosa*	Cimiracemoside A (**28**), C (**30**), E (**100**) and F (**109**), actein (**125**), 23-*epi*-26-deoxyactein (**117**), 26-deoxycimifigoside (**120**), **epi-actein**, acetyl shengmanol xyloside (**73**), cimicifugoside (**2**)	**
Lichrospher 100 RP-18 (125 mm×4.6 mm, 3 μm)	Acetonitrile in water: 40%, 0-10 min; 40-70%, 10-20 min; 70%, 20-30 min at a flow rate of 1.0 ml/min.	UV at 210 nm	*C. racemosa*	Actein (**125**)	*92*

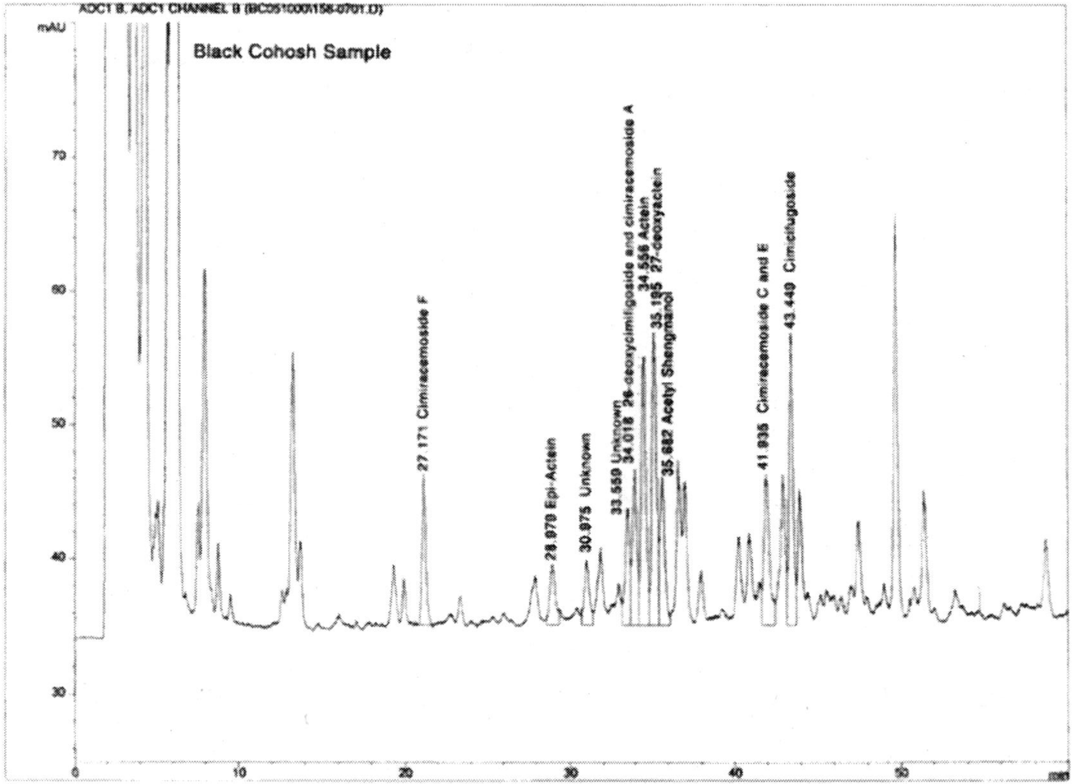

Figure 3. ELSD chromatogram of black cohosh (C. racemosa) as adopted from the following Website: http://www.nsfina.org/methods/blackcohoshset.html.

validated a TLC-fluorometry method for the quantitative analysis of formononetin and GC-MS for that of isoferulic acid (**169**).

Evaluation of Stability

Phytochemical complexity of botanicals presents a major challenge in our understanding on the chemical stability of individual compounds, the complex of compounds and the changes that may occur over storage and in particular once consumed by living organisms. As a consequence, the simple monitoring of the stability of one or two marker compounds does not sufficiently represent the global stability profile of a herbal medicinal product nor does it provide the full story of the efficacy of the botanical product and/or preparation. Rather, it is preferable to also determine the other major ingredients within the formulation matrix using proper analytical methods, e.g. chemical fingerprinting, and to demonstrate the proportional content remaining constant (see the Guideline on quality of herbal medicinal products/traditional herbal medicinal products, CPMP/QWP/2819/00 Rev 1). Although the stability studies on formulations of *Cimicifuga* plant derived products appears not to be well documented in the literature, recently Jiang and others (*88*) compared the chemical profiles of triterpene glycosidic and phenolic constituents in a 85-year-old plant sample with those in a newly collected sample of black cohosh using HPLC-PDA and LC-MS assays. Surprisingly, they observed that in the chromatograms from both plant samples (Figures 4 and 5) there was qualitatively great similarity, broadly indicative of the stability profile of the old sample. In addition, quantitative analysis revealed that the proportional amounts of each triterpene glycoside and phenolic ingredient in the extracts of old and new samples were also broadly similar except for the contents of caffeic acid (**167**) and ferulic acid (**168**) in the old sample being markedly lower than those in the newly collected one. Considering that no difference in the major peaks of the phenolics was detected in the chromatograms of two samples, the authors concluded that raw black cohosh materials might possess good storage stability and allow a long period of storage time. Nonetheless, such a conclusion must be tempered and limited, because caffeic acid and its derivatives are generally susceptible to chemical instability such as oxidation, leading to decreases in the contents (*93,94*), and a direct comparison difficult given the origin of the old root versus the new fresh sample and lack of detailed information on storage conditions. However, given that a root of 85-years old (assuming correct age determination was made) black cohosh was so similar to that of a younge fresh one (assuming 4-5 years of age), does indicate stablility within the root matrix. Further studies on the impact of processing technologies and storage on the bioactive compounds in black cohosh would contribute to the quality control of this medicinal, and few studies have focused on this area.

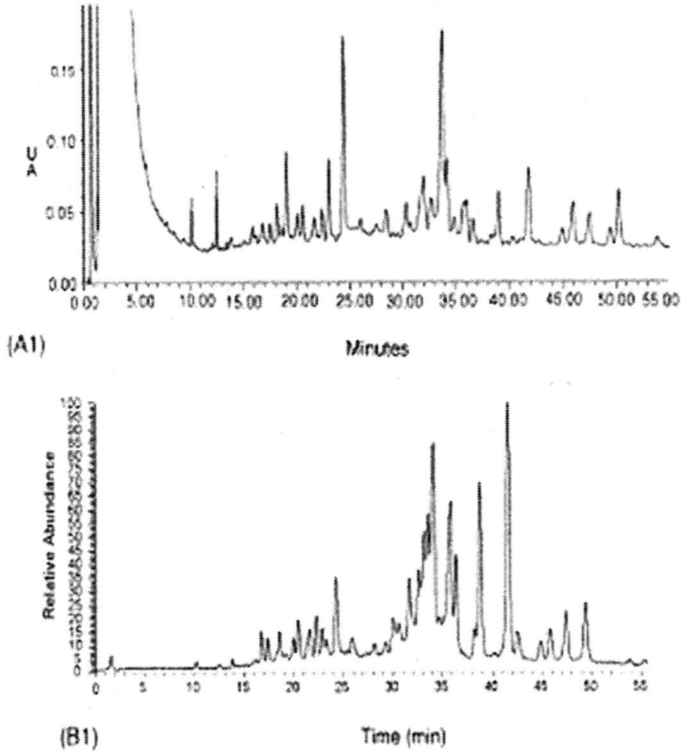

*Figure 4. HPLC–PDA and LC–MS spectra for the triterpene glycosides of two black cohosh samples. Spectra A: HPLC–PDA at 203 nm; spectra B: LC–MS TIC. Spectra 1 for the 85-year-old black cohosh plant material; spectra 2 for the modern plant material. Condition: Waters C-18 (150×3.9 mm, 5 μm) column, mobile phase gradient from 5% v/v acetonitrile (**A**) in water (**B**) to 75% v/v within 55 min with a flow rate of 1 ml/min (5-18% **A** at 0-18 min, 28-35% **A** at 28-35 min, 35-55% **A** at 36-45 min and 55-75% **A** at 45-55 min). Adopted from Jiang et al., 2006.*

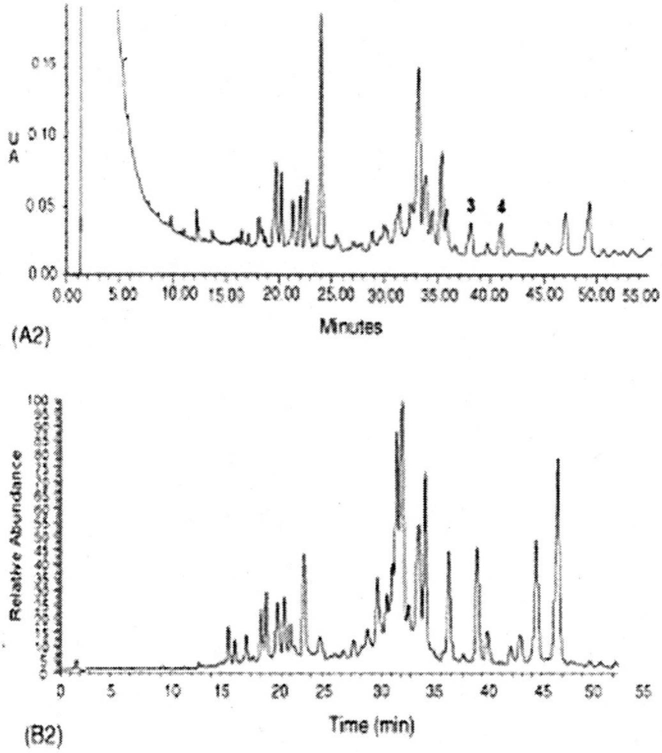

Figure 4. Continued.

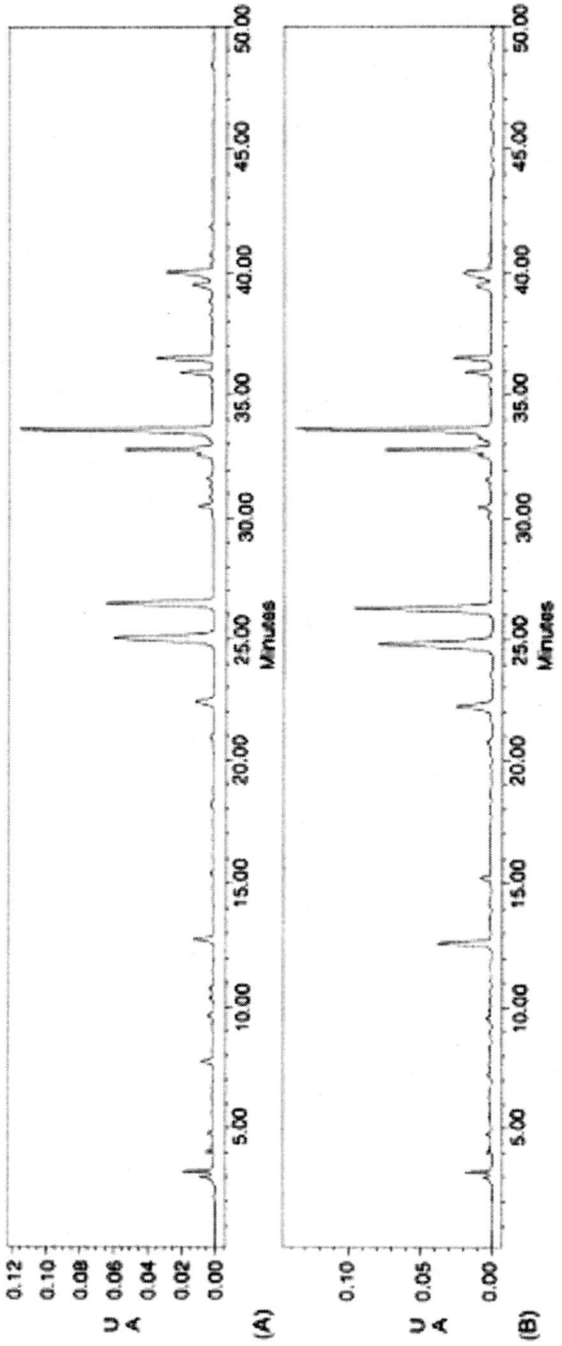

AU
0.12
0.10
0.08
0.06
0.04
0.02
0.00
(A)
AU
0.10
0.05
0.00
(B)
5.00
10.00
15.00
20.00
25.00
30.00
35.00
40.00
45.00
50.00
Minutes

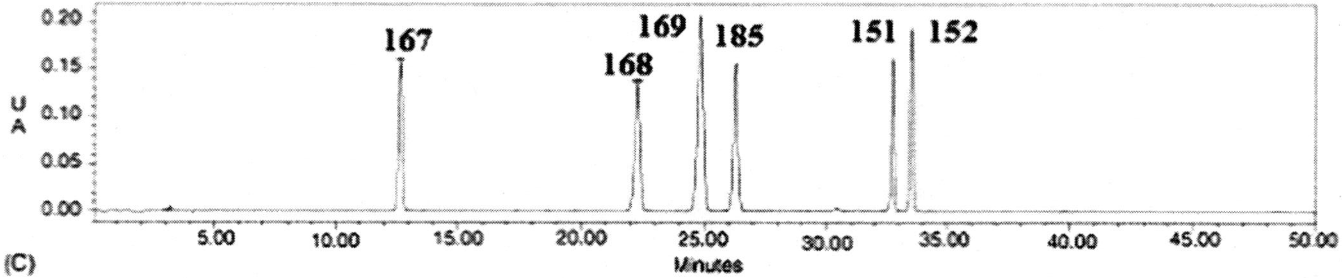

*Figure 5. HPLC–PDA (320 nm) spectra for the phenolic constituents. A: the 85-year-old black cohosh plant material; B: the modern plant material; C: the mixture of six phenolic reference compounds, caffeic acid (**167**), ferulic acid (**168**), isoferulic acid (**169**), fukinolic acid (**185**), cimicifugic acid A (**151**), and cimicifugic acid B (**152**). Condition: Phenomenex Aqua C-18 (250×4.6 mm, 5 μm) column, mobile phase using a gradient of acetonitrile (**A**) in 10% aqueous formic acid water (**B**) 5-15% **A** at 0-15 min, 15% **A** at 15-20 min, 15-50% **A** at 20-50 min and 50-100% **A** at 50-55 min with a flow rate of 1 mL/min. Adopted from Jiang et al., 2006.*

Conclusion

Phytochemical studies have revealed the presence of over 180 triterpene glycosides and phenolics in the medicinal *Cimicifuga* species. Important constituents present in these species include triterpene glycosides, i.e., 23-epi-26-deoxyactein (**117**), actein (**125**), cimigenol xyloside (**2**), cimiracemoside C (**30**) etc., aromatic acids, e.g. caffeic acid (**167**), ferulic acid (**168**) and isoferulic acid (**169**), and others, e.g. cimifugin (**175**) and cimipronidine (**181**).

The chemical complexity, and the difficulty in analysis of individual tripterpene glycosides in black cohosh still presents an enormous challenge in the botanical standardization and quality control of *Cimicifuga* herbs derived products. Many of the commercially available black cohosh preparations, if standardized, are standardized based on the content of total triterpene glycosides, while other compounds and classes of compounds are ignored. As a result, the chemical profiles of black cohosh derived products may not be identical to those from the originated plants, and qualitative differences in the individual compounds and the ratios of the individual compounds may occur even if the total tripterpene glycoside content is similar. The chemical fingerprinting of triterpene glycosides varies from products to products between different producers and processors due not only to different processing technologies, but also due to different analytical techniques used. While analytical methods applied to black cohosh derived products may permit the quantitation of triterpene glycoside content, improved and standardized analytical methods that are robust, validated, and associated with the pharmacological efficacy of the intended application of the product, would not only improve the ability to confirm the botanical identity and address issues of compound stability, but provide an increased quality of commercially available products.

From the perspective of quality control, a fully validated method(s) capable of identifying the correct species, quantifying the contents of chemical markers and ensuring the stability is highly desired. In addition, quantitative analysis of chemical ingredients does not necessarily confer to good quality control for the potency of *Cimicifuga* derived products unless a relationship of chemical quantity and biological activity is well characterized. As a consequence, *in vitro* or *in vivo* bioassays capable of quantitatively determining relevant biological activities are also overdue.

Acknowledgements

Authors wish to express their thanks and appreciation to our respective institutions, including the Institute of Medicinal Plant Development (IMPLAD), Chinese Academy of Medical Sciences & Peking Union Medical College and the New Jersey Agricultural Experiment Station and Rutgers University. Part of this work was conducted also in concert with our Purdue University and University of Alabama NIH-funded Botanical Center for Age-Related Diseases (Grant Number: OD-00-004) .

References

1. The Eighth Chinese Pharmacopoeia Commission. Pharmacopoeia of the People's Republic of China. Chinese Medical Science Publishers, Beijing, **2005**.
2. Takemoto, T.; Kusano, G. *Yakugaku Zasshi* **1967**, *87*, 1569-1572.
3. Sakurai, N.; Inoue, T.; Nagai, M. *Yakugaku Zasshi* **1972**, *92*, 724-728.
4. Zhang, Q.W.; Ye, W.C.; Hsiao, W.W.L.; Zhao, S.X.; Che, C.T. *Chem. Pharm. Bull.* **2001**, *49*, 1468-1470.
5. Li, C.J.; Li, Y.H.; Chen, S.F.; Xiao, P.G. *Acta Pharm. Sin.* **1994**, *29*, 449-453.
6. Li, C.J.; Chen, D.H.; Xiao, P.G. *Acta Pharm. Sin.* **1993**, *28*, 777-781.
7. Chen, S.N.; Fabricant, D.S.; Lu, Z.Z.; Fong, H.H.S.; Farnsworth, N.R. *J. Nat. Prod.* **2002**, *65*, 1391-1397.
8. Takemoto, T.; Kusano, G.; Kawahara, M. *Yakugaku Zasshi* **1970**, *90*, 64-67.
9. Kusano, A.; Shibano, M.; Kitagawa, S.; Kusano, G.; Nozoe, S.; Fushiya, S. *Chem. Pharm. Bull.* **1994**, *42*, 1940-1943.
10. Pan, R.L.; Chen, D.H.; Si, J.Y.; Zhao, X.H.; Shen, L.G. *Acta Pharm. Sin.* **2003**, *38*, 272-275.
11. Kusano, A.; Shibano, M.; Kusano, G. *Chem. Pharm. Bull.* **1995**, *43*, 1167-1170.
12. Watanabe, K.; Mimaki, Y.; Sakagami, H.; Sashida, Y. *Chem. Pharm. Bull.* **2002**, *50*, 121-125.
13. Kusano, A.; Shibano, M.; Tsukamoto, D.; Kusano, G. *Chem. Pharm. Bull.* **2001**, *49*, 437-441.
14. Kusano, A.; Takahira, M.; Shibano, M.; Miyase, T.; Kusano, G. *Chem. Pharm. Bull.* **1999**, *47*, 511-516.
15. Li, J.X.; Kadota, S.; Hattori, M.; Yoshimachi, S.; Shiro, M.; Oogami, N.; Mizuno, H.; Namba, T. *Chem. Pharm. Bull.* **1993**, *41*, 832-841.
16. Kusano, A.; Shibano, M.; Kusano, G.; Miyase, T. *Chem. Pharm. Bull.* **1996**, *44*, 2078-2085.
17. Kusano, A.; Shimizu, K.; Idoji, M.; Shibano, M.; Minoura, K.; Gusano, G. *Chem. Pharm. Bull.* **1995**, *43*, 279-283.
18. Sakurai, N.;, Kimura, O.; Inoue, T. *Chem. Pharm. Bull.* **1981**, *29*, 955-960.
19. Bedir, E.; Khan I.A. *Chem. Pharm. Bull.* **2000**, *48*, 425-427.
20. Bedir, E.; Khan, I.A. *Pharmazie* **2001**, *56*, 268-269.

21. Shao, Y.; Harris, A.; Wang, M.F.; Zhang, H.J.; Cordell, G.; Bowman, M.; Lemmo, E. *J. Nat. Prod.* **2000**, *63*, 905-910.
22. Pan, R.L. Ph.D. Dissertation, Peking Union Medical College, 2002.
23. Pan, R.L.; Chen, D.H.; Si, J.Y.; Zhao, X.H.; Shen, L.G. *Zhongguo Zhongyao Zazhi* **2003**, *28*, 230-232.
24. Kusano, G.; Takemoto, T. *Yakugaku Zasshi* **1975**, *95*, 1133-1137.
25. Kadota, S.; Li, J.X.; Tanaka, K.; Namba, T. *Tetrahedron* **1995**, *51*, 1143-1166.
26. Kusano, A.; Shibano, M.; Kusano, G. *Chem. Pharm. Bull.* **1996**, *44*, 167-172.
27. Kusano, A.; Shibano, M.; Kusano, G. *Chem. Pharm. Bull.* **1999**, *47*, 1175-1179.
28. Sakurai, N.; Koeda, M.; Inoue, T.; Nagai, M. *Chem. Pharm. Bull.* **1994**, *42*, 48-51.
29. Li, C.J.; Chen, D.H.; Xiao, P.G. *Acta Chim. Sin.* **1994,** *52*, 722-726.
30. Pan, R.L.; Chen, D.H.; Si, J.Y.; Zhao, X.H.; Shen, L.G. *J. Asian Nat. Prod. Res.* **2004**, *6*, 63-67.
31. Kusano, G.; Idoji, M.; Sogoh, Y.; Shibano, M.; Kusano, A.; Iwashita, T. *Chem. Pharm. Bull.* **1994**, *42*, 1106-1110.
32. Koeda, M.; Aoki, Y.; Sakurai, N.; Nagai, M. *Chem. Pharm. Bull.* **1995**, *43*, 771-776.
33. Liu, Y.; Chen, D.H.; Si, J.Y.; Tu, G.Z.; An, D.G. *Nat. Prod. Res.* **2003**, *17*, 243-246.
34. Sakurai, N.; Koeda, M.; Aoki, Y.; Nagai, M. *Chem. Pharm. Bull.* **1995**, *43*, 1475-1482.
35. Kimura, O.; Sakurai, N.; Nagai, M. *Yakugaku Zasshi* **1982**, *102*, 538-545.
36. Kusano, G.; Murakami, Y.; Sakurai, Y. *Yakugaku Zasshi* **1976**, *96*, 82-85.
37. Kusano, G.; Uchida, H.; Murakami, Y. *Yakugaku Zasshi* **1976b**, *96*, 321-325.
38. Kusano, A.; Takahira, M.; Shibano, M.; Miyase, T.; Okuyama, T.; Kusano, G. *Heterocycles* **1998c**, *48*, 1003-1013.
39. Nishida, M.; Yoshimitsu, H.; Nohara, T. *Chem. Pharm. Bull.* **2003**, *51*, 354-356.
40. Chen, S.N.; Li, W.K.; Fabricant, D.S.; Santarsiero, B.D.; Mesecar, A.; Fitzloff, J.F.; Fong, H.H.S.; Farnsworth, N.R. *J. Nat. Prod.* **2002**, *65*, 601-605.
41. Zhu, N.Q.; Jiang, Y.; Wang, M.F.; Ho, C.-T. *J. Nat. Prod.* **2001**, *64*, 627-629.
42. Kusano, A.; Takahira, M.; Shibano, M.; In, Y.; Ishida, T.; Miyase, T.; Kusano, G. *Chem. Pharm. Bull.* **1998a**, *46*, 467-472.
43. Kusano, A.; Takahira, M.; Shibano, M.; Miyase, T.; Kusano, G. *Chem. Pharm. Bull.* **1998b**, *46*, 1001-1007.

44. Kusano, G.; Hojo, S.; Kondo, Y. *Chem. Pharm. Bull.* **1977**, *25*, 3182-3189.
45. Li, J.X.; Kadota, S.; Pu, X.F.; Namba, T. *Tetrahedron Lett.* **1994**, *35*, 4575-4576.
46. Li, C.J.; Li, Y.H.; Xiao, P.G.; Mabry, T.J.; Watson, W.H.; Krawiec, M. *Phytochemistry* **1996**, *42*, 489-494.
47. Liu, Y.; Chen, D.H.; Si, J.Y.; Tu, G.Z.; An, D.G. *J. Nat.Prod.* **2002**, *65*, 1486-1488.
48. Liu, Y.; Chen, D.H.; Si, J.Y.; Pan, R.L.; Tu, G.Z.; An, D.G. *Yaoxue Xuebao*, **2003**, *38*, 763-766.
49. Wende, K.; Muegge, C.; Thurow, K.; Schoepke, T.; Lindequist, U. *J. Nat. Prod.* **2001**, *64*, 986-989.
50. Borrelli, F.; Izzo, A.A.; Ernst, E. *Life Sci.* **2003**, *73*, 1215-1229.
51. Jarry, H.; Harnischfeger, G. *Planta Med.* **1985**, *51*, 46-49.
52. Sakurai, N.; Kozuka, M.; Tokuda, H.; Nobukuni, Y.; Takayasu, J.; Nishino, H.; Kusano, A.; Kusano, G.; Nagai, M.; Sakurai, Y.; Lee, K.H. *Bioorg. Med. Chem. Lett.* **2003**, *11*, 1137-1140.
53. Einbond, L.S.; Shimizu, M.; Xiao, D.H.; Nuntanakorn, P.; Lim, J.T.E.; Suzui, M.; Seter, C.; Pertel, T.; Kennelly, E.J.; Kronenberg, F.; Weinstein, I.B. *Breast Cancer Res. Treat.* **2004**, *83*, 221-231.
54. Tian, Z.; Yang, M.S.; Huang, F.; Li, K.G.; Si, J.Y.; Chen, S.L.; Chen, S.B.; Xiao, P.G. *Cancer Lett.* **2005**, *226*, 65-75.
55. Sakurai, N.; Wu, J.H.; Sashida, Y.; Mimaki, Y.; Nikaido, T.; Koike, K.; Itokawa, H.; Lee, K.H. *Bioorg. Med. Chem. Lett.* **2004**, *14*, 1329-1332.
56. Takahira, M.; Kusano, A.; Shibano, M.; Kusano, G.; Koizumi, K.; Suzuki, R.; Kim, H.S.; Wataya, Y. *Biol. Pharm. Bull.* **1998**, *21*, 823-828.
57. Takahira, M.; Kusano, A.; Shibano, M.; Kusano, G.; Sakurai, N.; Nagai, M.; Miyase, T. *Chem. Pharm. Bull.* **1998**, *46*, 362-365.
58. Takahira, M.; Kusano, A.; Shibano, M.; Kusano, G.; Miyase, T. *Phytochemistry* **1998**, *49*, 2115-2119.
59. Chen, S.N.; Fabricant, D.S.; Lu, Z.Z.; Zhang, H.J.; Fong, H.H.S.; Farnsworth, N.R. *Phytochemistry* **2002**, *61*, 409-413.
60. Graf, E. *Free Rad. Biol. Med.* **1992**, *13*, 435-448.
61. Burdette, J.E.; Chen, S.N.; Lu, Z.Z.; Xu, H.Y.; White, B.E.P.; Fabricant, D.S.; Liu, J.H.; Fong, H.H.S.; Farnsworth, N.R.; Constantinou, A.I.; Breemen, R.B.; Pezzuto, J.M.; Bolton, J.L. *J. Agric. Food Chem.* **2002**, *50*, 7022-7028.
62. Hirabayashi, T.; Ochiai, H.; Sakai, S.; Nakajima, K.; Terasawa, K. *Planta Med.* **1995**, *61*, 221-226.
63. Noguchi, M.; Nagai, M.; Koeda, M.; Nakayama, S.; Sakurai, N.; Takahira, M.; Kusano, G. *Biol. Pharm. Bull.* **1998**, *21*, 1163-1168.
64. Kruse, S.O.; Lohning, A.; Pauli, G.F.; Winterhoff, H.; Nahrstedt, A. *Planta Med.* **1999**, *65*, 763-764.

65. Liu, I.M.; Chi, T.C.; Hsu, F.L.; Chen, C.F.; Cheng, J.T. *Planta Med.* **1999**, *65*, 712-714.
66. Loser, B.; Kruse, S.O.; Melzig, M.F.; Nahrstedt, A. *Planta Med.* **2000**, *66*, 751-753.
67. Kusano, A.; Seyama, Y.; Nagai, M.; Shibano, M.; Kusano, G. *Biol. Pharm. Bull.* **2001**, *24*, 1198-1201.
68. Kondo, Y.; Takemoto, T. *Chem. Pharm. Bull.* **1972**, *20*, 1940-1942.
69. Ito, M.; Kondo, Y.; Takemoto, T. *Chem. Pharm. Bull.* **1976**, *24*, 580-582.
70. Li, C.J.; Chen, D.H.; Xiao, P.G. *Zhong Cao Yao.* **1995**, *26*, 288-290.
71. Cai, P.; Pu, X.F.; Peng, S.L.; Zhang, X.R.; Ding, L.S. *J. Asian Nat. Prod. Res.* **2005**, *7*, 145-149.
72. Lai, B.; Kansal, V.K.; Singh, R.; Sankar, C.; Kulkarni, A.; Gund, V. *Indian J. Chem.* **1998**, *37B*, 881-893.
73. Fabricant, D.S.; Nikolic, D.; Lankin, D.C.; Chen, S.N.; Jaki, B.U.; Krunic, A.; van Breemen, R.B.; Fong, H.H.S.; Farnsworth, N.R.; Pauli, G.F. *J. Nat. Prod.* **2005**, *68*, 1266-1270.
74. Baba, K.; Kozawa, M.; Hata, K.; Ishida, T.; Inoue, M. *Chem. Pharm. Bull.* **1981**, *29*, 2182-2187.
75. Kusano, G.; Shibano, M.; Idoji, M.; Minoura, K. *Heterocycles* **1993**, *36*, 2367-2371.
76. Banker, G.S. In *Modern Pharmaceutics*, Banker, G.S.; Rhodes, C.T., Eds.; Marcel Dekker Inc.: New York, 2002, pp 1-22.
77. Applequist, W.L. *Flora* **2003**, *198*, 358–365,.
78. Xu, H.; Fabricant, D.S.; Piersen, C.E.; Bolton, J.L.; Pezzuto, J.A.; Fong, H.; Totura, S.; Farnsworth, N.R.; Constantinou, A.I. *Phytomedicine* **2002**, *9*, 757-762.
79. Zerega, N.J.C.; Mori, S.; Lindqvist, C.; Zheng, Q.Y.; Motley, T.J. *Econom. Bot.* **2002**, *56*, 154-164.
80. Wang, Z.; Sun, S.Q.; Li, X.B.; Zhou, Q.; Lin, L.; Du, D.G. *Spectro. Spectral. Anal.* **2001**, *21*, 311-313.
81. Wang, H.K.; Sakurai, N.; Shih, C.Y.; Lee, K.H. *J. Agri. Food Chem.* **2005**, *53*, 1379-1386.
82. Jiang, B.; Yang, H.; Nuntanakorn, P.; Balick, M.J.; Kronenberg, F.; Kennelly, E.J. *J. Ethnopharmacol.* **2000**, *96*, 521-528.
83. He, K.; Zheng, B.L.; Kim, C.H.; Rogers, L.; Zheng, Q.Y. *Planta Med.* **2000**, *66*, 635-640.
84. Struck, D.; Tegtmeier, M.; Harnischfeger, G. *Planta Med.* **1997**, *63*, 289-289.
85. Ganzera, M.; Bedir, E.; Khan, I.A. *Chromatographia* **2000**, *52*, 301-304.
86. Kong, L.; Li, X.; Zou, H.F.; Wang, H.L.; Mao, X.Q.; Zhang, Q.; Ni, J.Y. *J. Chromatogr. A* **2001**, *936*, 111-118.

87. Li, W.K.; Chen, S.N.; Fabricant, D.; Angerhofer, C.K.; Fong, H.H.S.; Farnsworth, N.R.; Fitzloff, J.F. *Anal. Chim. Acta* **2002**, *471*, 61-75.
88. Jiang, B.; Kronenberg, F.; Nuntanakorn, P.; Qiu, M.H.; Kennelly, E.J. *J. Agric. Food Chem.* **2006**, *54*, 3242-3253.
89. He, K.; Zheng, B.L.; Kim, C.H.; Rogers, L.L.; Zheng, Q.Y. In *Quality Management of Nutraceuticals*, Ho, C.-T.; Zheng, Q.Y., Eds., ACS Symp. Ser. 803, American Chemical Society: Washington, D.C., 2002, pp. 90-116.
90. Pan, R.L.; Chen, D.H.; Shen, L.G.; Si, J.Y.; Hu, Z.L.; Zhu, C.D. *Chin. J. Pharm. Anal.* **2000**, *20*, 396-398.
91. Schmidt, A.H. *J. Liquid Chromatogr. Tech.* **2005**, *28*, 871-881.
92. Panossian, A.; Danielyan, A.; Mamikonyan, G.; Wikman, G. *Phytochem. Anal.* **2004**, *15*, 100-108.
93. Bergeron, C.; Gafner, S.; Batcha, L.L.; Angerhofer, C.K. *J. Agric. Food Chem.* **2002**, *50*, 3967-3970.
94. Li, W.K.; Sun, Y.K.; van Breemen, R.B.; Fitzloff J.F. *J. Liquid Chromatogr. Tech.* **2004**, *27*, 2507-2529.

Chapter 6

Chemistry, Quality, and Functional Properties of Grains of Paradise (*Aframomum melegueta*), a Rediscovered Spice

H. Rodolfo Juliani[1], Cara Welch[1], Juliana Asante-Dartey[2], Dan Acquaye[2], Mingfu Wang[1], and James E. Simon[1]

[1]New Use Agriculture and Natural Plant Products Program, Cook College, and the New Jersey Agricultural Experiment Station (NJAES), Rutgers, The State University of New Jersey, 59 Dudley Road, New Brunswick, NJ 08901–8520
[2]Agribusiness in Sustainable Natural African Plant Products, H/NO C 205/29, Mempeasem, East Legon, Accra, Ghana

This study evaluates the variation of quality, essential oil and polyphenol profile and antioxidant activity of Grains of Paradise (*Aframomum melegueta* (Roscol) K. Schum) from Ghana and reviews the pharmacological properties of this species. This work has shown the potential of Grains of Paradise, a forgotten spice to provide the western world with new flavors and taste, as it contains higher levels of pungent principles ([6]-gingerol) than ginger and offers a distinct flavor and aroma and functional properties.

Zingiberaceae is the largest family of the Order Zingiberales and is found through the tropics This family, predominantly Asian, with 40 genera and ca. 900 species has provided economically important species which have established themselves as aromatic spices. The genus *Zingiber* (ginger), *Curcuma* (turmeric), *Alpinia* (galanga) and *Kaempferia* all represent rhizomatous spices, while the "cardamoms" include seed spices of the genus *Elettaria* (small cardamom), *Amomum* and *Aframomum* are known as large cardamoms (*1*). The genus *Aframomum* includes 50 species that are restricted to tropical Africa. The seeds of *A. melegueta* (Roscol) K.Schum are known as the grains of paradise, melegueta pepper or guinea peppers.

Spices are not only valuable to add flavor to foods, but their functional properties have been recognized since ancient times (*2,3*). Some spices includeing ginger were used as food preservatives (*4*), herbal medicines to cure urinary complaints, piles, jaundice, to increase flow of saliva and aid digestion (*3*).

In western Africa, grains of paradise have been extensively used as condiments (*5,6*), but also in popular medicine for the treatment of coughs, bronchitis, rheumatitis, digestive complaints (*7*), as aphrodisiac and antiparasitic (*6*).

Five centuries ago grains of paradise was very popular in Europe as a substitute of true pepper (*3,8*) though today, it is almost unknown in the modern western world. The global spice trade is expected to increase with the growing consumer demand in importing countries for more exotic, ethnic tastes in food (*2*), and this tendency will create new commercial opportunities for the developing world.

Although the essential oil and phenolic composition in *A. melegueta* has been studied, the variability of their components and quality aspects are less understood. Such lack of scientific information on the chemical composition, quality parameters and functional properties of natural products, may limit the access to markets (*9,10,11*). Thus, by providing users and the international community with consistent and defined products new commercial opportunities for local communities can be developed and a wider range of consumers could enjoy and gain access to this old yet "new spice".

In view of the commercial potential of grains of paradise, this study evaluates the variation of quality, essential oil and polyphenol profile and antioxidant activity of *A. melegueta* from Ghana and reviews the pharmacological properties of this species.

Material and Methods

Plant material

As part of our quality control program on African natural products, we conducted quality control and chemical analyses of grains of paradise from Akante (Ghana). The sun dried seed were received from 2003 to 2005 (samples 2003, 2004 and 2005a-b). The 2003 sample was grounded and separated in the seed coat (known as chaffs) and the whitish endosperm (plus the embryo). A commercial sample of ginger rhizome (New Jersey, local store) was included for comparison purposes. Each procedure was run at least in duplicate.

Quality Control Analysis

Moisture, total ashes, total insoluble ashes and essential oil content were assessed for each sample using methods described by the Food Chemical Codex (*12*).

Chemical and Antioxidant Activity Analysis

The ground seeds (200 mg) were extracted with 25 mL of methanol through sonication for one hour and subjected to HPLC, total phenol content, and antioxidant activity analyses. Two commercial ginger samples (1.0 g of fresh ginger) was also extracted with 25 mL of methanol and sonicated for one hour as a comparison and the dry weight was factored in at the time of calculations. The total phenols were measured using the Folin Ciocalteu's reagent (*13*) and the results were expressed as a percent of gallic acid equivalents on a dry weight basis (m/m). The antioxidant activity was evaluated using the ABTS method3 (*14*) and the results were expressed as a percent of Trolox (a water soluble analog of vitamin E) on a dry weight basis (m/m).

Quantitative HPLC Analysis

The analysis was performed on a Waters 2695 separations module with photodiode array detector. The HPLC was run on a Phenomenex Luna C18 column (5 μm, 250 x 4.6 mm) and the detection wavelength was 282 nm, the injection volume was 20 μL and flow rate was 1.0mL/min. The mobile phase was water (solvent A) and acetonitrile (solvent B) in the following gradient system: 0 min (55% A and 45% B), 8 min (50% A and 50% B), 15 min (45% A and 55% B), 40 min (10% A and 90% B), 45 min (55% A and 45% B), 55 min (55% A and 45% B). The total running time was 55 minutes with no post running time. Dihydrocapsaicin was used for the quantification of [6]-gingerol and the results were expressed as a percent of [6]-gingerol on a dry weight basis (m/m). [6]-Gingerol was tentatively analyzed by HPLC-MS by comparing the UV and MS spectra with the reference standards and by their $[M+1]^+$ and $[M+Na]^+$ ions.

Essential Oil Analysis

The essential oils were analyzed by a gas chromatograph (GC) coupled to a mass spectrometer (MS) (Agilent GC System 6890 Series, Mass Selective Detector, Agilent 5973 Network, FID detector) (*15*).

Results and Discussion

Quality Standards

The grains of paradise is a small seed (around 2 mm), the testa is brown or brown-reddish and the endosperm is white. The aroma is spicy and woody and the taste is sharp and pungent (Table I). Most of the samples were characterized by low levels of fine particles though the samples from the year 2004 showed higher levels. The foreign matter ranged from 1 to 1.45% (Table I).

The moisture percent ranged from 9.6% whole seeds (2003) and higher amount was observed for the 2004 sample, the required maximum standards for moisture is 11% according to international standards for spices (e.g. pepper).

The bulk density, an important measure in filling retail containers, was highest in the whole seeds (2003) and lower amount in the seeds of the years 2004-2005, the endosperm and chaffs were less dense (412 and 260, respectively).

The highest ash content was observed in the chaffs (6%), while for the endosperm and the whole seeds ranged from 2 to 4%. The total insoluble ashes, a classic determination of cleanliness (contamination by sand and earth) showed values lower than 1%. The international standard values for black pepper and ginger are 1.5% and 2%, respectively (*16*).

The moisture (8.4%) and total ashes (3.9%) has also been described for *A. melegueta* from Nigeria (*17*).

The seed showed low levels of essential oils ranging from 0.25 to 0.6%. The essential oil of the seed coat and endosperm showed no significant differences (Table I). The 2003 and 2004 samples showed the highest levels (0.5-0.6%, respectively), while the 2005a-b samples showed lower levels (0.25%). Other reports showed also lower essential oil levels (0.3-.75%, 0.21%, 0.8%) (*8, 18, 6*). Ginger rhizomes exhibited higher levels of essential oils (1%) when compared with *A. melegueta* seeds.

Phenolic Content

The highest amount of phenols were found in the seed coat while the whitish endosperm showed lower levels (2.2%), the total phenols in the whole seeds varied from 2 to 2.8% (Table I).

The HPLC analysis showed that *A. melegueta* seeds are dominated by high levels of [6]-gingerol, a component also found in ginger as a major component (Figure 1). Our analysis showed that grains of paradise yielded higher levels of [6]-gingerol, ranging from 1 to 1.6% while the initial content on two local

Table I. Quality Control Parameters of Grains of Paradise (*Aframomum melegueta*) from Ghana

Type	*Endosperm* 2003	*Chaff* 2003	*Whole kernels* 2003	2004	2005a	2005b
Color	Cream	Brown	Dark brown reddish		Dark brown	
Taste			Hot and spicy			
Fine particles (%, m/m)	-[1]	-	0.3[2]	1.0[2]	0.1[2]	0.1[2]
Foreign matter (%, m/m)	-	-	1.4 ± 0.1	1.0[2]	1.5 ± 0.1	1.3[2]
Moisture (%, m/m)	8.8[2]	9.9[2]	9.6[2]	10.4[2]	9.7[2]	9.3[2]
Bulk Density (%, m/m)	412.3 ± 8	259.7± 1	717.8±2	685.0 ± 5	656.7 ±2	668.3 ± 5
Total Ashes (%, m/m)	3.3[2]	6.2[2]	3.8[2]	3.2[2]	2.0[2]	2.1[2]
Insoluble Ashes (%, m/m)	0.6[2]	0.4[2]	0.7[2]	0.36[2]	0.9[2]	0.8[2]
EO (%, m/v)	0.6[2]	0.47[2]	0.5[2]	0.6[2]	0.25[2]	0.25[2]
Total Phenols (%, m/m)	2.22[2]	3.33[2]	2.02[2]	2.1 ± 0.2	2.8[2]	2.56[2]
Antioxidant activity (g Trolox/100 g)	-	-	-	4.9± 0.2	5.8[2]	5.7±0.2
[6]-gingerol (%, m/m)	0.95[2]	1.49[2]	1.03[2]	1.05[2]	1.62[2]	1.50[2]

1-Not tested, 2 – Standard Error less than 0.1

commercial ginger rhizome were only 0.45% and 0.75% (Figure 1). The major pungent principle in ginger has been reported to be [6]-gingerol (*19*).

Other researchers have reported that the phenolic composition of grains of paradise from Ghana showed high levels of [6]- and [7]-paradol and [6]-shogaol (*20*). The crude methanolic extract of grains of paradise from Nigeria yielded four major components, gingerdione (1.7%), [6]-gingerol (1.3%), [6]-paradol (0.9%) and shogaol (0.8%) (*21*). However, another study reported that shogaol is produced by the thermal dehydration of gingerols and not naturally present in the plant (*22*).

Although [6]-gingerol and total phenols in general are important indicators of quality, they are not usually considered nor listed in international standards for spices. Our results showed that the total phenols using the Folin Ciocalteu's

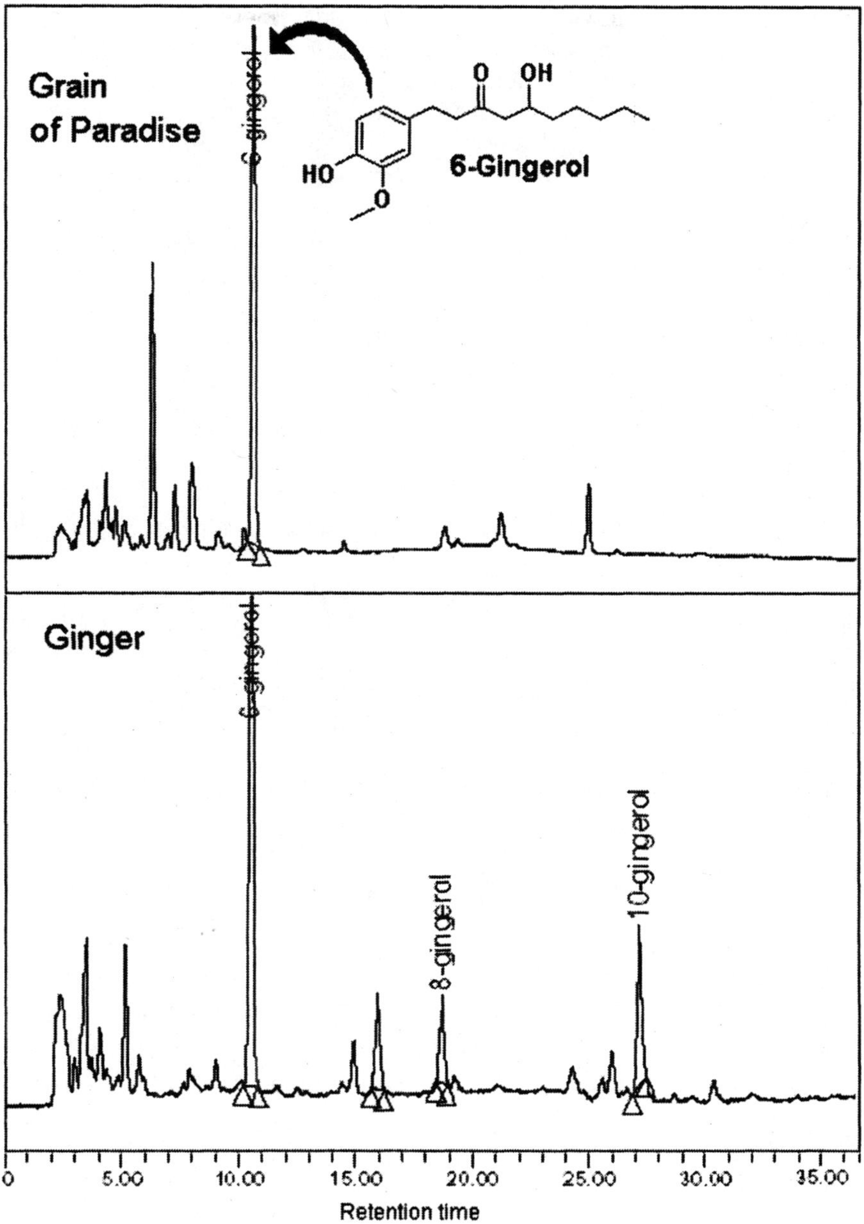

Figure 1. HPLC profile of the methanolic extract of Grains of Paradise (top) and Ginger (bottom), showing the main phenol [6]-gingerol (upper right).

reagent (*13*) (and possible antioxidant activity) (*14*), is a simple colorimetric method (in comparison with HPLC) to estimate the amount of [6]-gingerol in quality control programs, as there was a good correlation between the total phenols and [6]-gingerol content in *A. melegueta* (R^2=0.63). Our results suggest that the minimum amount of total phenols and [6]-gingerol in grains of paradise should be 2% and 1%, respectively.

In addition, we also found that the grains of paradise extracts were able to scavenge preformed free radical monocation(ABTS$^{\bullet+}$) (*14*), thus showing antioxidant activity. In the samples 2004-2005, the antioxidant activity of 100 g of grains of paradise were equivalent to 5 to 6 g of vitamin E (Table I).

Essential Oil Composition

The majority of the *A. melegueta* samples were characterized by high levels of sesquiterpenes (79-93%) (Table II). Almost all the samples were dominated by high levels of α-humulene (45-53%) and (*E*)-caryophyllene (23-29%), and lower levels of linalool (5-1.6%) and an unidentified monoterpene (0.3-8%). The 2005a seeds showed higher levels of monoterpenes, that were dominated by limonene/1,8 cineole (14%), 3-carene, α– and β- pinenes (8%, respectively) and lower levels of (*E*)-caryophyllene (10%) and α-humulene (26%). These results showed a variation in the chemical composition of essential of *A. melegueta.* The essential oil of ginger was also characterized by higher levels of monoterpenes that were dominated by neral (11%) and geranial (19%), components that give ginger the characteristic lemony aroma. Ginger was also characterized by its typical sesquiterpenes, a-zingiberene (17%), b-bisabolene (7%), b-sesquiphell-andrene (4%) and Ar-curcumene (4%) (Table II).

Another report also described essential oils of *Aframomum melegueta* to contain high levels of (*E*)-caryophyllene (22%) and α-humulene (61%) (*18*) . A sample from Cameroon showed a different essential oil profile, with the oil being dominated by α-humulene (31%), humulene oxide (28%), caryophyllene oxide (18%) and (E)-caryophyllene (9%) (*6*), while a sample from Central African Republic was dominated by α-pinene (>30%) (*23*). These reports confirmed the intraspecific variation of grains of paradise essential oils from Ghana (Table II).

Functional Properties

Although the literature on *A. melegueta* is scarce, the functional properties of such botanicals will depend on its active principles, e.g. essential oils (*24*) and polyphenols (*11, 25*).

The methanolic extract of *A. melegueta* fruits exhibited significant antimicrobial properties against Gram (+), Gram (-) bacteria and fungi (*26*). The ethanolic extracts of the seeds also showed activities against *Escherichia coli* and *Bacillus cereus*. The extract also showed antioxidant properties that inihibited the formation of peroxides in groundnut oil (*17*). These results suggest that the oleoresin of grains of paradise could be used as food preservatives, in the same way of ginger (e.g. to extend shelf life of cheese) (*4*).

Other studies has demonstrated that *A. melegueta* phenols (gingerdione, [6]-gingerol, [6]-paradol and shogaol) showed strong antifeedant activities against termite workers. The results suggested a protective role for these components against seed predators in their natural habitats. These components could be used as alternative forms of insect control (*21*). This observation is supported by the fact that the seed coat (chaffs) accumulated higher amount of total phenols and [6]-gingerol (Table I), thus protecting the endosperm and embryo against insect attack.

Others have observed that [6]-gingerol and other ginger phenols exhibited antitumor-promoting and antiproliferative activities (*27,28*). [6]-Gingerol inhibited angiogenesis that may be useful in the treatment of tumors and other angiogenesis-dependent diseases by selective inhibition of neovessel formation at the tumor site (*28*).

Gingerol related compounds from ginger had inhibitory effect on oxidation of methyl linoleate and induced-oxidation of liposomes (*29*). As lipid peroxidation may play a very important role in cell proliferation especially those of tumors (*30,31*), thus lipid peroxidation control could be a mechanism of action of antioxidants as anti-tumoral. Our results on the antioxidant activity of *A. melegueta* (Table I), support such observations.

Cyclooxygenase-2 (COX-2) a key enzyme in the prostaglandin biosynthesis is considered as a molecular target of many chemopreventive as well as anti-inflammatory agents. [6]-gingerol has been reported to inhibit tumor promotion through the inhibition the COX-2 expression (*32*). Topical application of [6]-gingerol significantly lower the incidence of initiated papilloma formation in mouse skin and significantly suppressed tumor promotion induced by inflammation (*33*).

[6]-Gingerol possesses antiinflamatory and analgesic activities. The antinociceptive effect of [6]-gingerol may be attributed to inhibition of prostaglandin release and other mediators inhibition of paw edema induced by carrageenin (*34*). Other reports has also shown that ginger extracts had high anti-inflammatory activities, fractions containing gingerols showed potent inhibition of LPS-stimulated PGE2 production, comparable to indomethacin (*19*). Gingerol components and their derivatives showed a more potent anti-platelet action than aspirin (*35*). A patent on the use of grains of paradise as antiinflamatory has been filled and also a cosmetic application to improve aesthetic appearance of skin using *A. melegueta* as one of the ingredients (*36, 37*).

Table II. Essential oil composition of Grains of Paradise and Ginger

	Endosp	Chaff	Whole seeds				Ginger
	2003	2003	2003	2004	2005a	2005b	2005
α-Pinene	-[1]	-	-	-	8.3 [2]	-	1.7
Sabinene	-	-	-	-	4.0	-	0.0
β-Pinene	-	-	-	-	7.9	-	0.2
Myrcene	-	-	-	-	2.9	-	1.2
α-Phellandrene	-	-	-	0.0	3.7	-	0.4
(+)-3-Carene	-	-	-	-	9.3	-	-
α-Terpinene	-	-	-	-	1.3	-	-
p-Cymene	0.0	-	-	-	1.3	-	-
Limonene+1,8-Cineole	-	-	-	-	13.8	-	-
β-Phellandrene+1,8-Cineole	-	-	-	-	-	-	12.3
(Z)-b-Ocimene	0.0	-	-	0.1	0.3	0.3	-
Unknown	1.2	0.3	1.1	1.6	3.1	7.6	-
(E)-β-Ocimene	0.4	-	0.1	2.1	3.3	8.1	-
Terpinolene	0.2	0.03	0.1	0.0		0.0	0.2
Linalool	3.3	2.1	1.8	1.6	1.3	4.7	0.7
α-Terpineol	-	-	-	-	-	-	1.1
Unknown	0.4	0.2	0.6	0.2	-	0.2	-
Neral	-	-	-	-	-	-	11.2
Geranial	-	-	-	-	-	-	19.1
Bornyl-acetate	0.6	0.5	-	0.2	-	0.0	0.0
β-Elemene	1.7	0.8	0.3	0.08	-	-	0.2
(E)-Caryophyllene	24.9	21.6	29.3	27.2	10.1	22.5	0.2
α-Humulene	45.4	51.0	53.9	56.5	25.6	53.3	1.0
Allo-Aromadendrene	2.2	2.2			-	-	-
α-Muurolene	3.3	2.4	0.2	0.0	-	-	0.1
Germacrene-D	0.8	0.6	0.0	0.3	-	0.0	-
AR-Curcumene	-	-	-	-	-	-	3.5
α-Selinene	1.2	0.9	0.3	-	-	-	-
α-Zingiberene	-	-	-	-	-	-	16.6
α-Bisabolene	-	-	-	2.1	0.1	1.4	-
α-Bulnesene	0.4	0.3	-	-	-	-	-
β-Bisabolene	0.7	0.6	0.1	0.2	-	-	6.6
Unknown	1.0	1.1	0.2	-	-	-	0.4

Table II. *Continued.*

	Endosp	Chaff	Whole seeds				Ginger
	2003	2003	2003	2004	2005a	2005b	2005
β-Sesquiphellandrene	-	-	-	-	-	-	4.2
Nerolidol	1.2	2.1	1.1	1.1	-	0.1	0.9
α-Caryophyllene alcohol	0.1	0.0	0.0	-	-	-	-
Caryophyllene-oxide	0.7	1.3	1.3	2.0	0.9	0.2	-
Unknown	1.2	2.5	2.2	3.1	0.2	1.2	-
Monoterpenes	6.2	3.1	3.7	5.9	60.5	20.9	48.2
Sesquiterpenes	84.8	87.5	89	92.6	36.9	78.7	33.7
Total-analyzed	91	90.6	92.7	98.5	97.4	99.7	81.9

1- not detected or in traces amounts (<0.01%), 2–Relative percent of the total oil

The popular uses of *A. meleguta* for digestive complaints suggest that its essential oil may also exert digestive properties, since other essential oils containing α-pinene. caryophyllene and caryophyllene oxide reported by literature (*38, 39*), showed antispasmodic properties. Ginger rhizomes containing the pungent principles [6]-gingerol and [10]-gingerol increased bile secretion in rats (*40*). [6]-gingerol also exerted hepatoprotective accion against carbon tetrachloride and galactosamine induced cytotoxicity in rat hepatocytes (*41*). All these observation support the uses of *A. melegueta* for digestive complaints.

The high content of Zn in *A. melegueta*, may be of interest in improving human nutrition under situations of Zn deficiency and its elevated concentrations of Se, could exert cancer prevention properties. No toxic heavy metals such as Cd, As, Pb, Hg were detected in those Nigerian samples, and the spice was reported as safe to consume (*42*).

This work has shown the potential of grains of paradise, a forgotten spice in the western world, yet an old and traditionally popular spice in the countries of its origin, should be re-examined for its spice and functional food value. Grains of Paradise can provide the western world with yet a new flavor and taste, and it contains higher levels of pungent principles than ginger. This species could also have other additional and important health benefits due to the antioxidant properties of their constituents, these activities could include antiinflamatory, antitumor, analgesic, and digestive. Its essential oil and oleoresin also shows promise for industrial applications. Further research is warranted to support the safety and efficacy of grains of paradise and its functional food applications despite its long use and history in the countries of origin.

Table III. Proposed quality standards for whole seeds of *Aframomum melegueta* from Ghana

Characteristic	*Requirement*
Color	Brown, brown/reddish
Aroma	Spicy, woody, free from foreign odors
Taste	Sharp pungent
Excreta (mammalian and others)	Practically free
Extraneous foreign matter (%, m/m) maximum	1
Fine particles	1
Moisture (%, m/m) maximum	10
Bulk Density (g/l) minimum	650
Total Ashes (%, m/m) maximum	4
Insoluble Ashes (%, m/m) maximum	1
Essential oil content (%, m/m) minimum	0.2
Total Phenols (%, m/m) minimum	2
[6]-gingerol (%, m/m) minimum	1

Since the lack of quality standards may limit access to markets, we propose the following initial standards for grains of paradise to provide a framework to the users and international community for a consistent and defined grains of paradise product (Table III). The development of clear grades and standards for grains of paradise should provide a foundation upon which processors, producers as well as buyers and users can objectively define this product to the enjoyment and health benefit of consumers

Acknowledgements

We thank the Ghanaian communities involved in these domestication and commercialization studies, who, with assistance from the ASNAPP project, were among the first to actually begin to export this specialized tea collected from the wild. In particular, we thank Kodzo Gbewonyo, BioResources of Ghana and New Jesey for his interest and support of this work. We thank Carol Wilson, USAID Chief Technical Officer of our Partnership for Food and Industry in Natural Products (PFID/NP) project supported by the Office of Economic Growth, Agriculture and Trade (EGAT/AG) of the USAID (Contract Award No. AEG-A-00-04-00012-00) in support of their global economic development programs. We also thank Jerry Brown, USAID project officer, for his support and encouragement as this work originally began as part of the Agri-Business in

Sustainable African Natural Plant Products Program (ASNAPP) with funding from the USAID (Contract Award No. HFM-O-00-01-00116). Finally, we recognize and thank the New Use Agriculture and Natural Plant Products Program (NUANPP) and the New Jersey Agricultural Experiment Station, Rutgers University.

References

1. Madhusoodanan, K.J.; Saideswara Rao, Y. In *Handbook of Herbs and Spices*; Peter, K.V., Ed.; CRC Press: Boca Raton, FL, 2001; pp. 134-142.
2. Peter, K.V. In *Handbook of Herbs and Spices*; Peter, K.V., Ed.; CRC Press: Boca Raton, FL, 2001; pp. 1-12.
3. Rosengarten, F. *The Book of Spices*. Livinsgton Publishing Company: Philadelphia, PA, 1969.
4. Belewu, M.A.; Belewu, K.Y.B.; Nkwunonwo, C.C. *African Journal of Biotechnology* **2005**, *4*, 1076-1079.
5. Dahlgre, R.M.T.; Clifford, H.T.; Yeo, P.F. *The Families of the Monocotyledons*, Springer Verlag: Berlin, Germany, 1985.
6. Menut, C.; Lamaty, G.; Amvam Zollo, P.H.; Atogho, B.M.; Abondo, R.; Bessiere, J.M. *Flavour Fragrance J.* **1991**, *6*, 183-186.
7. Ayiku, M.N.B. *Ghana Herbal Pharmacopeia*. Technology Transfer Center: Accra, Ghana, 1992, pp. 86-88.
8. Guenther E. *The Essential Oils, Vol. 5*, Van Nostrand Reinhold: New York, 1949, p 104.
9. Juliani, H.R.; Simon, J.E.; Ramboatiana, M.M.R.; Behra, O.; Garvey, A.; Raskin, I. *Acta Horticulturae* **2004**, *629*, 77-81.
10. Juliani, H.R.; Behra, O.; Moharram, H.; Ranarivelo, L.; Ralijerson, B.; Andriantsiferana, M.; Ranjatoson, N.; Rasoarahona, J.; Ramanoelina, P.; Wang, M.; Simon, J.E. *Perfumer and Flavorist* **2005**, *30*, 60-65.
11. Juliani, H.R.; Wang, M.; Asante-Dartey, J.; Acquaye, D.; Moharram, H.; Koroch, A.R.; Simon, J.E. In: *Herbs: Challenges in Chemistry and Biology*, Wang, M.; Sang, S.; Hwang, L.S.; Ho, C.-T., Eds.; ACS Symp. Ser. 925, American Chemical Society: Washington, D.C., 2006, pp. 126-142.
12. Committee on Food Chemical Codex. Food Chemical Codex. National Academy Press: Washington, DC, 1996.
13. Gao, X.; Bjork, L.; Trajkovski, V.; Uggla, M. *J. Sci. Food Agric.* **2000**, *80*, 2021-2027.
14. Re, R.; Pellegrini, N.; Proteggente, A.; Pannala, A.; Yang, M.; Rice Evans, C. *Free Rad. Biol. Med.* **1999**, *26*, 1231-1237.
15. Juliani, H.R.; Zygadlo, J.A.; Scrivanti, R.; de La Sota, E.; Simon, J.E. *Flavour Fragr. J.* **2004**, *19*, 541-543.

16. Muggeridge, M.; Foods, L.; Clay, M. In *Handbook of Herbs and Spices*, Peter, K.V., Ed.; CRC Press: Boca Raton, FL, 2001.
17. Adegoke, G.O.; Makinde, O.; Falade, K.O.; Peters, P.I.U. *Eur. Food Res. Technol.* **2003**, *216*, 526-528.
18. Ajaiyeoba, E.O.; Ekundayo, O. *Flavour Fragrance J.* **1999**, *14*, 109-111.
19. Jolad, S.D.; Lantz, R.C.; Chen, G.J.; Bates, R.B.; Timmermann, B.N. *Phytochem.* **2005**, *66*, 1614-1635.
20. Tackie, A.N.; Dwuma, B.D.; Ayim, J.S.K.; Dabra, T.T.; Knapp, J.E.; Slatkin, D.J.; Schiff, P.L., Jr. *Phytochem.* **1975**, *14*, 853-854.
21. Escoubas, P.; Lajide, L.; Mizutani, J. *Phytochem.* **1995**, *40*, 4, 1097-1099.
22. Chen, C.C.; Rosen, R.T.; Ho, C.-T. *J. Chromatogr.* **1986**, *360*, 175-184.
23. Lamaty, G.; Menut, C.; Koudou, J.; Regnier, P.; Bessiere, J.M. *J. Essential Oil Res.* **1993**, *5*, 81-83.
24. Koroch, A.R.; Zygadlo, J.A.; Juliani, H.R. In *Flavours-Chemistry, Technology and Resources*, Berger, R.G., Ed.; 2006, in press.
25. Wang, M.; Juliani, R.; Simon, J.E.; Ekanem, A.; Liang, C.P.; Ho, C.-T. In *Phenolic Compounds in Foods and Natural Health Products*, Shahidi, F.; Ho, C.-T., Eds.; ACS Symposium Series 909. American Chemical Society: Washington DC., 2005, pp. 118-142.
26. Konning, G.H.; Agyare, C.; Ennison, B. *Fitoterapia* **2004**, *75*, 65-67.
27. Surh, Y.J. *Mutation Res.* **1999**, *428*, 305-327.
28. Kim, E.C.; Min, J.K.; Kim, T.Y.; Lee, S.J.; Yang, H.O.; Han, S.H.; Kim, Y.M.; Kwon, Y.G. *Biochem. Biophys. Res. Commun.* **2005**, *335*, 2, 300-308.
29. Masuda, Y.; Kikuzaki, H.; Hisamoto, M.; Nakatani, N. *BioFactors* **2004**, *21*, 293-296.
30. Udilova, N.; Jurek, D.; Marian, B.; Gille, L.; Schulte, H.R.; Nohl, H. *Food Chem. Toxicol.* **2003**, *41*, 1481.
31. Gonzalez, M.J. *Medical Hypotheses* **1992**, *38*, 106.
32. Kim, S.O.; Chun, K.S.; Kundu, J.K.; Surh, Y.J. *BioFactors* **2004**, *21*, 27-31.
33. Park, K.K.; Chun, K.S.; Lee, J.M.; Lee, S.S.; Surh, Y.J. *Cancer Lett.* **1998**, *129*, 139-144
34. Young, H.Y.; Luo, Y.L.; Cheng, H.Y.; Hsieh, W.C.; Liao, J.C.; Peng, W.H. *J. Ethnopharmacology* **2005**, *96*, 207-210.
35. Nurtjahja, T.E.; Ammit, A.J.; Roufogalis, B.D.; Tran, V.H.; Duke, C.C. *Thrombosis Res.* **2003**, *111*, 259-265.
36. Raskin, I.; Ilic, N. U.S. Pat. Appl. Publ. 2005-134195, 2005.
37. Mahalingam, H.; Kyrou, C.D.; Traudt, M.; Ptchelintsev, D. U.S. Pat. Appl. Publ. 741,383, 2004.
38. Camara, C.C.; Nascimento, N.R.F.; Macedo-Filho, C.L.; Almeida, F.B.S.; Fonteles, M.C. *Planta Med.* **2003**, *69*, 1080.
39. Sadraei, H.; Asghari, G.; Naddafi, A. *Phytother. Res.* **2003**, *17*, 645.

40. Yamahara, J.; Miki, K.; Chisaka, T.; Sawada, T.; Fujimura, H.; Tomimatsu, T.; Nakano, K.; Nohara, T. J. *Ethnopharmacology* **1985**, *13*, 217-225.
41. Hikino, H.; Kiso, Y.; Kato, N.; Hamada, Y.; Shioiri, T.; Aiyama, R.; Itokawa, H.; Kiuchi, F.; Sankawa, U. *J. Ethnopharmacology* **1985**, *14*, 31-39.
42. Obiajunwa, E.I.; Adeleke, C.A.; Olanrewaju, R.O. *J. Rad. Nuclear Chem.* **2002**, *252*, 473-476.

Chapter 7

Xylopia aethiopia (*Annonaceae*): Chemistry, Traditional Uses and Functional Properties of an "African Pepper"

H. Rodolfo Juliani[1], TaeOh Kwon[1,3], Adolfina R. Koroch[1], Julie Asante-Dartey[2], Dan Acquaye[2], and James E. Simon[1]

[1]New Use Agriculture and Natural Plant Products Program, Cook College, and the New Jersey Agricultural Experiment Station (NJAES), Rutgers, The State University of New Jersey, 59 Dudley Road, New Brunswick, NJ 08901–8520
[2]Agribusiness in Sustainable Natural African Plant Products, H/NO C 205/29, Mempeasem, East Legon, Accra, Ghana
[3]Joosan Scholarship Foundation (2005), Wonkwang University, Chunbuk 570–749, Korea

This review examines the natural products in *Xylopia aethiopia*, an African pepper, whose fruits have long been used as a spice and in traditional medicine. This paper discusses the botany, chemistry and pharmacological properties and presents current and potential uses and application of this spice and its natural products.

The *Annonaceae* family includes tropical and subtropical plants that are widespread in Africa, Australia, Asia, South and Central America (*1*). There are over 2230 species distributed over 124 genera estimated (*2*). Annonaceae consists in numerous fruit trees, fragrant and spicy plants. Numerous *Annonaceae* plants used extensively in the ethnomedicine in some areas where the plants are part of the flora (*1*).

The genus Xylopia is distributed throughout the tropics, particularly in Africa. *Xylopia aethiopica* (Dun) A. Rich. is an aromatic evergreen tree that grows wild and is also cultivated in Ghana. This species is largely found in West, Central and Southern Africa, and can reach up to 15m high, with bark slightly ridged, and leaves oblong-lanceolate (*3*). The slender fruits are red when ripe (fresh) turn black upon drying, and typically are 5 to 8 cm long containing 4-9 peppery seeds.

Fresh and dried fruits are commonly used in traditional medicine in Africa (*3, 4*). They are known as "spice tree", "African pepper", "Ethiopian pepper", " Negro pepper", "Senegal pepper" or "Guinean pepper". Both fruits and seeds are used as condiments and spices as substitute of pepper because of their spicy properties (*5*), normally in cooked foods or in the spicing of beverages. In herbal medicine the fruits are used as a carminative, stimulant and as additive to other remedies for the treatment of skin infections, as digestive, appetizer antiemetic agents (*3*). Other medicinal uses reported are for the treatment of cholera, diarrhea, dysentery and wound sepsis, to reduce fever, "clean system" of newly delivered mothers, against coughs and to cure stomach problems (*4*). Also, due to their attractive aroma, the crushed powdered fruits are also used in mixtures with shea butter fat and cosmetics products (*6*).

Many studies have reported on the composition of the essential oils of fruits (*1, 7-11*), seeds (*12, 13*), stem bark (*14*), root bark and leaves (*10*). Tauri *et al.*, (*15*) reported on the identification of the most odor-active compounds in the dried fruits. However, there is limited information about the functional properties of the natural products of the dried fruits and even less on the quality of the consumed product. Lack of standardization and quality control are the main disadvantages of traditional medicine (*16*). Products with defined quality standards may have better access to more markets and capture a premium in the market when introduced in a consistent and reliable manner.

This work aims to provide information on the composition of essential oils, antioxidant activity, quality control standards, and to identify the pharmacological uses of the dried fruits of *X. aethiopica* using the product of Ghana as an illustrative case study.

Material and Methods

Plant Material

As part of our quality control program on African natural products, we conducted quality control and chemical analyses of dried fruits of *X. aethiopica* (Ghana). The sun dried fruits were received from 2004 to 2005. Each procedure was run at least in duplicate.

Quality Control Analysis

Moisture, total ashes, total insoluble ashes and essential oil content were assessed for each sample using methods described by the Food Chemical Codex (*17*).

Chemical and Antioxidant Activity Analysis

The ground dried fruits (200 mg) were extracted with different concentrations of methanol and 60% of methanol was the best concentrations for extracting total phenols (Figure 1), through sonication for total phenol content, and antioxidant activity analyses. The total phenols were measured using the Folin Ciocalteu's reagent (*18*) and the results were expressed as a percent of gallic acid equivalents on a dry weight basis (m/m). The antioxidant activity was evaluated using the ABTS method (*19*) and the results were expressed as a percent of Trolox (a water soluble analog of vitamin E) on a dry weight basis (m/m).

Essential Oil Analysis

The volatile oils were analyzed by a gas chromatograph (GC) coupled to a mass spectrometer (MS) (Agilent GC System 6890 Series, Mass Selective Detector, Agilent 5973 Network, FID detector) (*20*).

Results

Quality Parameters

The dried fruit of *X. aethiopica* is slightly curved with constriction, the color is brown, the aroma is resinous and fresh, without off odors (Table I). The presence of fine particles was low in both samples, while the foreign matter was very high in the 2004 sample (2.75%), but it showed acceptable levels (0.04%) in sample 2005. Procuring samples and products without such foreign matter can be accomplished by implementating GACP and training in the gathering, harvesting and storage of the product.

The moisture percent ranged from 8-9%, which more than meets the required maximum standards for moisture (@ 10%) according to international standards for botanicals (*21*).

Table I. Quality Control Parameters of Dried Fruits of (*X. aethiopica*) from Ghana

Year	2004	2005
Color	Brown	Brown
Taste	Slightly spicy	Slightly spicy
Aroma	Fresh, floral slightly fruity notes	Stronger fresh floral slightly fruity notes
Fine particles (%, m/m)	0.15±0.05	0.13±0.04
Foreign matter (%, m/m)	2.75±0.25	0.04±0.02
Moisture (%, m/m)	7.65 ± 0.19	8.84 ± 0.07
Total Ashes (%, m/m)	3.66 ± 0.05	3.25 ± 0.04
Insoluble Ashes (%, m/m)	0.09±0.01	0.05 ± 0.03
EO (%, m/v)	2.00 ± 0.1	4.5 ± 0.3
Total Phenols (%, m/m)	0.7 ± 0.05	0.81± 0.07
Antioxidant activity (g Trolox/100g)	1.29±0.01	3.21±0.22

Total Phenols and Antioxidant Activity

The extraction solvent is of critical importance for extracting the phenols from botanicals (*22*), and we found that 60% of methanol in water was the optimal for the extraction of *X. aethiopica* total phenols (Figure 1). Our results showed that the amount of phenols in this spices was very low (0.8%) (Table I) in both samples.

Ground dried fruits of our samples exhibited poor antioxidant activity which contrasts the antioxidant activity previously reported for this same fruit (*23*). This could be accounted by the differences among the test systems employed, the origin and genetics of the plant, or differing quality of the samples.

Phenolic compounds constitute the largest proportion of known natural antioxidants because they may scavenge free radicals involved in the lipid peroxidation (*24*). Plant phenolics are multifunctional and can act as reducing agents, hydrogen donating antioxidants and singlet oxygen quenchers (*25*). Thus, the low antioxidant activity might be associated with the low amount of phenols reported.

The potential antioxidant activity of essential oils obtained from dried fruits was observed (*10*). The activity was related with the high content of Germacrene

D (25.1%) and α– (5%) and β–pinenes (21.6%). Furthermore, 1,8-cineole have been demonstrated to possess in vitro antioxidant properties (*26-28*) while α–pinene showed low activity (*29*).

Essential Oils

The yields of the essential oils from dried fruits are shown in Table I. The content of essential oil from samples collected in 2005 is significant higher than those collected in 2004.

The essential oils from both samples are a complex mixtures of compounds characterized by high levels of monoterpenes (80.62, 76.33%) respectively (Table II, Figure 2). The main constituents of both samples were β- pinene, 1,8 cineole, α-pinene and myrtenol. Also sabinene, (*E*)-sabinol, terpin-4-ol and α–terpineol were detected but in lower levels.

Previous reports of essential oils of dried fruits of *X. aethiopica* showed that β- pinene was present in high proportions in samples from Benin (41.9%) (*7*) Egypt (14.6%) (*11*), Mali (19.1%) (*8*), Cameroon (18.3%) (*30*). High levels of 1,8 cineole (15.15%) were reported in samples from Nigeria (*31*), Benin (12.8%) (*9*) and Egypt (16.3%) (*11*). Sabinene has been found as the main components (36%) in essential oils from Benin (*9*) and from Cameroon (23.9%) (*32*) in contrast to our samples that was only a minor component (3.73 and 5.05% respectively).

Our findings agree with the low levels of Germacrene D that has been reported (7, 8), however there is a report with high levels (25.1%) from dried fruits from Ghana (*10*). The differences observed in the chemical composition of the essential oils likely reflect the influence of the geographic origin.

Functional Properties

Many functional properties of botanicals are attributable to their active components such as essential oils (*33-35*) and phenols (*22, 36-37*).

Antimicrobial and Antifungal Activity

Essential oils of *X. aethiopica* were reported to have antibacterial and antifungal properties. Essential oil characterized by β-pinene (18.3%), terpinen-4-ol (8.9%), α- phellandrene (7.1%), was found to inhibit the growth of Gram (+) and Gram (-) strains of bacteria and some human pathogenic fungi including species of the genus *Aspergillus and Candida* (*30, 38*). Toxicity against several

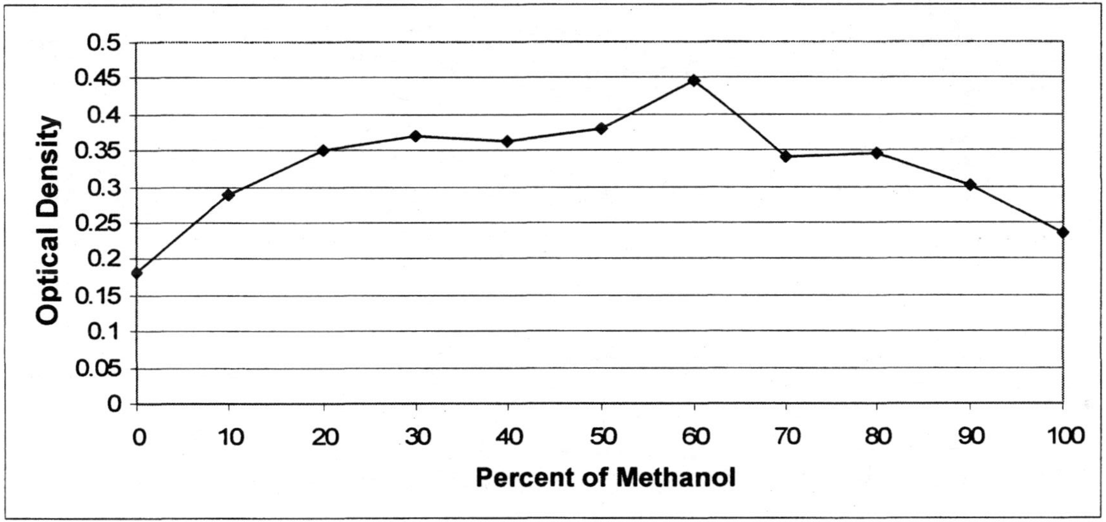

Figure 1. Recovery of total phenols of X. aethiopica in different concentrations of methanol and water.

Table II . Chemical Composition of Essential Oils of *X. aethiopica* from Ghana

RT	Components	2004	2005
5.21	α-Thujene	0.53[1]	0.6
5.39	α-Pinene	7.73	8.56
5.74	Camphene	0.2	0.25
6.35	Sabinene	3.73	5.05
6.47	β–Pinene	19.78	19.07
6.81	Myrcene	1.4	1.68
6.88	Unidentified	2.06	2.53
7.15	α - Phellandrene	0.17	0.25
7.49	α - Terpinene	0.48	0.51
7.73	*p* – Cymene	1.64	1.57
8.2	1,8 –Cineole	12.74	11.72
8.73	γ - Terpinene	0.86	0.89
8.98	*cis*-Sabinene hydrate	0.16	0.22
9.1	*cis*-Linalol oxide	0.07	0.07
9.3	Terpinolene	0.38	0.39
9.98	Linalool	0.79	0.71
10.62	Dehydro-sabinene ketone	0.13	0.1
10.83	α-Campholenal	0.48	0.43
11.27	*trans*-Sabinol	5.49	4.39
11.45	*cis*-Verbenol	1.49	1.15
12.02	Pinocarvone	1.87	1.48
12.19	*p*-Mentha-1,5-dien-8-ol	1.47	1.28
12.54	Terpin-4-ol	4.3	3.27
12.81	*p*-Cymenol	0.53	0.47
12.99	α -Terpineol	2.01	1.65
13.17	Myrtenol	7.75	5.75
13.56	Verbenone	1.28	0.97
13.91	(*E*)-Carveol	0.37	0.38
14.21	Dihydro carveol	0.28	0.36
14.57	Cumin aldehydes	0.16	0.18
16.02	α-Terpinen-7-al	0.08	0.15
16.56	Perilla alcohol	0.21	0.25
17.75	δ-Elemene	0.81	0.87
18.14	α-Cubenene	0.07	0.16
19	α -Copaene	1.71	1.82
19.49	β-Cubebene	0.1	0.21

Table II . *Continued.*

RT	Components	2004	2005
19.76	Cyperene	0.26	0.37
20.36	(*E*)-Caryophyllene	0.06	0.13
22.13	Germacrene D	0.25	0.34
22.68	Epi-Cubebol	0.42	0.57
23.29	Cubebol	0.18	0.32
24.5	Germacrene B	0.45	0.65
25.28	Caryophyllene Oxide	0.16	0.4
26.6	Epi-Cubenol[2]	2.56	2.68
	Monoterpenes	80.62	76.33
	Sesquiterpenes	6.54	8.03
	Total analyzed	87.65	84.85

[1]Values are expressed as relative percent of the total oil. 2 tentative identification

fungi and brine shrimps was also reported with essential oils dominated by 1,8 cineole (15.15%) and terpinen-4-ol (6.6%) (*31*). Also, extracts obtained of dried fruits have shown to inhibit spore germination of *Colletotrichum lindemuthianum*, a fungal pathogen of cowpea and bean, that causes anthracnose and can reduce grain yield by 35 to 50 % (*39*).

Also, aqueous fractions of *X. aethiopica* were reported as antimicrobial agent against *Escherichia coli, Pseudomonas aeruginosa, Salmonella typhy Staphylococcus aureus, Streptococcus fecalis, and Lactobacilli*. This finding may explain the use in traditional medicine in hot decoctions of these plants in the treatment of cough, dysentery and female sterilization, and in the management of stomach aches and gastroenteritis (*40-41*).

The low antimicrobial activity of some essential oils of *X. aethiopica* was correlated with large amounts of monoterpene hydrocarbons or sesquiterpenes (*32*). Several reports correlated the antimicrobial activity of essential oils with the bioactivities of their components (*33, 34, 35, 42, 43*).

Antimalarial Activity

Malaria is one of the most prevalent infections in the world, and is caused by parasites of the genus *Plasmodium* (the most severe *P. falciparum*), after transmission by *Anopheles* mosquitoes. The search for newer drugs intensifies as more resistant strains of the disease emerge. Traditional herbal medicine has many potential advantages. They are inexpensive and easily available,

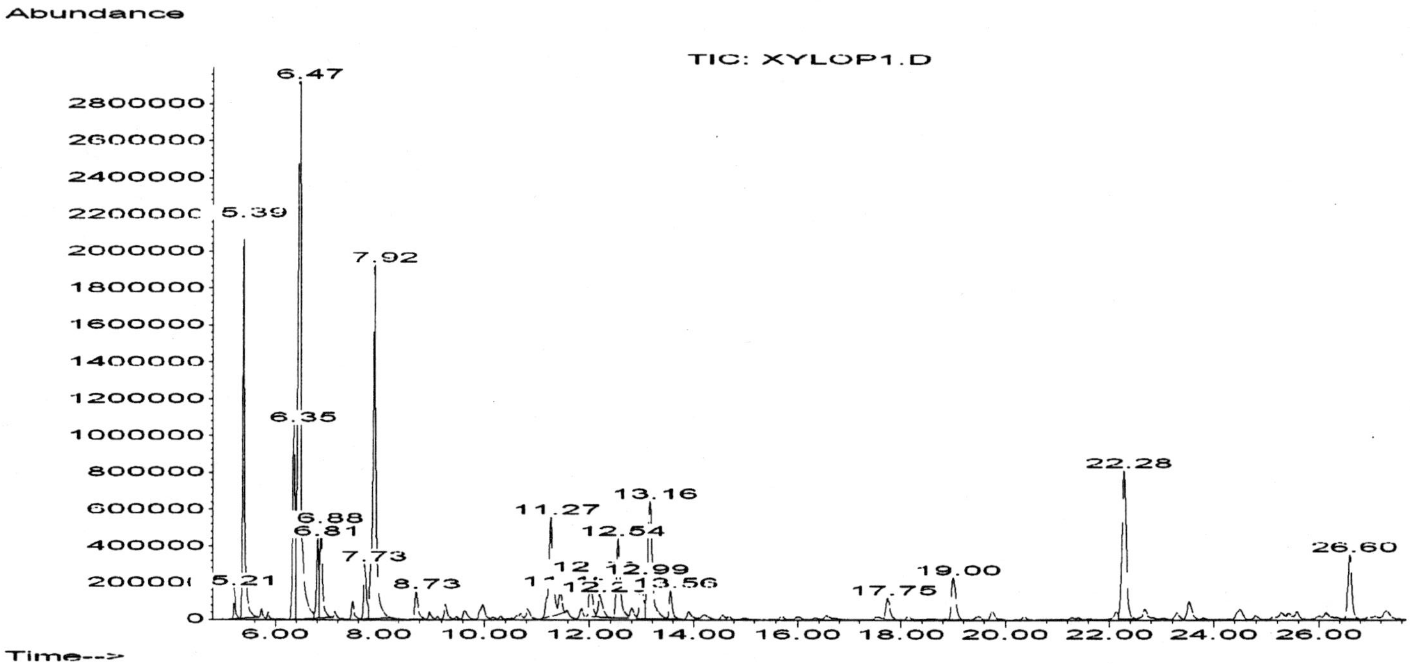

Figure 2. Chromatogram of the essential oils of X. aethiopica identifying the peaks according the retention times listed in Table II.

particularly if people grow them themselves, however the danger is the formulations used are not standardized, the products often not properly and rigously tested, and the difficulty in ensuring any plant grown will contain the needed pharmacological profile that would suggest effectiveness. Nevertheless, for many patients the use of herbal traditional medicine for malaria and other infectious diseases may be the only course of treatment available (*44, 45*).

Moreover, plant essential oils or their active components can be used as alternatives or adjuncts to current antiparasitic therapies (46). The difficulties in obtaining and administering these products may explain their uncommon use in traditional medicine, but they may offer a unique means of discovering new effective antimalarial drugs from plants. (*45*). The antimalarial properties of essential oils underline either their potential biological activity on the malaria parasites (47-*49*) or on the mosquito vector as insecticides (*50-52*). In a recent study, essential oils from stem bark of *X. aethiopica* were reported active against *P. falciparum* in culture (*14*), however the authors were not able to identify the specific components that elicit anti-plasmodial activity.

Although, plant products may be effective further collaborative research into their usefulness, pharmacognosy and toxicity needs to be evaluated. The potential variability of active ingredient(s) must also be addressed (*46*), particularly prior to any recommendations relative to growing one's own medicine. Insect control is becoming difficult because of the development of strains resistant against insecticides, and this has led to significant interest in finding less toxic, environmentally friendly yet effective insect control substances (*33, 35*). It was reported that hexane extracts of fruits of *X. aethiopica* exhibited strong antifeedant activity against wood destroying organisms such as the subterranean termites (*53*), crop pests (*Spodoptera litura*) and medically important insects such as the yellow fever mosquito *Aedes aegyptii* (*54*).

In folk medicine *X. aethiopica* is employed as carminative and as digestive. *In vitro* studies, have shown that essential oils containing α– and β–pinene produced the inhibition of gastric motility, that is the base of the treatment of some gastrointestinal disorders (*55*), being β–pinene more active than α–pinene (*56*). Other main component of the essential oils is 1, 8 cineole, and it was reported to possess anti-inflammatory, analgesic, gastroprotective and hepatoprotective activities (*57-59*). Moreover, its potential value as a dietary flavoring agent in the prevention of gastrointestinal inflammation and ulceration was observed (*60*). Thus, *X. aethiopica* could be used in combination with other aromatic plants as an ingredient in herbal teas and other food and beverage products.

In traditional folk medicine *X. aethiopica* is widely used as condiment in cooked foods and in the spicing of beverages. The primary reason for this widespread use is the lack of substitute spices (*3-5*).

There are different methods of fruit conservation. Vendors, specialized in trading medicinal plants, first boil or smoke the fresh fruits before drying them in the sun. Consequently, differences in the chemical composition of the resulting essential oils has been reported (*7*). Properly dried spices could be expected to store for a year or more without significant deterioration (*4*). Moreover, the use of spices is nutritionally important, since they are good alternative source of essential fatty acids (*13*) and can also contribute to mineral and protein consumption (*4, 13*). Considering, the term spice is used to cover the use of spices/herbs and certain aromatic vegetables to impart odor and flavor to foods (*61*), dried fruits from *X. aethiopica* could be used as a mild/aromatic spice, mixes of spices, sauces, condiments and seasonings.

Another application for this spice is for its antioxidant activity, for the Xylopia that exhibits higher levels of such activity. Food processors often look to new natural anti-oxidants for use in possessed foods as an alternative to synthetic antioxidants (*24*), and also as food preservatives, since essential oils would retard microbial contamination and therefore reduce the onset of spoilage (*43*).

Seeds oils of *X. aethiopica* were described as semi-drying and characterized by the relatively high saponification values. These physico-chemical properties, made it possible to be used for the preparation of shampoo but also because of its iodine value it was used in the preparation of resin and paint (*6*).

Several diterpenes such as kovalene, kaurene and trachylobanes, have also been described from stem bark for *X. aethiopica* (*62-65*). Recently, it was reported cardiovascular and diuretic effects of diterpenoids isolated from *X. aethiopica.* These were mainly attributed to presence of xylopic acid, kaurenoic/ dihydro-kaurenoic acids and some of their derivates (*66*).

Since the lack of standardization and quality controls may limit the access to more variable markets, we propose the following initial standards for *X. aethiopica,* to provide the users, buyers and the international community with consistent and defined products (Table III). Our results showed that the sample we examined in 2004 is of lower quality because it contained higher levels of foreign matter (Table I) and lower levels of active principles, e.g. essential oils, total phenols and antioxidant activities. Clearly, standards need to cover a range of expected values from low to high to include the expected range available in the marketplace. These standards reflect a conservative approach to this spice product and should allow many producers to more than meet the proposed levels.

In conclusion, *Xylopia aethiopica* is a flavorful spice, well recognized in the countries where the plant is indigenous, and with a long history of safe food use and consumption. While the plant has been long used in traditional medicine, the purposeful introduction of this spice into new markets seeking to diversity into ethnic spices and specialty foods, may find this "new African pepper" an excellent candidate.

Acknowledgements

We thank the Ghanaian communities involved in the ASNAPP domestication and commercialization of natural products studies. We thank Hanson Arthur, ASNAPP-Ghana for collecting the samples and shipping them to Rutgers. We thank Carol Wilson, USAID Chief Technical Officer of our Partnership for Food and Industry in Natural Products (PFID/NP) project supported by the Office of Economic Growth, Agriculture and Trade (EGAT/AG) of the USAID (Contract Award No. AEG-A-00-04-00012-00) in support of their global economic development programs. We also thank Jerry Brown, USAID project officer, for his support and encouragement as this work originally began as part of the Agri-Business in Sustainable African Natural Plant Products Program (ASNAPP) with funding from the USAID (Contract Award No. HFM-O-00-01-00116). This research was also supported in part by NIH grant OD-00-004 (Botanical Center for Age-Related Diseases) in our search for specific polyphenols. The authors also acknowledge support by the New Jersey Agricultural Experiment Station and by Cook College, Rutgers University We thank the Joosan Scholarship Foundation for providing a fellowship to TaoOh Kwon which funded his sabbatical study at Rutgers from Wonkwang University, Korea.

Table III. Proposed quality standards dried fruits of (*X. aethiopica*) from Ghana

Characteristic	*Requirement*
Color	Brown, brown/reddish
Aroma	Spicy, woody, free from foreign odors
Taste	Slightly spicy
Excreta (mammalian and others)	Practically free
Extraneous foreign matter (%, m/m) maximum	1
Fine particles (%, m/m) maximum	1
Moisture (%, m/m) maximum	10
Total ashes (%, m/m) maximum	4
Acid Insoluble ashes (%, m/m) maximum	1
Essential oil content (%, m/m) minimum	3
Total phenols (%, m/m) minimum	0.8
Antioxidant activity (%, g Trolox/100 g) minimum	3

References

1. Ekundayo, O. *J. Ess. Oil Res.* **1989,** *1,* 223-245.
2. Leboeuf, M.; Cave, A.; Bhaumik, P.K.; Mukherjee, B. *Phytochemistry* **1982,** *21,* 2783.
3. Ayiku, M.N.B. *Ghana Herbal Pharmacopeia.* Technology Transfer Center. Accra. Ghana, 1992, pp. 86-88.
4. Okeke, E.C. *J. Herbs, Spices, & Med. Plants* **1998,** *5 (4),* 51-63.
5. Claus, E.P. 1956. Volatiles Oils. In *Pharmacognosy* by Gathercoal, E.N.; Wirth, E.H. Third Edition. Lea & Febiger. Philadelphia, 1956, Chapter 7. pp. 263-362.
6. Ajiwe, V.I.E.; Okeke, C.A.; Ogbuagu, J.O.; Ojukwu, U.; Onwukeme, V.I. Bioresource Technol **1998,** *64,* 249-252
7. Ayedoun, A.M.; Adeoti, B.S.; Sossou, P.V.; Leclercq, P.A. *Flavour Fragrance J.* **1996,** *11,* 245-250.
8. Keita, B.; Sidibe, L.; Figueredo, G.; Chalchat, J.C. *J. Essent. Oil Res.* **2003,** *15,* 267-269.
9. Poitu, F. *J. Essent. Oil Res* **1996,** *3,* 329-330.
10. Karioti, A.; Hadjipavlou-Litina, D.; Mensah, M.L.K.; Fleischer, T.C.; Skaltsa, H. *J. Agric. Food Chem.* **2004,** *52,* 8094-8098.
11. Karawya, M.S.;Abdel Wahab, S.M.; Hifnawy MS. Planta Med. **1979,** *37,* 57-59.
12. Tomi, F.; Casanova, J. *J. Essent. Oil Res.* **1996,** *8,* 429-431.
13. Barminas, J.T.; James, M.K.; Abubakar, U.M. *Plant Foods For Human Nutrition* **1999,** *53,* 193-198.
14. Boyom, F.F.; Ngouana, V.; Zollo, P.H.A.; Menut, C.; Bessiere, J.M.; Gut, J.; Rosenthal, P.J. *Phytochem.* **2003,** *64,* 1269-1275.
15. Tauri, A.O.; Hofmann, T.;Schieberle, P. *J. Agric. Food Chem.* **1999,** *47,* 3285-3287.
16. Asase, A.; Oteng-Yeboah, A.A.; Odamtten, G.T.; Simmonds, M.S.J. *J. Ethnopharmacology* **2005,** *99,* 273-279.
17. Committee on Food Chemical Codex. 1996. Food Chemical Codex. National Academy Press. Washington DC.
18. Gao, X.; Bjork, L.; Trajkovski, V.; Uggla, M. *J. Sci. Food Agric.* **2000,** *80,* 2021-2027.
19. Re, R.; Pellegrini, N.; Proteggente, A.; Pannala, A.; Yang, M.; Rice Evans, C. *Free Rad. Biol. Med.* **1999,** *26,* 1231-1237.
20. Juliani, H.R. ; Zygadlo, J.A. ; Scrivanti, R. ; de La Sota, E.; Simon, J.E. *Flavour Fragrance J.* **2004,** *19,* 541-543.
21. Muggeridge, M.; Foods, L.; Clay, M. In *Handbook of Herbs and Spices,* Peter, K.V. (Ed.), CRC Press, Boca Raton, Florida, 2001.

22. Wang, M.; Juliani, R.; Simon, J.E.; Ekanem, A.; Liang, C.P.; Ho, C.-T. In *Phenolic Compounds in Foods and Natural Health Products*, F. Shahidi and C.T. Ho (Eds), ACS Symp. Ser. 909, American Chemical Society. Washington DC. 2005, pp. 118-142.
23. Adegoke, G.O.; Makinde, O.; Falade, K.O.; Uzo-Peters P.I. *Eur. Food Res. Technol.* **2003**, *216*, 526-528.
24. Madsen, H.L.; Bertelsen, G.; Skibsted, L.H. In Spices: Flavor Chemistry and Antioxidant Properties, S. J. Rich and C.-T. Ho (Eds.), American Chemical Society, Washington, D.C. 1997, p.p. 176-187.
25. Proteggente, A.R.; Pannala, A.S.; Paganga, G.; Van Buren, L.; Wagner, E.; Wiseman, S.; Van De Put, F.; Dacombe, C.; Rice-Evans, C.A. *Free Rad. Res.* **2002**, *36*, 217-233.
26. Mimica-Dukic, N.; Bozin, B.; Sokovic, M.; Mihajlovic, B.; Matavulj, M. *Planta Med.* **2003**, *69*, 413-419.
27. Candan, F.; Unlu, M.; Tepe, B.; Daferera, D.; Polissiou, M.; Sokmen, A.; Akpulat, H.A. *J. Ethnopharmacol.* **2003**, *87*, 215-220.
28. Tepe, B.; Donmez, E.; Unlu, M.; Candan, F.; Daferera, D.; Vardar-Unlu, G.; Polissiou, M.; Sokmen, A. *Food Chem.* **2004**, *84*, 519-525.
29. Kim, H.J.; Chen, F.; Wu, C.Q.; Wang, X.; Chung, H.Y.; Jin, Z.Y. *J. Agric. Food Chem.* **2004**, *52*, 2849-2854.
30. Tatsadjicu, L.N. *Fitoterapia* **2003**, *74*, 469-472.
31. Asekun, O.T.; Adeniyi, B.A. *Fitoterapia* **2004**, *75 (3-4)*, 368-370.
32. Chalchat, J.C.; Garry, R.P.; Menut, C.; Lamaty, G.; Malhuret, R.; Chopineau, J. *J. Essent. Oil Res.* **1997**, *9*, 67-75.
33. Zygadlo, J.A.; Juliani, H.R. *Minireview: Current Topics in Phytochemistry* **2000**, *3*, 203-214.
34. Kalemba, D.; Kunicka, A. *Curr. Med. Chem.* **2003**, *10*, 813-829.
35. Koroch, A.R.; Zygadlo, J.A.; Juliani, H.R. In *Flavours-Chemistry, Technology and Resources*, R.G. Berger, (Ed.) in press.
36. Pietta, P.G. *Journal of Natural Products* **2000**, *63*, 1035-1042
37. Lule, S.U.; Xia, W.S. *Food Rev. Internat.* **2005**, *21*, 367-388.
38. Konning, G.H.; Agyare, C.; Ennison, B. *Fitoterapia* **2004**, *75*, 65-67.
39. Amadioha, A.C.; Obi, V.I. *J. Herbs, Spices, & Med. Plants* **1998**, *6*, 33-40.
40. Okeke, M.I.; Iroegbu, C.U.; Jideofor, C.O.; Okoli, A.S.; Esimone, C.O. *J. Herbs, Spices, & Med. Plants* **2001**, *8*, 39-46.
41. Ijeh, I.I.; Omodamiro, O.D.; Nwanna, I.J. *African J. Biotechnol.* **2005**, *4*, 953-956
42. Griffin, S.G.; Wyllie, S.G.; Markham, J.L.; Leach, D.N. *Flavour Frag. J.* **1999**, *14*, 322-332.
43. Dorman, H.J.D.; Deans, S.G. *J. Appl. Microbiol.* **2000**, *88*, 308-316.
44. Willcox, M.; Bodeker, G. *Parasitology Today* **2000**, *16*, 220-221.

45. Benoit-Vical, F. *Idrugs* **2005**, *8*, 45-52.
46. Anthony, J.P.; Fyfe, L.; Smith, H. *Trends in Parasitology* **2005**, *21*, 462-468.
47. Valentin, A. ; Pelissier, Y. ; Benoit, F. ; Marion, C. ; Kone, D. ; Mallie, M. ; Bastide, Jm.; Bessiere, Jm. *Phytochemistry* **1995**, *40*, 1439-1442.
48. Lopes, N.P.; Kato, M.J.; Andrade, E.H.A.; Maia, J.G.S.; Yoshida, M.; Planchart, A.R.; Katzin, A.M. *J. Ethnopharmacol* **1999**, *67*, 313-319.
49. Tchoumbougnang, F.; Amvam Zollo, P.H.; Dagne, E.; Mekonnen, Y. *Planta Med* **2005**, *71*, 20-23.
50. Moore, S.J.; Lenglet, A.; Hill, N. *Journal of the American Mosquito Control Association* **2002**, *18*, 107-110.
51. Samarasekera, R.; Kalhari, K.S. *Journal of Essential Oil Research* **2005**, *17*, 301-303.
52. Prajapati, V.; Tripathi, A.K.; Aggarwal, K.K.; Khanuja, S.P.S. *Bioresource Technology* **2005**, *96*, 1749-1757.
53. Lajide, L. ; Escoubas, P. ; Mizutani, J. *Phytochem.* **1995**, *40*, 1105-1112.
54. Escoubas, P. ; Lajide, L. ; Mizutani, J. In *Natural and Engineered Pest Management Agents, ACS Symposium Series* **1994**, *551*, 162-171.
55. Camara, C.C.; Nascimento, N.R.F.; Macedo-Filho, C.L.; Almeida, F.B.S.; Fonteles, M.C. *Planta Med.* **2003**, *69*, 1080-1085.
56. Sadraei, H.; Asghari, G.R.; Hajhashemi, V.; Kolagar, A.; Ebrahimi, M. *Phytomedicine* **2001**, *8*, 370-376.
57. Santos, F.A.; Rao, V.S.N. *Eur. J. Pharmacol.* **1997**, *331*, 253–258.
58. Santos, F.A.; Rao, V.S.N. *Phytother. Res.* **2000**, *14*, 240-244.
59. Santos, F.A.; Silva, R.M.; Tomé, A.R.; Rao, V.S.N.; Pompeu, M.M.L.; Teixeira, M.J.; De Freitas, L.A.R.; De Souza, V.L. *J. Pharm. Pharmacol.* **2001**, *53*, 505-511.
60. Santos, F.A.; Silva, R.M.; Campos, A.R.; de Araujo, R.P.; Junior, R.C.P.L.; Rao, V.S.N. *Food Chem. Toxicol.* **2004**, *42*, 579-584.
61. Peter, K.V. 2001. Introduction. In Handbook of Herbs and Spices (K.V. Peter, Ed.). CRC Press, Boca Raton. Florida, 2001, pp 1-12.
62. Hasan, C.M.; Healey, T.M.; Waterman, P.G. *Phytochem.* **1982**, *21*, 1365-1368.
63. Faulkner, D.F.; Lebby, V.; Waterman, P.G. *Planta Med.* **1985**, *4*, 354-355.
64. Harrigan, G.G.; Bolzani, V. da S.; Gunatilaka, A.A.L.; Kingston, D.G.I. *Phytochem.* **1994**, *36*, 109-113.
65. Hanson, J.R. *Nat. Prod. Reports* **2005**, *22*, 594-602.
66. Somova, L.I.; Shode, F.O.; Moodley, K.; Govender, Y. *J. Ethnopharmacology* **2001**, *77*, 165-174

Chapter 8

Basil: A Source of Rosmarinic Acid

H. Rodolfo Juliani, Adolfina R. Koroch, and James E. Simon

New Use Agriculture and Natural Plant Products Program, Cook College, and the New Jersey Agricultural Experiment Station (NJAES), Rutgers, The State University of New Jersey, 59 Dudley Road, New Brunswick, NJ 08901–8520

In recent years natural compounds have gained much attention and therefore the production of these compounds and the identification of new natural sources of bioactive and antioxidant compounds have gained scientific and industrial important. This review examines new sources of a rosmarinic acid, a molecule with multiple biological activities. This paper discusses the extraction methods and accumulation of rosmarinic acid in different organs of basil, and in different basil varieties. A review of the pharmacological properties of rosmarinic acid are also presented.

Phenolic acids constitute a large group of naturally occurring organic compounds with a broad spectrum of pharmacological activities. One of the most abundant phenolic compounds is rosmarinic acid (RA) and in past years it has gained much attention as a molecule with multiple biological activities with low toxicity (*1, 2*). Therefore, studies of the production of a quantifiable natural source of this potent and versatile compound became important.

Rosmarinic acid is a well known constituent found in several families all over the higher plants Kingdom (*2, 3*). The Lamiaceae family has a cosmopolitan distribution and highly variable morphological characters. The occurrence of RA in Lamiaceae is limited to the subfamily Nepetoideae which comprises the majority of the essential oil-rich culinary and medicinal herbs such as basil, mint, sage, rosemary and thyme, but absent in the subfamily Lamioideae (*2, 4*).

Therefore, RA has been proposed as a strong chemotaxonomic marker for the subfamily Nepetoideae (*4-6*).

Rosmarinic acid (α-*O*-caffeoyl-3-4-dihydroxyphenyllactic acid) has two aromatic rings with two aromatic hydroxyl groups each and is one of the most abundant caffeic acid esters present in *Ocimum basilicum* (*7*). Rosmarinic acid is synthesized in basil in specialized structures (peltate glandular trichomes) that are located on the surface of the aerial parts of the plants (*8*). The enzymes involved in the biosynthesis of rosmarinic acid from its amino acid precursors have been studied in *Coleus blumei* (*2, 9*) and *Ocimum basilicum* (*8, 10*).

The RA content is largely dependent on the genotype and growing conditions of the plant and method of extractions. Thus the objective of this paper is to study different combinations of solvent for extraction of RA in order to maximize its recovery from basil and to study the RA content in different organs and varieties of basil.

Materials and Methods

Plant Material

Different basil varieties (Italian Large Leaf, Sweet Basil, Cinnamom Basil, Holy Basil, Dark Opal, Green Dark Opal, Osmin Purple, Green Osmin Purple, Purple Ruffles, Green Purple Ruffles and Red Rubin) were analyzed.

Extraction and Separation of Phenolic Compounds

Basil leaves were harvested from full blooming plants from the greenhouse (New Jersey Agricultural Experiment Station, Rutgers University). Samples were dried in oven at 38°C for six days. 50 mg of dry powdered were mixed with 25 ml of different concentrations (0-100%) of methanol-water. Samples were then sonicated for 20 minutes. The extracts was filtered and then analyzed by HPLC.

Total phenols were measured using the Folin Ciocalteu's reagent (*11*) and the results were expressed as gram of gallic acid equivalents on a dry basis (g gallic acid/100 g DW).

Separation of RA was achieved using an Agilent 1100 series HPLC with a phenyl hexyl column (3 µm, 150 x 4.6 mm, Phenomenex) and the detection wavelength was 280 nm, injection volume 20 µL and the flow rate 0.8 mL/min.. Mobile phase was solvent A: phosphoric acid (0.2%) + water and solvent B: acetonitrile (100%) injection volume was 5 µL. Quantitative determination of rosmarinic acid was made according to a reference curve using the peaks area and the results expressed as g RA/100g DW.

All standard phenolics and reagent chemicals were purchased from Sigma (St Louis)-Adrich (Milwaukee, WI). All determinations were run in triplicate and then averaged.

Results

The extracted total phenols and RA significantly varied with different concentration of methanol water. There was no significant difference in total phenolic content when extracted with 40 to 80% methanol/water. In contrast, RA content recovered reached the maximum with 70% of methanol/water (Figure 1). Thus, this was the concentration choosen for further analysis.

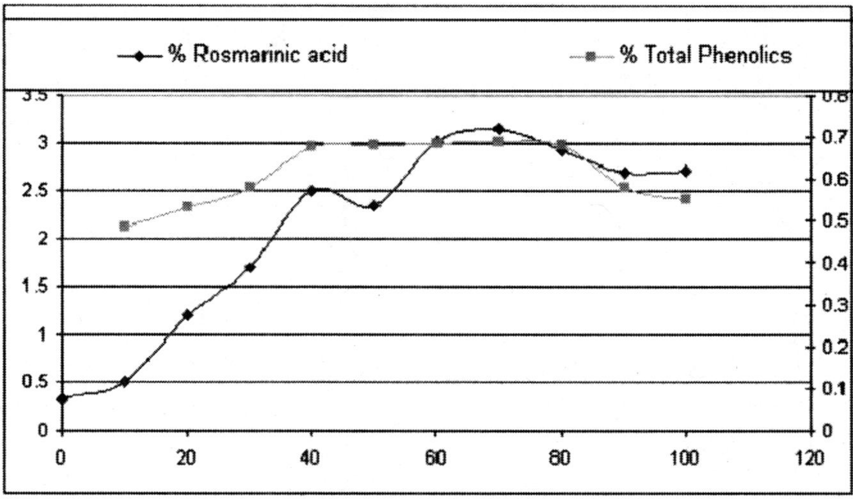

Figure 1. Effect of different percentage of methanol-water over the extraction of RA and total phenolics content.

In general, the analysis of phenolic acids in plants is performed by HPLC and is a relatively simple method. The majority of the published studies have applied the reversed-phase HPLC system with UV detection and organic water or organic buffer mobile phases (*5, 12*). Extraction of phenolic acid with methanol–water mixture has been reported (*12*), however, these authors reported the extraction with 60: 40 methanol-water in *Mellisa officinalis*. Also, the application of ultrasound bath with controlled temperature in comparison to classical extraction assay (shaking) has been also mentioned (*5, 12, 13*).

It has been reported that herbs (mainly culinary herbs) are the only source of rosmarinic acid (*3, 14*) and some example of the occurrence of RA in different herbs has been summarized (Table I).

Table I. Examples of Occurrence of RA in Different Culinary and Medicinal Plants

		Range of RA content	*Reference:*
Rosmarinus	*officinalis*	0.7-2.5%DW	*13-21*
		0.03%FW	*22*
Salvia	*officinalis* L.	0.07 -2.18% DW	*5, 13, 16, 19-23*
S.	*fructicosa*	0.03-1.4%PE	*5, 24*
S	*miltiorrhiza*	*	*25*
S.	*stenophylla*	*	*26*
S.	*repens*	*	*26*
S.	*runcinata*	*	*26*
Mentha	*arvensis*	0.5%DW	*26*
M.	*piperita*	0.01-0.4% DW	*13, 27*
M.	*spicata*	0.19-1.4% DW	*5, 20, 27*
Thymus	*vulgaris L.*	0.5-0.8% DW	*13, 16, 19*
		0.09% FW	*20, 21*
Origanum	*vulgare L*	1.27% PE	*23*
		0.12-2.5% DW	*5, 16, 19*
Satureja	*hortensis*	2.13% PE	*23*
		1.25% DW	*13*
Ocimum	*basilicum* L.	1.1-9.9%DW	*13, 16, 29, 30*
Melisa	*officinalis* L	0.9-2.7%DW	*5, 12, 13, 19, 20*

DW-dry weight

FW- fresh weight

PE- plant extract

RA was characterized in terms of their retention times, UV, mass spectra and commercial standard (rosmarinic acid) and among the various phenolic compounds; it was the major one in basil (Figure 2). Previous findings have reported that rosmarinic acid is the most abundant phenolic compound in other aromatics plants such as *Salvia officinalis* (*19, 22, 31*), *Melissa officinalis* (*12, 19*), *Thymus vulgaris*, *Rosmarinus officinalis* (*19*), *Origanum x majoricum* (*22*), *Origanum vulgare* (*5, 19*), *Satureja montana* (*5*).

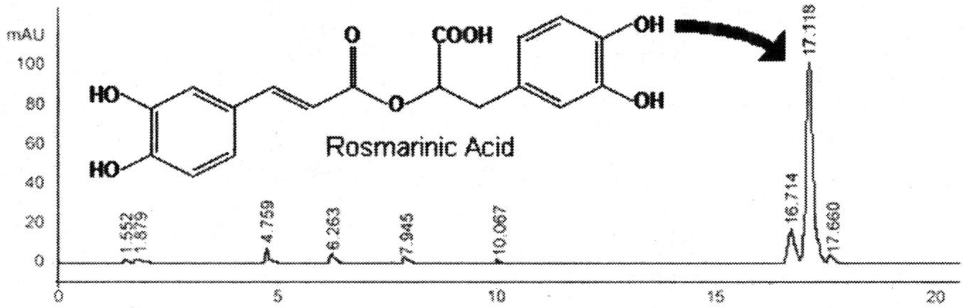

Figure 2. HPLC profile of Italian large leaf basil in methanolic (70%) extract showing rosmarinic acid as the main polyphenolic component.

Effect of Different Methods of Processing Rosmarinic Acid Content

To investigate if the sample drying processing has an effect on RA recovery, two different methods of drying processing in Italian Large Leaf basil (IIL) were compared: freezed dry (the sample is frozen and dry under vacuum conditions for 2 days) and oven dry (at 38°C in oven for 6 days). High amounts of total phenolics and RA were extracted with both methods of drying, showing there was no difference between freeze dry and oven dry to extract total phenolics and RA in Italian large leaf basil (Table II).

Table II. Effect of Different Methods of Processing the Basil Leaves on the Rosmarinic Acid and Total Phenolics Content.

Method	*Total phenolics (%) (g GA/100 g DW)*	*% Rosmarinic acid (ILL)*
Freeze dry (2 days)	7.5 ± 0.37*	3.75 ± 0.03*
Dry in oven at 38°C	8.8 ± 0.66	3.81 ± 0.03

mean ± standard error

IIL -Italian Large Leaf basil

The analysis of methanol/ water extracts from leaves, inflorescences and stems of Italian Large Leaves basil, indicated that content of RA is higher in leaves and lower in stems (Table III).

Table III. RA Content in Leaves, Flowers and Stems in Italian Large Leaf Basil

Organ	*% RA content (g RA/ 100 g DW)*
Leaves	4.7 ± 0.3
Flowers	2.6 ± 0.4
Stems	1.5 ± 0.06

RA Content in Different Basil Varieties

Under greenhouse conditions, the highest phenolic content was observed in dark opal (12.6%), green dark opal (10.2%) and Green Osmin Purple (10.5%), while in the other varieties the content ranged from 3.6% in cinnamon basil to 9.5% in Red Rubin (Table IV). The accumulation of RA showed a similar trend reaching higher values in dark opal (4.8%), green dark opal (4.4%), lower values were observed in the other varieties, the lowest accumulation was observed in cinnamon and sweet basil (0.7% and 1.2%, respectively).

From the greenhouse trial (Table IV), the varieties Dark opal and Italian Large Leaf were selected to measure the amount of RA in field-grown plants.

Table IV. Total Phenolic and Rosmarinic Acid Content in Leaves of Different Basil Varieties Grown Under Greenhouse Conditions

*Basil varieties**	*phenolic content (%) (g GA/100 gDW)*	*Rosmarinic acid content (g RA/100gDW)*
Italian Large Leaf	6.3	2.9 ± 0.1
Sweet	5.6	1.2 ± 0.2
Cinnamom	3.6	0.7 ± 0.1
Dark opal	12.6	4.8 ± 0.2
Green Dark Opal	10.2	4.4 ± 0.2
Osmin purple	8.2	2.9 ± 0.3
Green osmin purple	10.5	3.6 ± 0.1
Purple ruffles	9.3	2.3 ± 0.1
Green purple ruffles	7.1	2.7 ± 0.4
Red rubin	9.5	3.3 ± 0.1

Studies were conducted to determined the accumulation of RA during growth and development in these two varieties. In the preflowering stage the lowest amount of biomass were observed as well as the lowest amoung of RA (Figures 3, 4). The maximum amount of biomass was observed at full flowering and after that the biomass decresed probable due to the absition of leaves. This tendency was also observed in the Dark Opal basil but it was less marked. The amount of RA increased steadily in Italian Large Leaf, the highest amount (2%) was

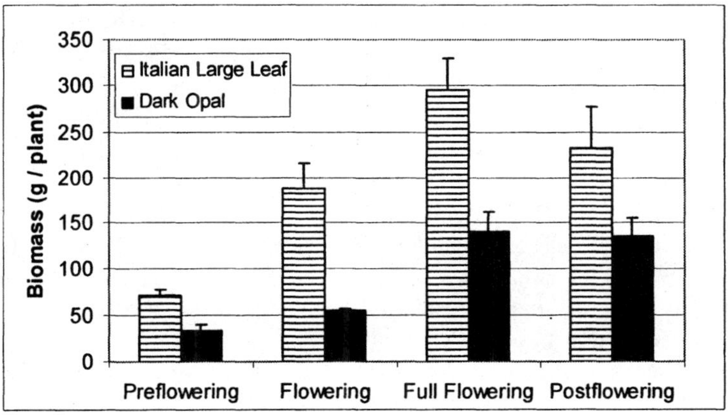

Figure 3. Biomass accumulation during growth in Dark Opal and Italian Large Leaf basil. The plus bars represent the standard error.

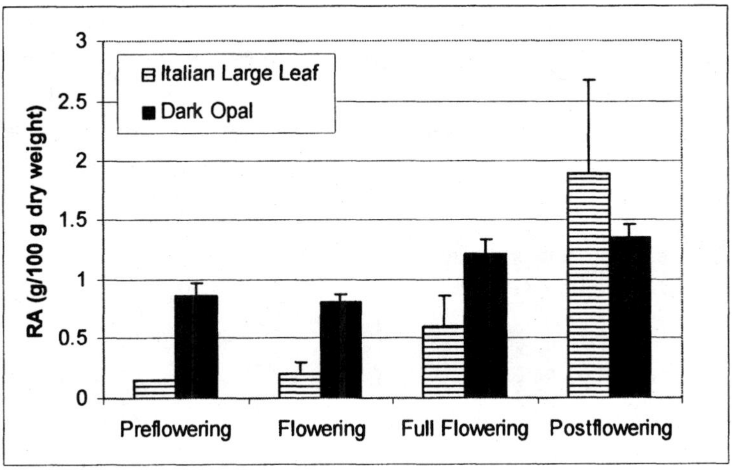

Figure 4. Rosmarinic acid (RA) accumulation during growth in Dark Opal and Italian Large Leaf basil. The plus bars represent the standard error.

observed at postflowering, while in Dark Opal highest accumulation were observed at full flowering and postflowering (1.2-1.3%) (Figure 4).

Biological Activities of Rosmarinic Acid

Antioxidant Activity

Spices are common food adjuncts, which have been used as flavoring, seasoning, and coloring agents and sometimes as preservatives throughout the world for thousand of years in many countries (*32-34*). Many spices have been recognized to have medicinal properties and possess many beneficial effects on health (*22, 35, 36*).

Phenolics exhibit a wide range of biological activities, and many of these activities have been attributed to their free radical scavenging, antioxidant properties, chelation of redox active metal ions, modulation of gene expression and interaction with the cell signaling pathways (*15, 16, 22, 37-39*). In the last years the interest of natural antioxidant such as those present in spices, has increased because of the increasing restrictions in the use of the synthetic antioxidans in foods (*38, 40, 41*).

The radical scavenging and antioxidant activities of phenolics are dependant upon the arrangement of functional groups about the nuclear structure (*38, 42-44*). Rosmarinic acid is a strong reducing scavenging agent and this is mainly attributable to the four phenolic hydroxyl groups separated in two pairs located in ortho-positions of the benzene rings (*21, 29, 44, 45*).

Extracts containing rosmarinic acid have been reported with high antioxidant activity in *Rosmarinus officinalis* (*15, 17, 18, 21*); *Salvia sp* (*21, 22, 26, 37*), *Thymus vulgaris* (*22, 43*), *Ocimum basilicum* (*29*), *Lepechinia graveolens* (*46*), *Origanum sp* (*22, 24, 36*), and *Prunella vulgaris* (*47*) among others.

Synergic effects of RA and other plants constituents has been suggested that might influence the differences in the antioxidant ability of the plants extracts (*22, 34, 48*). Extracts containing mixtures of lycopene and RA showed synergic antioxidant activity (*49*). Recently, Vattem *et al.*, (*50*) reported synergistic effects of cranberry phenolics and RA mixtures, functioning as antimutagen by modulating the redox environment of the mutagen and by modulating DNA repair systems.

Antiviral Ativity

Several reports demonstrated that RA effectively inhibited Human Immunodeficiency Virus (HIV) integrase, an enzyme essential for viral replication that integrates the viral reverse transcribed DNA into the host-cell DNA, with I50 values below 10 M (*51-54*).

Also, the inhibition of Herpes Simplex Virus type 1 (HSV-1) has been attributed to rosmarinic acid content (*2, 55, 56*).

Anti-inflammatory Ativity

Several studies indicate that RA has anti inflammatory activity and have been demonstrated that this characteristic can be used to treat various inflammatory disorders. Rosmarinic acid has been reported as an anti inflammatory agent, by inhibiting of TCR-signaling and subsequent T cell proliferation or by inhibiting of lipoxygenases and cyclooxygenases or by the interference of RA with the complement cascade (*1, 2, 57-61*). Moreover, Osakabe *et al.*, (*59*) concluded that part of the anticarcinogenic effects of *Perilla frutescens* extract is due to RA via two independent mechanisms: inhibition of the inflammatory response and scavenging of reactive oxygen radicals.

Antibacterial Ativity

RA showed antimicrobial activity against a range of soil-borne microorganisms, with its most deleterious effect against *Pseudomonas aeruginosa,* an opportunistic soil bacterium and human pathogen (*7, 62*). Similarly, these authors reported that RA showed potent antifungal activity against *Aspergillus niger*, one of the most prolific, widespread and invasive fungal plant pathogen known, and also can cause severe infection in humans. Extracts of *Prunella vulgaris* showed antibacterial activity against *Staphylococcus sp* (*47*).

Other Parmacological Ativities

Other interesting pharmacological activities of rosmarinic acid have been reported. Thus, beneficial effects of RA on suppression of collagen-induced arthritis in mice was demonstrated (*63*). RA was also found to show mild antithrombotic effect (*54*). RA in lower doses can produce anxiolytic-like effect without exerting locomotor alterations or DNA damage in brain tissue (*64*).

Recent reports reveals that RA has the potential to inhibit the lymphocyte cell- specific kinase, thus impairing T cell- restricted signaling and generating immunosuppression (*58, 65*).

RA is capable of inhibiting mesangial cell proliferation, one of the major histological findings in various renal diseases, and suppressing mRNA expressions of PDGF (platelet-derived growth factor) and c-myc in PDGF-stimulated mesagial cells (*66*).

Potent inhibitory action of RA against the toxic effect of mycotoxins has been reported (*67*), and these results were attributed to its antioxidants properties. Also, methanolic extracts of *Salvia sp* exhibited moderate antimalarial activity (*26*).

Biotechnological Production of RA

In vitro accumulation of Rosmarinic acid in suspensions of cells (*68-74*), hairy roots (*7, 75, 76*), and multiple shoots (*77-78*) have been reported. Thus, *in vitro* cultures can provide an attractive alternative of RA production since the concentration of the metabolites can sometimes be higher than the original plants, easier to extract and grow in a bioreactor with the possibility to up scaling the process (*79*).

Moreover, it has been also reported that RA accumulation can be increased by the addition of fungal elicitors (*7, 80, 81*), with methyl jasmonate (*73, 82*) or Ag^+ (*81*). These results suggest that rosmarinic acid could have a role in plants as a defense compound against pathogens (*2, 7*).

Conclusions

This review has shown that basil is a promising new source of rosmarinic acid since all organs were found to accumulate significant amounts of this component. A variation of rosmarinic acid content was observed in different basil varieties, both green and pigmented varieties accumulated significant amounts of RA. Agriculture practices such harvesting time were found to modulate the accumulation of rosmarinic in basil, while the extraction method would also affect the maximum recovery of this important bioactive component. Rosmarinic acid was found to be a component with different activities that include antioxidant, antiinflamatory as well as antimicrobial activities against important pathogens.

Acknowledgements

We thank the New Use Agriculture and Natural Plant Products Program (NUANPP) and the New Jersey Agricultural Experiment Station, Rutgers University, and the Rutgers Cooperative Extension Service, and the New Jersey Farms Bureau. We also thank Dr. Mingfu Wang, University of Hong Kong for his suggestions.

References

1. Parnharm, M.J.; Kesselring, K. *Drugs of the Future* **1985**, *10*, 756-757.
2. Petersen, M.; Simmonds, M.S.J. *Phytochem.* **2003**, *62*, 121-125.
3. Clifford, M.N. *J. Sci. Food Agr.* **1999**, *79*, 362-372.
4. Pedersen, J.A. *Biochem. Syst. Ecol.* **2000**, *28*, 229-253.
5. Janicsak, G.; Mathe, I.; Miklossy-Vari, V.; Blunden, G. *Biochem. Syst. Ecol.* **1999**, *27*, 733-738.
6. Grayer, R.J.; Eckert, M.R.; Veitch, N.C.; Kite, G.C.; Marin, P.D.; Kokubun, T.; Simmonds, M.S.J.; Paton, A.J. *Phytochem.* ***2003***, *64*, 519-528.
7. Bais, H.P.; Walker, T.S.; Schweizer, H.P.; Vivanco, J.A. *Plant Physiol. Bioch.* **2002**, *40*, 983-995.
8. Gang, D.R.; Wang, J.H.; Dudareva, N.; Nam, K.H.; Simon, J.E.; Lewinsohn, E.; Pichersky, E. *Plant Physiol.* **2001**, *125*, 539-555.
9. Petersen, M. *Phytochem.* **1997**, *45*, 1165-1172.
10. Gang, D.R.; Beuerle, T.; Ullmann, P.; Werck-Reichhart, D.; Pichersky, E. *Plant Physiol.* **2002**, *130*, 1536-1544.
11. Gao, X.; Bjork, L.; Trajkovski, V.; Uggla, M. *J. Sci. Food Agr.* **2000**, *80*, 2021-2027.
12. Caniova, A.; Brandsteterova, E. *J. liq. Chromatogr. R. T.* **2001**, *24*, 2647-2659.
13. Zgorka, G.; Glowniak, K. *J. Pharmaceut. Biomed.* **2001**, *26*, 79-87.
14. Clifford, M.N. *J. Sci. Food Agr.* **2000**, *80*, 1033-1043.
15. del Bano, M.J.; Lorente, J.; Castillo, J.; Benavente-Garcia, O.; del Rio, J.A.; Ortuno, A.; Quirin, K.W.; Gerard, D. *J. Agr. Food Chem.* **2003**, *51*, 4247-4253.
16. Shan, B.; Cai, Y.Z.; Sun, M.; Corke, H. *J. Agr. Food Chem.* **2005**, *53*, 7749-7759.
17. Frankel, E.N.; Huang, S.W.; Aeschbach, R.; Prior, E. *J. Agr. Food Chem.* **1996**, *44*, 131-135.
18. Cervellati, R.; Renzulli, C.; Guerra, M.C.; Speroni, E. *J. Agr. Food Chem.* **2002**, *50*, 7504-7509.
19. Ziakova, A.; Brandsteterova, E. *J. liq. Chromatogr. R. T.* **2003**, *26*, 443-453.
20. Wang, H.F.; Provan, G.J.; Helliwell, K.S. *Food Chemistry* **2004**, *87*, 307-311.
21. Cuvelier, M.E.; Richard, H.; Berset, C. *J. Am. Oil Chem. Soc.* **1996**, *73*, 645-652.
22. Zheng, W.; Wang, S.Y. *J. Agr. Food Chem.* **2001**, *49*, 5165-5170.
23. Areias, F.; Valentao, P.; Andrade, P.B.; Ferreres, F.; Seabra, R.M. *J. Agr. Food Chem.* **2000**, *48*, 6081-6084
24. Exarchou, V.; Nenadis, N.; Tsimidou, M.; Gerothanassis, I.P.; Troganis, A.; Boskou, D. *J. Agr. Food Chem.* **2002**, *50*, 5294-5299.

25. Liu, A.H.; Li, L.; Xu, M.; Lin, Y.H.; Guo, H.Z.; Guo, D.A. *J Pharmaceut Biomed.* **2006**, *41*, 48-56.
26. Kamatou, G.P.P.; Viljoen, A.M.; Gono-Bwalya, A.B.; van Zyl, R.L.; van Vuuren, S.F.; Lourens, A.C.U.; Baser, K.H.C.; Demirci, B.; Lindsey, K.L.; van Staden, J.; Steenkamp, P. *J. Ethnopharmacol.* **2005**, *102*, 382-390.
27. Kosar, M,; Dorman, H.J.D.; Baser, K.H.C.; Hiltunen, R. *J. Agr. Food Chem.* **2004**, *52*, 5004-5010.
28. Areias, F.M.; Valentao, P.; Andrade, P.B.; Ferreres, F.; Seabra, R.M. *Food Chem.* **2001**, *73*, 307-311.
29. Jayasinghe, C.; Gotoh, N.; Aoki, T.; Wada, S. *J. Agr. Food Chem.* **2003**, *51*, 4442-4449.
30. Javanmardi, J.; Khalighi, A.; Kashi, A.; Bais, H.P.; Vivanco, J.M. *J. Agr. Food Chem.* **2002**, *50*, 5878-5883.
31. Ollanketo, M.; Peltoketo, A.; Hartonen, K.; Hiltunen, R.; Riekkola, M.L. *Eur. Food Res. Technol.* **2002**, *215*, 158-163.
32. Platel, K.; Srinivasan, K. *Indian J. Med. Res.* **2004**, *119*, 167-179.
33. Srinivasan, K. *Food Rev. Int.* **2005**, *21*, 167-188.
34. Capecka, E.; Mareczek, A.; Leja, M. *Food Chem.* **2005**, *93*, 223-226.
35. Madsen, H.l.; Bertelsen, G. *Trends Food Sci. Tech.* **1995**, *6,* 271-277.
36. Pizzale, L.; Bortolomeazzi, R.; Vichi, S.; Uberegger, E.; Conte, L.S. *J. Sci. Food Agr.* **2002**, *82*, 1645-1651.
37. Dorman, H.J.D.; Peltoketo, A.; Hiltunen, R.; Tikkanen, M.J. *Food Chem.* **2003**, *83*, 255-262.
38. Soobrattee, M.A.; Neergheen, V.S.; Luximon-Ramma, A.; Aruoma, O.I.; Bahorun, T. *Mutat. Res-Fund. Mol. M.* **2005**, *579*, 200-213.
39. Hinneburg, I.; Dorman, H.J.D.; Hiltunen, R. *Food Chem.* **2006**, *97*, 122-129.
40. Schilderman, P.A.E.L.; Tenvaarwerk, F.J.; Lutgerink, J.T.; Vanderwurff, A.; Tenhoor, F.; Kleinjans, J.C.S. *Food Chem. Toxicol.* **1995**, *33*, 99-109.
41. Madsen, H.L.; Nielsen, B.R.; Bertelsen, G.; Skibsted, L.H. *Food Chem* **1996**, *57*, 331-337.
42. Cuvelier, M.E.; Richard, H.; Berset, C. *Biosci. Biotech. Biochem.* **1992**, *56*, 324-325.
43. Dapkevicius, A.; van Beek, T.A.; Lelyveld, G.P.; van Veldhuizen, A.; de Groot, A.; Linssen, J.P.H.; Venskutonis, R. *J. Nat. Prod.* **2002**, *65*, 892-896.
44. Chen, J.H.; Ho, C.T. *J. Agr. Food Chem.* **1997**, *45*, 2374-2378.
45. Cao, H.; Cheng, W.X.; Li, C.; Pan, X.L.; Xie, X.G.; Li, T.H. *J. Mol. Struc-Theochem.* **2005**, *719*, 177-183.
46. Parejo, I.; Caprai, E.; Bastida, J.B.; Viladomat, F.; Jauregui, O.; Codina, C. *J. Ethnopharmacol.* **2004**, *94*, 175-184.
47. Psotova, J.; Kolar, M.; Sousek, J.; Svagera, Z.; Vicar, J.; Ulrichova, J. *Phytother. Res.* **2003**, *17*, 1082-1087.

48. Milos, M.; Mastelic, J.; Jerkovic, I. *Food Chem.* **2000**, *71*, 79-83.
49. Shixian, Q.; Da, Y.; Kakuda, Y.; Shi, J.; Mittal, G.; Yeung, D.; Jiang, Y. *Food Rev. Int.* **2005**, *21*, 295-311.
50. Vattem, D.A.; Jang, H.D.; Levin, R.; Shetty, K. *J. Food Biochem.* **2006**, *30*, 98-116.
51. Mazumder, A.; Neamati, N.; Sunder, S.; Schulz, J.; Pertz, H.; Eich, E.; Pommier, Y. *J. Med. Chem.* **1997**, *40*, 3057-3063.
52. Tewtrakul, S.; Miyashiro, H.; Nakamura, N.; Hattori, M.; Kawahata, T.; Otake, T.; Yoshinaga, T.; Fujiwara, T.; Supavita, T; Yuenyongsawad, S.; Rattanasuwon, P.; Dej-Adisai, S. *Phytother. Res.* **2003**, *17*, 232-239.
53. Asres, K.; Seyoum, A.; Veeresham, C.; Bucar, F.; Gibbons, S. *Phytother. Res.* 2005, *19*, 557-581.
54. Jiang, R.W.; Lau, K.M.; Hon, P.M.; Mak, T.C.W.; Woo, K.S.; Fung, K.P. *Curr. Med. Chem.* **2005**, *12*, 237-246.
55. Jassim, S.A.A.; Naji, M.A. *J. Appl. Microbiol.* **2003**, *95*, 412-427.
56. Sokmen, M.; Serkedjieva, J.; Daferera, D.; Gulluce, M.; Polissiou, M.; Tepe, B.; Akpulat, H.A.; Sahin, F.; Sokmen, A. *J. Agr. Food Chem.* **2004**, *52*, 3309-3312.
57. Kang, M.A.; Yun, S.Y.; Won, J.W. *Blood* **2003**, *101*, 3534-3542.
58. Won, J.; Hur, Y.G.; Hur, E.M.; Park, S.H.; Kang, M.A.; Choi, Y.; Park, C.; Lee, K.H.; Yun, Y. *Eur. J. Immunol.* **2003**, *33*, 870-879.
59. Osakabe, N.; Yasuda, A.; Natsume, M.; Yoshikawa, T. *Carcinogenesis* **2004**, *25*, 549-557.
60. Hur, Y.G.; Yun, Y.D.; Won, J.W. *J. Immunol.* **2004**, *172*, 79-87.
61. Sanbongi, C.; Takano, H.; Osakabe, N.; Sasa, N.; Natsume, M.; Yanagisawa, R.; Inoue, K.; Sadakane, K.; Ichinose, T.; Yoshikawa, T. *Clin. Exp. Allergy* **2004**, *34*, 971-977.
62. Walker, T.S.; Bais, H.P.; Deziel, E.; Schweizer, H.P.; Rahme, L.G.; Fall, R.; Vivanco, J.M. *Plant Physiol.* **2004**, *134*, 320-331.
63. Youn, J.; Lee, K.H.; Won, J.; Huh, S.J.; Yun, H.S.; Cho, W.G.; Paik, D.J. *J. Rheumatol.* **2003**, *30*, 1203-1207.
64. Pereira, P.; Tysca, D.; Oliveira, P.; Brun, L.F.D.; Picada, J.N.; Ardenghi, P. *Pharmacol. Res.* **2005**, *52*, 199-203.
65. Ahn, S.C.; Oh, W.K.; Kim, B.Y.; Kang, D.O.; Kim, M.S.; Heo, G.Y.;
66. Makino, T.; Ono, T.; Muso, E.; Yoshida, H.; Honda, G.; Sasayama, S. *Nephrol. Dial. Transpl.* **2000**, *15*, 1140-1145.
67. Renzulli, C.; Galvano, F.; Pierdomenico, L.; Speroni, E.; Guerra, M.C. *J. Appl. Toxicol.* **2004**, *24*, 289-296.
68. Kovatcheva, E.; Pavlov, A.; Koleva, I.; Ilieva, M.; Mihneva, M. *Phytochem.* **1996**, *43*, 1243-1244.
69. Ilieva, M.; Pavlov, A. *Appl. Microbiol. Biot.* **1997**, *47*, 683-688.
70. Pavlov, A.I.; Ilieva, M.P.; Panchev, I.N. *Biotechnol. Progr.* **2000**, *16*, 668-670.

71. Kintzios, S.; Makri, O.; Panagiotopoulos, E.; Scapeti, M. *Biotechnol. Lett.* **2003**, *25*, 405-408.
72. Santos-Gomes, P.C.; Seabra, R.M.; Andrade, P.B.; Fernandes-Ferreira, M. *J. Plant Physiol.* **2003**, *160*, 1025-1032.
73. Nitzsche, A.; Tokalov, S.V.; Gutzeit, H.O.; Ludwig-Muller, J. *J. Agr. Food Chem.* **2004**, *52*, 2915-2923.
74. Fedoreyev, S.A.; Veselova, M.V.; Krivoschekova, O.E.; Mischenko, N.P.; `Denisenko, V.A.; Dmitrenok, P.S.; Glazunov, V.P.; Bulgakov, V.P.; Tchernoded, G.K.; Zhuravlev, Y.N. *Planta Med.* **2005**, *71*, 446-451.
75. Tada, H.; Murakami, Y.; Omoto, T.; Shimomura, K.; Ishimaru, K. *Phytochem.* **1996**, *42*, 431-434.
76. Chen, H.; Chen, F.; Chiu, F.C.K.; Lo, C.M.Y. *Enzyme Microb. Tech.* **2001**, *28*, 100-105.
77. Al-Amier, H.; Mansour, B.M.M.; Toaima, N.; Korus, R.A.; Shetty, K. *J. Agr. Food Chem.* **1999**, *47*, 2937-2943.
78. Al-Amier, H.; Mansour, B.M.M.; Toaima, N.; Korus, R.A.; Shetty, K. *Food Biotechnol.* **1999**, *13*, 227-253.
79. Kintzios, S.; Kollias, H.; Straitouris, E.; Makri, O. *Biotechnol. Lett.* **2004**, *26*, 521-523.
80. Kim, H.; Chen, F.; Wang, X.; Rajapakse, N.C. *J. Agr. Food Chem.* **2005**, *53*, 3696-3701.
81. Yan, Q.; Shi, M.; Ng, J.; Wu, H.Y. *Plant Science* **2006**, *170*, 853-858.
82. Kim, H.J.; Chen, F.; Wang, X.; Rajapakse, N.C. *J. Agr. Food Chem.* **2006**, *54*, 2327-2332.

Tea and Health

Chapter 9

A Review on the Laboratory Investigations and Epidemiological Studies of Black and Pu-Erh Tea

Priscilla Mok[1], Raymond Chuen-Chung Chang[1], Mingfu Wang[2], and Kwok-Fai So[1]

Departments of [1]Anatomy, Faculty of Medicine and [2]Botany, The University of Hong Kong, Pokfulam, Hong Kong, SAR China

There have been a great number of studies reporting the beneficial activities of green tea and its major catechin constituent, EGCG, but claims of black tea and its exclusive major polyphenols have surfaced recently, warranting further investigation into its active fractions and their abilities. Various methods have been used to analyze the phenolic compounds and caffeine in black tea with HPLC methods coupled with absorbance and diode detection as the dominant ones. Extraction and purification of the compounds have proven to be a difficult barrier, as the structures of these compounds are complex and some have yet to be characterized (such as thearubigins) or discovered. Also, many of the studies conducted have been *in vitro* investigations with supra-physiological dosages, giving rise to the need for a more detailed study on their bioavailability and biotrasnformation *in vivo*. A variety of mechanisms have been proposed to be responsible for the overall anticancer effect; this may involve tumor apoptosis, antiproliferative activity, attenuation of angiogenesis, and modulation of cell-cycle regulation or signal transduction pathways. Whilst anticarcinogenic activities usually involve polyphenols acting as prooxidants, these catechin derivatives are similar to catechins, most notably

EGCG, in that it is also capable of antioxidant activity. Black tea, in addition to its capabilities of suppressing ROSes and deterring inflammation, has also been shown to lower some cardiovascular risk factors. The few studies on the effects of tea consumption (particularly black tea) have had heterogeneous results. Some laboratory results have been promising, but the link between these investigations and confirmation through epidemiological evidence remains to be substantiated.

Tea is derived from the plant *Camellia sinensis*, its wide popularity due to its easy preparation and relatively low cost. It has been consumed as a beverage for nearly 5000 years; Chinese history relates that the legendary emperor Shen Nung discovered the drink c. 2737 BCE (*1*). Almost every region of the world has its own method of preparing the infusion, taking the leaves from the plant and then processing them in various ways to produce the different types of tea.

The variations in tea are attributed to the types of leaves used and the fermentation processing. Many studies have been conducted on green tea, where the leaves are picked and subsequently steamed or pan-fried to prevent them from undergoing oxidation. Thus, constituent analyses of green tea have reported the highest catechin concentrations out of all major types of tea, with the most significant components being (–)-epigallocatechin-3-gallate (EGCG), (–)-epicatechin (EC), (–)-epicatechin-3-gallate (ECG) and (–)-epigallotecatechin (EGC). EGCG has been widely studied because of its purported anticarcinogenic, neuroprotective, hypolipidemic and antimutagenic activities, among others. It also has the lowest caffeine content compared to other teas. Oolong tea is semi-oxidized—after a period of time the leaves are "panned" to terminate the oxidation process. Black tea resides on the other end of the fermentation spectrum, undergoing full oxidation. The leaves are laid out to wilt until they are fairly dried. Next, they are rolled so that exposure to the air encourages oxidation. Pu-erh tea is still subject to debate as to whether it is a subcategory of black tea or belongs in a category of its own. It is unique because it is processed by "double fermentation", and also can be aged for many years, acquiring increasing value as it matures. Pu-erh tea undergoes oxidation like black tea but in addition, true fermentation, cultivating microbes in the process that may be responsible for its alleged medicinal properties. Pu-erh tea itself can be divided into two categories: the green variety is picked and heated to stop the natural oxidation, but then shaped into cakes and put away to slowly age. Black pu-erh is different in that after the leaves are heated, they are dried and then rewetted and allowed to ferment in a temperate environment to oxidize. There have been a great number of studies reporting the beneficial activities of green tea and its major catechin constituent, EGCG, but claims of black tea and its exclusive major polyphenols have surfaced recently, warranting further investigation into its active fractions and their abilities. Therefore, this review will focus on the two types of fermented tea, black and pu-erh, and the studies that have been conducted thus far on their activities. Furthermore, as the second-most popular drink in the world, epidemiological studies will be evaluated to gain further insight into tea's relative ability to bestow health and wellness to consumers.

The Composition of Tea

Studies on the composition of pu-erh tea have been rare, but black tea has received some analysis. It is by knowing the chemical constituents that the effects observed in *in vitro* and *in vivo* studies can best be justified.

The Compounds of Black Tea

The fermentation process of black tea induces changes to the components that would usually be present in green tea, the major transformations involving the oxidation of catechins by polyphenol oxidase to quinone and then condensation, forming theflavins, thearubigins, theasinensins, and bisflavanols (*1*). More specifically, dimerization of the polyphenols produces theaflavins (~2-6% of dry mass), and polymerization produces thearubigins (~15-20%), whilst ring expansion to tropolones further alter the original leaves. Other constitutents found include nitrogenous materials (~17%) consisting of protein (~6%), amino and nucleic acids (~8%) and theanine (~3%), an amino acid exclusive to tea (*2*). Theflavins in black tea have beeen further characterized to include theaflavin (TF1), theaflavin-3-gallate (TF-2A), theaflavin-3′-gallate (TF-2B), and theaflavin-3,3′-digallate (TF3). Their abundance in black tea have been reported by Leung *et al.* to be 0.08%, 0.34%, 0.11% and 1.05%, respectively (*3*). Several studies have shown that the presence of gallate moieties are significant (*3-7*). Sachinidis *et al.* found that only catechins with a gallate moiety were able to inhibit the tyrosine-induced phosphorylation of PDGF beta-receptor, curtailing the proliferation of A172 human glioblastomas (*4*), whilst Chen *et al.* observed a stronger inhibitory effect in SARS-CoV 3C-like protease activity in theaflavins TF-2B and TF3, both compounds which contain a gallate moiety attached at the 3' position (*5*). Leung *et al*, in their study of the antioxidant abilities of green tea catechins and black tea theaflavins, found that theaflavins were at least tantamount to if not more potent than their catechin precursors in suppressing the $CuSO_4$-induced oxidation of LDL. The antioxidant ability was in the order of

TF3 > ECG > EGCG ≥ TF-2B ≥ TF-2A > TF1 ≥ EC > EGC, and the hypothesis for this observed trend was grounded in the presence of the gallate moiety—TF3 possessed two and performed better than theaflavin compounds with only one; TF-2A and TF-2B with only one moiety outperformed TF1 which had none. Consistent with this observation, EC and EGC were less effective inhibitors than EGCG and ECG, the latter two possessing one gallate each. Therefore it would appear that the gallate moiety is a major participant in the antioxidant activities of these compounds (*3*).

Caffeine (2-8%) is another significant component, and is present in the highest concentrations in black tea and pu-erh compared to other varieties. When evaluating the individual abilities of tea polyphenols, caffeine then presents itself as a potentially confounding factor as it is able to induce body fat loss and exhibits anticarcinogenic activities of its own (*8,9*). The use of decaffeinated teas allows studies to more conclusively explain whether an effect was caused by caffeine or other tea components. In UVB-induced carcinogenesis, mice who were administered p.o. green tea, black tea or caffeine experienced inhibition whilst the noncaffeinated groups had less of an inhibitory effect (*10,11*). Caffeine's recognized ability to decrease tissue fat levels may have caused the indirect effect of inhibition, since the dermal fat under a tumor has been proposed to be a source of energy for growth and proliferation (*11*).

However any observed abilities of polyphenols would not be relevant without studying its bioavailability and biotransformation *in vivo*. It appears that the bioavailability of black tea polyphenols as it is regularly consumed may not be entirely practical for any short-term benefits because of their transient bioavailability. Henning *et al.* examined the bioavailabilty and bioactivity of tea flavanols between green tea, black tea and green tea supplement. Looking at Trolox equivalents as a measure of antioxidant activity, total plasma flavanol concentrations, green tea initially had the highest flavanol concentration, however as the trial progressed, green tea supplement's concentration surpassed it, exhibiting a delayed but overall greater absorption. Furthermore, an increase of 7.5 μmol/L in flavanol concentration was prerequisite to observing an increase in antioxidant activity; the average total plasma concentration for all three interventions was 1.2 μmol/L, suggesting that a supplement would be more efficient if flavanols were to be used for pharmacological purposes (*12*). Although this study focused more on the bioavailability of catechins present, the same concern exists for theaflavins, their derivatives. Thus, this is an area that deserves further investigation. One point to consider is that bioavailability differs between species, as showed by Kim *et al.* who observed the presence of green tea catechins in rats and mice after chronic consumption for varying periods of time. EGCG levels were greater in rats than in mice, and these findings implicate that more pronounced differences may exist between the absorption of polyphenols in animals and humans (*13*). Similarly, when aromatase activity was observed in rat ovarian and human placental microsomes, there was a discrepancy in the most effective inhibitory compound. The most potent was TF1 in the rat ovary, with an IC_{50} of 5.72±0.61 μM. Conversely for human placenta, the lowest IC_{50} belonged to TF2 with 3.23±0.08 μM, illustrating that caution should be taken when extrapolating results from animal to man (*7*).

Methods of Analysis

Various methods have been used to analyze the phenolic compounds and caffeine in black tea with HPLC methods coupled with absorbance and diode detection as the dominant ones (*14,15*). Capillary electrophoresis is an alternative method for analysis of black tea phenolic compounds (*15*). However, there are very few published methods that can simultaneously separate catechins, theaflavins, and other phenolic compounds. Lee and Ong published the first HPLC method that can separate catechins and theaflavins simultaneously in 2000 (*15*). The separation was performed on a guard and analytical cartridge system (Whatman PartiSphere 5 C18, 5 μM, 110*4.6 mm) with column temperature at 32 ^{O}C. The mobile phases were water with 5% acetonitrile and 0.035% trifluoroacetic acid (v/v) (A) and acetonitrile containing 0.025% trifluoracetic acid (v/v) (B) gradient which started from A:B (90:10) and graduated increased to 20% B in 10 min, to 40% at 16 min, to 50% at 20 min and back to 40% from 25 to 27 min and then balanced at 10% B for 3 min, the total running time was 30 minutess and the flow rate was set at 1 mL/min, the injection volume was 10 μL and the detection wavelengths were set at 205, 275 and 375nm. Rio *et al.* published two high resolution, gradient elution reverse-phase HPLC systems for the separation of 30 compounds in black tea, which is by far is the best HPLC methods for analysis of phenolic compounds and purine alkaloids in green and black tea (*16*). The separation were carried out using a Phenomenex Synergie RP-MAX column (4 μM, 250*4.6 mm, C_{12} reverse phase) with column temperature set at 40 ^{O}C, eluted with a 60 min gradient of either a 4 to 25% (for analysis of catechins and hydrycinnamates) or a 10 to 30% gradients (for separation of flavonols and theaflavins) of acetonitrile in water containing 1% formic acid. The flow rate was set at 1 mL/min and the detection wavelength was at 280 nm. LC/MS was also utilized for rapid identification purpose in this publication. Several methods also have been published to analyze black tea compounds in biological fluids including HPLC analysis of catechins and theaflavin with coulometric array detection (*17*) and analysis of theaflavins using HPLC coupled with electrospray mass spectrometry (*18*). Phenolic compounds also exist in black tea in the polymer formats (thearubigins), which can not be accurately analyzed so far.

Activities of Black and Pu-Erh Tea

Many studies have been conducted on green tea, yet there have been sparse investigations on the activities of black tea. Conventionally assumed to be bioactively inferior to green tea because of its lower catechin content, black tea is beginning to pique interest in its own exclusive polyphenol compounds, especially theaflavins. Extraction and purification of the compounds have proven to be a difficult barrier, as the structures of these compounds are complex and some have yet to be characterized (such as thearubigins) or discovered. Also, many of the studies conducted have been *in vitro* investigations with supraphysiological dosages, giving rise to the need for a more detailed study on their bioavailability and biotrasnformation *in vivo* to validate the results.

Anticarcinogenic Activities

The preventive and alleviative actions of black tea have been studied with varying types of cancer including that of the lung, skin, gastrointestinal tract, prostate, liver and mammary glands. A variety of mechanisms have been proposed to be responsible for the overall anticancer effect; this may involve tumor apoptosis, antiproliferative activity, attenuation of angiogenesis, and modulation of cell-cycle regulation or signal transduction pathways. Lu *et al.* experimented on mice pre-treated with ultraviolet B light and examined its implications on tumor proliferation, size and multiplicity. The mice were hyperplastic, considered to be at high risk for developing tumors, and were administered p.o. either green tea, black tea or caffeine as the intervention. The results they obtained suggested that tumor multiplicity was strongly associated (P=0.0001) with the amount of fat away from tumors, and that it was the treatments containing caffeine that negatively affected multiplicity via reduction of dermal fat levels. Decaffeinated teas were less effective, concurring with the view that caffeine can influence adipose tissue. It is also due to its presence that a direct causal relationship cannot be drawn between tea polyphenols and dermal fat levels. Tumor size appeared to be regulated by a different method due to their weak correlation (P=0.034), although the observation that the dermal fat layer beneath tumors was thinner indicates it may be used as an energy source for enlargement. In fact, administration of caffeine alone decreased the dermal fat layer by 36% under small tumors, and 97% under large tumors, showing that the reduction effect was amplified in large tumors. The decrease in dermal fat and tumor multiplicity may be attributed to a synergistic effect between the polyphenols and caffeine (*11*). Black tea, with its higher caffeine content, may enhance in the long-term the otherwise similar inhibitory effects exhibited by both teas in the aforementioned study. In another skin tumorgenesis study, epidermal growth factor (EGF) and platelet-derived growth factors (PDGF) were both inductors of EGF-receptor and PDGF-receptor, respectively. A study was carried out by Liang *et al.* to investigate the potential antiproliferative effects exerted by green tea and black tea in human A431 epidermoid cells and NIH3T3 mouse fibroblast cells; EGCG and TF3 were equally efficient in preventing proliferation but TF3 exceeded EGCG in reducing autophosphorylation of the receptors. In the same order, the IC50 values of TF3 were 15 and 18 μM and for EGCG, 26 and 28 μM for A431 and NIH3T3 cells. TF3 was also the only compound able to continue inhibition of autophosphorylation in the presence of EGF (*19*).

Black tea polyphenols have also been shown to modulate aromatase activity in a study conducted by Way *et al.* Aromatase converts androgens to estrogen, and is the final step in estrogen synthesis. Elevated aromatase activity is associated with tumor proliferation in estrogen-dependent breast cancers. Furthermore, the overexpression of the HER2/neu, a receptor of the tyrosine kinase family, endows cells with enhanced resistance to anti-estrogen therapies such as tamoxifen. When theaflavins were extracted from black tea and used as the treatment, aromatase activity was found to decrease in a dose-dependent manner, dehydroepiandrosterone (DHEA)-treated MCF-7 breast cancer cell proliferation was inhibited by 10 μM theaflavins and EGCG, and in HER2/neu-transfected cells, antiproliferation was not only controlled but the abilities of tamoxifen, an estrogen blocker, were enhanced rather than suppressed. The mechanism behind this reduction in resistance to tamoxifen was found to be due to decreasing the level of phosphorylation of tyrosine. In this study, the strongest inhibition of aromatase activity in rat ovarian microsomes was exerted by TF1 (*7*). This demonstrates that a galloyl group is not always necessary for an observed anticarcinogenic effect—however, it is also possible that the observation was from indirect causes, such as the metabolism of estrogen or decrease in body fat levels (*20*).

In many studies, it has been suggested that the efficacy of tea polyphenols is derived from the presence of the gallate moieties in theaflavins and catechins. The chemical structure of a gallate group is composed of an aromatic ring with three hydroxyl radicals that may confer to it its observed therapeutic properties. This is especially pronounced in experiments where the anticarcinogenic effect is observed roughly in the order TF3>TF2>TF1. Addition of green tea catechins will find themselves similarly ranked in the same fashion amongst the theaflavins in decreasing number of gallate groups. In an *in vitro* study of prostrate cancer, the NADPH-dependent enzyme steroid 5α-reductase was treated with green and black tea polyphenols to evaluate their roles in abrogating the conversion of testosterone to the more active dihydrotestosterone (DHT). Subsequently, this ligand joins with androgen receptor (AR) to mediate androgen-regulated gene transcription via androgen-responsive elements (AREs). Elevated activity of 5α-reductase is associated with prostate cancer, and thus 5α-reductase inhibitors may prove to be potentially therapeutic. 1,2,3,4,6,-Penta-*O*-galloyl-β-D-glucose (5GG) was also tested, possessing five galloyl moieties and being structurally similar to EGCG. 5α-reductase activity assay indicated that TF3 was the most effective (TF3 > 5GG > TF2B > TF2A > EGCG >> TF1), however treatment with other gallate group-containing compounds such as gallic acid and n-propyl gallate showed no effect, implying that a gallate group was not the singular factor in modulating the activity of the enzyme. Proliferation of the lymph node carcinoma of the prostrate (LNCaP) cells were considerably inhibited by TF3 and EGCG; the mechanism for this action was twofold: 1) there was a significant down-regulation of AR expression and 2) via the suppression of fatty-acid synthase (FAS), a lipogenic enzyme whose activity may be directed by androgens and serve a supporting role in facilitating LNCaP cells (*6*).

Yang *et al.* have found in their interventions with transformed and non-transformed human bronchial cell lines that TF3 was the most potent (IC_{50}=22 μM), comparable to that of EGCG's IC_{50} of 22-24 μM. However, cell viability was not affected by any compound for concentrations up to 30 μM. An apoptosis assay showed that with TF3, the percentage of apoptotic cells increased sharply and stabilized after 12 hours, whilst EGCG experienced a lag phase and then peaked in apoptotic acitivity after 24 hours. It was also found that EGCG's apoptotic activity was dependent on the induction of H_2O_2 production, which ceased after addition of catalase. TF3's mechanism appears to be regulated differently as the morphological results show an irregularly-shaped cell and cytoplasmic granules. The underlying mechanism behind this may be due to down-regulating the phosphorylation of c-jun protein, decreasing AP-1 activity and modulating gene transcription favoring cellular termination. Similar to the results from Lee *et al.* the galloyl structures were significant here, with the percentage of apoptotic cells caused by each intervention in the order TF3 > EGC > EGCG > EC (*21*). Other studies have also reported the importance of the gallate group to theaflavins in which they investigated the apoptosis of BEL-7402 cells from human liver cancer and MKN-28 stomach cancer cells treated with tea polyphenols (*22*). Lee *et al.* further analyzed the significance and possible action of the galloyl structures by comparing it with the structure of the androgen testosterone. They reported that: 1) the A, B, C rings of catechins and theaflavins laid in a planar arrangement, similar to testosterone's rings and possibly encouraging hydrophobic attraction, 2) the galloyls in all theaflavins and EGCG are rotatable and 3) TF3's B ring is fixed, whilst EGCG's ring is rotatable, perhaps enhancing its competitive inhibition with NADPH for the 5α-reductase binding site (*6*).

This selection of studies demonstrates the potential of black tea in a wide range of anticarcinogenic activities. The results have shown how theaflavins are able to modulate *in vitro* the mechanisms that regulate signal transduction pathways, modulating the expression of transcription factors that down-regulate cellular growth factors and up-regulate those that encourage tumor apoptosis. It has been proposed that since tea is first and foremost a beverage, its ingestion would naturally result in greater contact with the gastrointestinal system, where it is expected to exert the strongest anticarcinogenic effects (*20*). However, due to the artificial nature of these studies, more *in vivo* experiments are required to conclusively determine the bioavailability and consequently the practical applications of these findings.

Anti-inflammatory Activities / Antioxidant Activities

Whilst anticarcinogenic activities usually involve polyphenols acting as prooxidants, these catechin derivatives are similar to catechins, most notably EGCG, in that it is also capable of antioxidant activity. The scavenging of reactive oxygen species (ROS), reactive nitrogen species (RNS) and free transition metals have been associated with anti-inflammatory effects (*23,24*). It has not been elucidated in any studies whether theaflavins exhibit the same dose-dependent biphasic quality as EGCG. Lin *et al.* examined the factors that may affect the quantity of tea polyphenols using reverse-phase HPLC. After the extracts were obtained, they also investigated the effects of adding these extracts to nitric oxide (NO), acute promyelocytic leukemia HL-60 cells, and plasmid DNA damaged by the Fenton reaction to induce hydroxyl radical production. The suppression of LPS-induced NO was in the order: pu-erh tea (PT) $\geq$ black tea (BT) > green tea (GT) > ooloong tea (OT). The abrogative effects of pu-erh tea may be due to their lower levels of catechin monomers, since they are more extensively fermented. This reduced permeability in cellular membranes has been postulated to be beneficial in blocking LPS from attaching to its receptor and initiating transduction pathways downstream towards increased inducible nitric oxide synthase (iNOS). Pu-erh tea and also showed the strongest antoxidative activities; hydroxyl radical scavenging was in the order PT > GT > BT > OT. HL-60 cell apoptosis was led by GT and OT, whose lanes in the agarose gels clearly indicated DNA ladder formation, an evident sign of apoptosis (*25*). Pu-erh tea's constituents have been consistently difficult to quantify and characterize, owing to their complexity and the double fermentation that further oxidizes present structures and cultivates a number of microbes. It may be precisely due to these bioactive microbes that are responsible for pu-erh's unique capacities. More sophisticated techniques will need to be developed in order to determine the exact mechanism of pu-erh's bioactivity.

In a comparative study between green tea and black tea, it was reported that black tea surpassed green tea in chemopreventive abilities such as in the containment of NO production and another ROS, superoxide anion. The combination of both species results in formation of peroxynitrite, a highly reactive radical (*26-28*). A 10 μg total catechin treatment of black tea was found to abrogate NO production by 70%, and a reconstitution experiment demonstrated that removal of all catechin and theaflavin components relieved black tea of its inhibitive faculties. Step-wise addition of some active fractions restored the abrogative activity to 63%, excluding thearubigins, EGC and ECG. Reverse-transcription PCR revealed that the suppressive activity of BT was due to modifying iNOS expression (*29*). A previous study has ascribed this to the down-regulation of nuclear factor-κB (NF-κB), the associated transcription factor of iNOS (*30*). The NF-κB pathway also played a main role in Aneja *et al.*'s study on the anti-inflammatory potential of theflavin in reducing the expression of interleukin-8 (IL-8) in human respiratory cells *in vitro*. The experimental results show that theaflavin counteracts nearly every step in the pathway leading down to the expression of the IL-8 gene. Tumor necrosis factor-α (TNF-α) is a cytokine released by tissues under stress and damage; this event activates IκB-kinase (IKK), which has the potential to phosphorylate the inhibitory protein IκB, liberating NF-κB. Using an enzyme-linked immunosorbent assay, it was found that TNF-α was inhibited completely at high concentrations (>30 μM). The activation of IKK was also affected by theaflavin—in an IKK assay using an anti-IKK-γ antibody for immunoprecipitation, treatments with theaflavin concentrations at 10 and 30 μg/mL elicited inhibition. The next step following activation of IKK is phosphorylation of IκB. Western blotting revealed a concentration-dependent protective role by theaflavin in preventing the degradation of IκB. An electrophoretic mobility shift assay exhibited impedance of the activation of NF-κB, however again at the fairly high concentrations of 10 and 30 μg/mL. It is important to note that below 10 μg/mL concentration, some of these outcomes would not have occurred while others would have diminished efficacy. Following successful activation of NF-κB would result in binding with the oligonucleotide in the IL-8 promoter region. The observations from a Northern blot assay showed that IL-8 mRNA expression was decreased in a concentration-dependent manner, warranting further investigation into its underlying mechanism. It was determined that theflavin only modulated the NF-κB pathway only at high concentrations. However, there was also a significant effect on the AP-1 pathway even at concentrations as low as 1 and 3 μg/mL, suggesting that theaflavin regulates IL-8 expression primarily via the AP-1 pathway. These observations may be of particular relevance for patients of acute lung injury as there are high levels of IL-8 in the serum (*31*).

Human red blood cells (RBC) were the focus of another study looking at the antioxidant properties of tea. Halder *et al.* induced oxidative damage using a variety of agents such as phenylhydrazine, Cu^{2+} with ascorbic acid, resulting in the typical indicators such as lipid peroxidation, deterioration of membranous proteins and increased membrane viscosity. Addition of black tea extracts to each incubation system resulted in abrogation of oxidative damage. Malondialdehyde is produced during the process of lipid peroxidation, but addition of 50 μl of extract in a system with phenylhydrazine produced a 1% decrease from the control. Furthermore, the microviscosity of the membrane was also preserved. Addition of oxidative stressors xanthine and xanthine oxidase to the control system elevated the microviscosity value (η) from 1.53±0.02 to 2.04±0.04. However, the addition of 50 μL of black tea extract reinstated the value to 1.55±0.05 (*32*). Although results from these studies appear promising, it is crucial to note that the dosages applied in each treatment were most likely supraphysiological concentrations. More studies will need to be done to see whether these results can be replicated *in vivo.*

Hypolipidemic Activities

A few mechanisms are associated with hypercholestrolemic effects, such as increased LDL oxidation, greater serum triglyceride levels, and greater fibrinogen levels (*33-36*). Black tea, in addition to its capabilities of suppressing ROSes and deterring inflammation, has also been shown to lower these cardiovascular risk factors. A comparative study by Kuo *et al.* was conducted to observe the hypolipidemic abilities of four types of tea (GT, OT, PT and BT). After feeding Sprague-Dawley rats one of four types of tea leaves for 30 weeks, it was found that the amount of suppressed body weight was in the order OT > PT > BT > GT. For serum triglyceride levels, PT and OT were the most effective, but PT and GT were more consistent in reducing overall cholesterol, even though 4% OT only showed a dramatic 35% drop after 30 weeks. In addition, PT was most efficient in lowering LDL (PT > GT > OT > BT) and was the only tea that concurrently raised HDL levels as well. The authors postulate that PT's pronounced capacities in this study may be attributed to the high molecular weight compounds produced during fermentation, as was studied by Hayakawa *et al.* (*37,38*). However, these compounds have yet to be covered extensively. LDL oxidation was used as a model in Leung *et al.*'s experiment to compare the antioxidant abilities of green tea and black tea (*3*). Its oxidation has been shown in various studies to exacerbate the risk of developing atherosclerosis, promoting cholesterol accumulation and intoxicating endothelial cells (*34,39*). It was observed that black tea's antioxidant abilities were comparable and in some cases, greater than that of green tea through the measurement of the thiobarbituric acid-reactive substances (TBARS) produced during LDL oxidation. The inhibition of TBARS was in the order of TF3 > ECG ≥ EGCG ≥ TF2B ≥ TF2A > TF1 ≥ EC > EGC. At the lowest concentration of polyphenols used, 5 μmol/L, the lag times in production of TBARS was 4 (TF1), 7 (TF2A), 7.5 (TF2B), and 16 (TF3) hours. At greater concentrations of 20 μmol/L and 40 μmol/L, there was complete abrogation of TBARS production, indicating that all theaflavin compounds had successfully deterred oxidation. It was noted that the theaflavins were consistently more effective inhibitory agents than their one of their precursors. TF2B, derived from EGC and ECG was stronger than EGC but not as effective as ECG. Similar to previously covered studies on black tea and the significance of the galloyls in anticarcinogenesis, an analogous trend was observed here, possibly due to the extra hydroxyl groups on the gallate moieties.

It appears that the full fermentation of pu-erh tea and black tea may have useful implications in controlling the deleterious effects of hypercholestrolemia. However, long-term epidemiological studies are needed to corroborate with these findings in confirming the cardiovascular benefits of black tea.

Epidemiological Studies

The few studies on the effects of tea consumption (particularly black tea) have had heterogeneous results. They are usually not as conclusive as laboratory findings, mentioning the various confounding factors and issues that may have occluded the study's results. In a meta-analysis exploring the relationship between tea-drinking and cardiovascular risk, the studies that it attempted to combine also displayed mixed results. This was mostly due to differing geographical locations in which the studies were conducted, giving rise to regional differences in preparation of the beverage (such as tea strength) and also differing etiological factors that cause incidences of cardiovascular disease in each population. Another issue noted by the authors was that of publication bias, in which studies that appear to challenge the popularly-accepted health benefits of tea are usually not accepted (*40*).

However, there have been some studies that have concluded a possibly beneficial role of black tea. Duffy *et al.* found in a crossover study that individuals with coronary artery disease (CAD) experienced a significant improvement in the dilation of the brachial artery ($P<0.001$) after consuming black tea. Prior to the intervention, the mean baseline dilation was 6.0±3.4%. After the short-term tea intervention, dilation had increased to 9.3±3.9%. This was proposed to be due to a reversal of any endothelial vasomotor dysfunction in patients with CAD. The duration of the study was 8 weeks, indicating that although tea can exert benefits in the short-term, it remains to be seen whether this effect can continue with extended consumption habits (*41*). Another study examined the incidence of squamous cell carcinoma in a case-control study of elders in Arizona, USA. It was found that there was an inverse association between consumption of hot black tea and squamous cell carcinoma ($P=0.05$). However, the same relationship was not observed for ice tea, demonstrating that variations in tea preparation can have significant discrepancies. The lack of association with iced tea may be that the beverage is usually consumed in a relatively more diluted state and that the lower temperatures promotes formation of "tea cream", which is the complexation of theaflavins and thearubigins with caffeine. This complex precipitates out of the tea, settling at the bottom and thus these polyphenols would not be consumed (*42*). In contrast, Goldbohm *et al.* investigated the possible relationship between cancer incidence and consumption of black tea. In a cohort study with a follow-up period of 4.3 years, they concluded that their study showed no significant relationship in favor of the anticarcinogenic abilities of black tea. Nevertheless, the authors do point out that cancer has a long induction period, so dietary habits from the past year may not have influenced the development of cancer a period of time later. In addition, subclinical cases of cancer were not taken into account. This may have affected the results because dietary habits (such as increased tea consumption) may have changed due to this without the subject's knowing (*43*).

As demonstrated, epidemiological studies have the potential to provide substantial evidence for the results acquired in laboratory investigations on black tea. There are a few general considerations that can be accrued from the majority of these surveys. First, the inaccuracy of the respondent's answers may be potentially influential, especially in a small-scale study. Dietary recall is potentially susceptible to errors since individuals do not tend to pay particular attention to such a natural and daily activity as eating and drinking. Second, questionnaires tend to not be extremely specific as to the method of tea consumption. However it has already been demonstrated by Hakim *et al.* that variables such as tea temperature and brewing method have significant ramifications on a study's drawn conclusions. Questionnaires that do not specify the type of tea consumed are expected to be even more susceptible to inaccurate results since levels of tea polyphenols and caffeine would pose as confounding factors.

Concluding Remarks

Black tea has long been known to possess curative qualities, but only recently have they attracted enough interest to necessitate a preliminary inquiry into those exact qualities and their mechanisms. Some laboratory results have been promising, but the link between these investigations and confirmation through epidemiological evidence remains to be substantiated.

First, the bioavailability of theaflavins and other black tea constituents merit a deeper examination. These also include the microbes present in black tea and further characterization of some fractions, such as thearubigins, as they are present in significant amounts. By knowing this, the dosages required to achieve the desired effect can thus be determined. To improve laboratory results, the types of tea may need to be standardized, since there are many factors that can considerably influence the character of the tea. These may include plucking, withering, fermenting, storage and brewing methods, all which can drastically vary the amounts of polyphenols and caffeine (*1*). *In vivo* studies will also need to be conducted, to verify *in vitro* results, and more sophisticated extraction and purification techniques are also in order. Epidemiological evidence on black tea has been sparse, but positive findings from the laboratory can assist in determining the direction of research.

References

1. Balentine, D.A.; Harbowy M.E.; Graham H.N. In *Caffeine*; Spiller, G.A., Ed.; CRC Press, Boca Raton, FL 2004, pp. 38-72.
2. Harbowy M.E.; Balentine D.A. *Crit. Rev. Plan. Sci.* **1997**, *16*, 415-480.
3. Leung, L.K.; Su, Y.; Chen, R.; Zhang, Z.; Huang, Y.; Chen, Z.Y. *J. Nutr.* **2002**, *131*, 2248-2251.
4. Sachinidis, A.; Seul, C.; Seewald, S.; Ahn, H.; Ko, Y.; Vetter, H. *FEBS Lett.* **2000**, *471*, 51-55.
5. Chen, C.N.; Lin, C. P. C.; Huang, K.K.; Chen, W.C.; Hsieh, H.P.; Liang, P.H.; Hsu, J.T.A. *Evid. Based Complement. Alternat. Med.* **2005**, *2*, 209-215.
6. Lee, H.H.; Ho, C.-T.; Lin, J.K. *Carcinogenesis.* **2004**, *25*, 1109-1118.
7. Way, T.D.; Lee, H.H.; Kao, M.C.; Lin, J.K. *Eur. J. Cancer* **2004**, *40,* 2165-2174.
8. Westerterp-Plantenga, M.S,; Lejeune, M.P.; Kovacs, E.M. *Obes. Res.* **2005**, *13*, 1195-1204.
9. Hashimoto, T.; He, Z.; Ma, W.Y.; Schmid, P.C.; Bode, A.M.; Yang, C.S.; Dong, Z. *Cancer Res.* **2004**, *64*, 3344-3349.
10. Huang, M.T.; Xie, J.G.; Wang, Z.Y.; Ho, C.-T.; Lou, Y.R.; Wang, C.X.; Hard, G.C.; Conney, A.H. *Cancer Res.* **1997**, *57*, 2623-2629.
11. Lu, P.; Lou, Y.R.; Lin, Y.; Shih, W.J.; Huang, M.T.; Yang, C.S.; Conney, AH. *Cancer Res.* **2001**, *61*, 5002-5009.
12. Henning, S.; Niu, Y.; Lee, N.; Thames, G.D.; Minutti, R.R.; Wang, H.; Go, V. L. W.; Heber, D. *Am. J. Clin. Nutr.* **2004**, *80*, 1558-1564.
13. Kim, S.; Lee, M.J.; Hong, J.; Li, C.; Smith, T.J.; Yang, G.Y.; Seril, D.N.; Yang, C.S. Nutr. Cancer 2000, 37, 41-48.
14. Temple, C.; Clifford, M. N. J. *J. Sci. Food Agric.* **1992**, *21*, 351.
15. Lee B.L.; Ong, C. N. J. *Chromatography* A. **2000**, *881*, 439.
16. Rio, D.D.; Stewart, A.J.; Mullen, W.; Burns, J.; Lean, M. E.J.; Brighenti, F.; Crozier, A.J. *Agric Food Chem*. **2004**, *52*, 2807.
17. Lee, M.J.; Prabhu, S.; Meng, X.; Li, C.; Yang, C.*S. Analytical Biochemistry*. **2000**, *279*, 164.

18. Mulder T. P. J.; Platerink C. J. van; Schuyl P. J. W.; Amelsvoort J. M. M. Van. *J. Chromatography* B. **2001**, *760*, 271.
19. Liang, Y.C.; Chen, Y.C.; Lin, Y.L.; Lin-Shiau, S.Y.; Ho, C.-T.; Lin, J.K. *Carinogenesis* **1999**, *20*, 733-736.
20. Yang, C.S.; Maliakal, P.; Meng, X. *Annu. Rev. Pharmacol. Toxicol.* **2002**, *42*, 25-54.
21. Yang, G.Y.; Liao, J.; Li, C.; Chung, J.; Yurkow, E.J.; Ho, C.-T.; Yang, C.S *Carcinogenesis* **2000**, *21*, 2035-2039.
22. Tu, Y.Y.; Tang, A.B.; Watanabe, N. *Acta Biochim. Biophys. Sin.* **2004**, *36*, 508-512.
23. Pan, M.H.; Lin-Shiau, S.Y.; Ho, C.-T.; Lin, J.H.; Lin, J.K. *Biochem. Pharmacol.* **2000**, *59*, 357-367.
24. Nomura M. *Carcinogenesis* **2000**, *21*, 1885-1890.
25. Lin, Y.S.; Tsai, Y.J.; Tsay, J.S.; Lin, J.K. *J. Agric. Food Chem.* **2003**, *51*, 1864-1873.
26. Herold, S.; Fago, A. *Comp. Biochem. Physiol. A Mol. Integr. Physiol.* **2005** In press.
27. Li, J.; Baud, O.; Vartanian, T.; Volpe, J.J.; Rosenberg, P.A. *Proc. Natl. Acad. Sci.* **2005**, *102*, 9936-9941.
28. Hayashi Y.; Sawa Y.; Nishimura M.; Fukuyama N.; *et al. Eur. J. Cardiothorac. Surg.* **2004**, *26*, 276-280.
29. Sarkar A,; Bhaduri A. *Biochem. Biophys. Res. Commun.* **2001**, *284*, 173-178.
30. Lin Y.L.; Tsai S.H.; Lin-Shaiu, S.Y.; Ho, C.-T.; Lin, J.K. *Eur. J. Pharmacol.* **1999**, *367*, 379-388.
31. Aneja, R.; Odoms, K.; Denenberg, A.; Wong, H.R.; *Crit. Care Med.* **2004**, *32*, 2097-2103.
32. Halder, J.; Bhaduri, A.N. *Biochem. Biophys. Res. Commun.* **1998**, *244*, 903-907.
33. Paramo, J.A.; Beloqui, O.; Roncal, C.; Benito, A.; Orbe, J. *Haematologica* **2004**, *89*, 1226-1231.
34. Palinski, W.; Rosenfeld, M.E.; Yla-Herttuala, S.; Gurtner, G.C.; Socher, S.S.; Butler, S.W.; Parthasarathy, S.; Carew, T.E.; Steinberg, D.; Witztum, J.L. *Proc. Natl. Acad. Sci.* **1989**, *86*, 1372-1376.
35. Gordon; T.; Castelli; W.P.; Hjortland; M.C.; Kannel; W.B.; Dawber, T.R. *J. Am. Med. Assoc.* **1997**, *238*, 497-499.
36. Neaton, J.D.; Blackburn, H.; Jacobs, D.; Kuller, L.; Lee, D.J.; Sherwin, R.; Shih, J.; Stamler, J.; Wentworth, D. *Arch. Intern. Med.* **1992**, *152*, 1490-1500.
37. Kuo, K.L.; Weng, M.S.; Chiang, C.T.; Tsai, Y.J.; Lin-Shiau, S.Y.; Lin, J.K. *J. Agric. Food Chem.* **2005**, *53*, 480-489.

38. Hakayawa, S.; Kimura, T.; Saeki, K.; Koyama, Y.; Aoyagi, Y.; Noro, T.; Nakamura, Y.; Isemura, M. *Biosci. Biotechnol. Biochem.* **2001**, *65*, 459-462.
39. Steinberg D.; Parthasarathy S.; Carew T.W.; Knoo J.C.; Witztum, J.L. *N. Engl. J. Med.* **1989**, *320*, 915-924.
40. Peters, U.; Poole, C.; Arab, L. *Am. J. Epidem.* **2001**, *154*, 495-503.
41. Duffy, S.J.; Keaney, J.F.; Holbrook, M.; Gokce, N.; Swerdloff, P.L.; Frei, B.; Vita, J.A. *Circulation* **2001**, *104*, 151-156.
42. Hakim, I.A.; Harris, R.B.; Weisgerber, U.M. *Cancer Epidemiol. Biomarkers Prev.* **2000**, *9*, 727-731.
43. Goldbohm, R.A.; Hertog, M. G. L.; Brants, H. A. M.; Poppel, G.; van den Brandt, P.A. *J. Natl. Cancer Inst.* **1996**, *88*, 93-100.

Chapter 10

Black Tea Polyphenols Theaflavins Inhibit the Growth of LNCaP Prostate Cancer Cells through Suppressing Androgen Receptor and 5α-Reductase Activity

Jen-Kun Lin[1], Hung-Hsiao Lee[1], and Chi-Tang Ho[2]

[1]Institute of Biochemistry, College of Medicine, National Taiwan University, Taipei, Taiwan
[2]Department of Food Science, Rutgers, The State University of New Jersey, 65 Dudley Road, New Brunswick, NJ 08901–8520

Tea and tea polyphenols have been considered as potential cancer chemopreventive and anti-obesity agents in the general population. Investigation into tea polyphenols have showed many profound biochemical and pharmacological activities including anti-oxidant and pro-oxidant effects, induction of apoptosis and cell cycle arrest in cancer cells, inhibition of cellular proliferation and tumor progression through suppressing epidermal growth factor receptor signaling pathway and inducible nitric oxide synthase expression through down-regulation of NFκB activation, inhibition of tumor promotion through suppressing AP-1 activation and fatty acid synthase expression. We recently found that the black tea polyphenols TF3, TF-2A, and TF-2B inhibited rat liver microsomal 5α-reductase activity and significantly reduced androgen responsive LNCaP prostate cancer cell growth, suppressed expression of the androgen receptor and lowered androgen-induced prostate specific antigen and fatty acid synthase protein level. Similarly, the green tea polyphenol EGCG also showed the inhibitory effect, but to a lesser extent, on the 5α-reductase activity and LNCaP prostate cancer cell growth. These findings suggest that tea polyphenols might function as chemopreventive agents for prostate carcinogenesis through suppressing the function of androgen and its receptor.

Prostate cancer is the most frequently diagnosed malignancy and second leading cause of death due to cancer in males in the western world (*1*). In the human prostate, androgens mediate critical processes involved in the normal development, organizational structure and mature function of the gland (*2*). Evidence shows that androgens can also be risk factors for prostate cancer development (*3*). Other risk factors include modifiable factors such as diet, diet supplements, obesity, and non-modifiable factors such as age, race, family history and presence of certain genetic polymorphisms (*4*). Strategies for developing prevention clinical trials for prostate cancer have focused primarily on prevention by hormonal modulation and through the use of natural and synthetic bioactive food components and dietary supplements. It appeared that clinical trials of androgen-insensitive prostate cancer cell lines have been conducted investigating bioactive food components such as tea polyphenols (*5*).

The androgenic hormones testosterone (T) and dihydrotestosterone (DHT) exert their cellular effects by means of interaction with the androgen receptor (AR), a ligand-dependent transcriptional factor. AR belongs to the steroid/nuclear receptor superfamily, and possesses higher affinity for DHT than for T. In the prostatic cell, T is transformed into DHT by the enzyme 5α-reductase. After T or DHT binding to the AR , ligand-activated AR complexed with co-activator proteins and general transcription factors bind to androgen-response elements (AREs) located promoter regions of androgen-regulated genes which then serves to activate or to repress transcription (*6,7*).

Many reports have described the cancer chemopreventive effects of green tea polyphenols, especially, (-)-epigallocatechin-3-gallate (EGCG) in prostate carcinogenesis (*8-11*). EGCG the major catechin present in green tea is a promising cancer chemopreventive agent. Green tea polyphenols exert remarkable preventive effects against prostate cancer in a mouse model and many of these effects are mediated by the ability of polyphenols to induce apoptosis in cancer cells (*5*). Using a cDNA microarray, it has been found that EGCG treatment of LNCaP cells results in induction of genes that functionally exhibit growth-inhibitory effects and repression of genes that belong to the G-protein signaling network (*9*). The chemopreventive action of catechins in the TRAMP mouse model of prostate carcinogenesis is accompanied by clusterin over-expression (*10*). Immunohistochemistry experiments confirmed clusterin down-regulation during prostate cancer onset and progression, and clusterin sustained expression in chemoprevented TRAMP mice. A possible role for clusterin as a novel tumor suppressor gene in the prostate is thus suggested (*10*).

Published investigations have shown that green tea polyphenols such as EGCG are capable reducing tumor burden in animal models and inhibiting proliferation and inducing apoptosis in cell culture models (*12-15*). Although green tea polyphenols are generally considered to be more potent in influencing these processes, several studies have reported that polyphenols from black tea are similarly (or even more) effective in suppressing these processes (*16,17,18*).

As black tea is more readily consumed in Western nations, it is worth investigating the effects of black tea polyphenols theaflavins, on mechanisms involved in prostate cancer development. The major black tea polyphenols theaflavins include theaflavin (TF-1), theaflavin-3 -gallate (TF-2A), theaflavin-3′-gallate (TF-2B) and theaflavin-3,3′-digallate (TF-3). Among black tea theaflavins, TF-3 is generally considered to be the more effective components for the inhibition of carcinogenesis. In our previous studies, TF-3 has been found to possess anti-proliferative activity on several tumor cell lines including A431 and NIH3T3 through blocking the growth factor binding to its receptor (*18*). TF-3 may also exert its anti-inflammatory and cancer chemopreventive actions by suppressing the activation of nuclear factor κB (NFκB) through inhibition of inhibitor κB kinase (IKK) activity (*19*). In addition, TF-3 may inhibit TPA-induced protein kinase C (PKC) and transcription activator protein-1 (AP-1) binding activities (*20*).

In this study, we examined whether theaflavins have inhibitory effects on androgen production and action of prostate cancer. We used rat liver microsomes as 5α-reductase enzyme source. We found that theaflavins inhibit 5α-reductase activities. Furthermore, they significantly reduced androgen responsive LNCaP prostate cancer cell growth, suppressed expression of the AR and lowered the secretion of prostate specific antigen (PSA). Our results suggest that theaflavins can attenuate the function of androgen and AR, which may be useful for the prevention of prostate cancer.

Materials and Methods

Materials

EGCG, gallic acid, and n-propyl gallate were purchased from Sigma (St. Louis, MO). TF-1, TF-2A. TF-2B and TF-3 were isolated from black tea (*21*). Testosterone (T), dihydrotestosterone (DHT) and androst-4-ene-3,17-dione were purchased from Sigma. [^{14}C-]-Testosterone (57 mCi/mmol) was purchased from Amersham (Arlington Height, IL).

Isolation of Rat Liver Microsomes

Rat liver microsomes were isolated by the method of Liang *et al.* (*22*). One female Sprague-Dawley rat (body weight 400 g), that was made to fast overnight to decrease the concentration of liver glycogen was then put to death by carbon dioxide asphyxiation. The liver was then removed for microsome isolation as described (*22*).

5α-reductase Assay

The 5α-reductase assay was performed as described (*23*). Briefly, the reaction solution contains 1 mM dithiothreitol, 40 mM potassium phosphate, pH 6.5, 100 μM NADPH, [^{14}C-]-testosterone (3.5 μM) and liver microsome (20 μg protein) in a total volume of 0.5 mL, with or without designated concentrations of polyphenols. The reaction was started with the addition of the enzyme preparation. After incubation at 37 °C for 20 min, the reaction mixture was extracted with 2 ml ethylacetate. The ethyl acetate phase (upper phase) was transferred to a tube and evaporated to dryness with nitrogen gas. The steroids were taken up in 50 μL ethylacetate and chromatographed on a TLC plate (Merck silica gel $60F_{254}$ type), using chloroform- methanol (96:4) as the developing solvent system at 4 °C. For comparison, standards of unlabeled steroids (10 μg each of T, DHT, and androst-4-ene-3,17-dione) were also submitted to TLC. The plate was viewed under a UV lamp (254 nm) to locate T and androst-4-ene-3,17-dione. Since DHT can not be visualized under UV, it was located with iodine vapors. Detection of the radiolabeled steroids was performed by autoradiography. [^{14}C-]-T and [^{14}C-]-DHT were measured by densitometer (IS-100 Digital Imaging System). The conversion of T to DHT was calculated from the ratio of the radioactivity of DHT to the sum of the radioactivities of T and DHT (*22, 23*).

Cell Cultures and Treatments

The human prostate cancer cell line LNCaP was obtained from The American Type Culture Collection (Rockville , MD) and propagated in 24-well, 60 or 100 mm culture dishes at the desired density in RPMI 1640 medium supplemented with 10% fetal calf serum (FCS) (Gibco, Grand Island, NY), 1% penicillin/streptomycin (P-S) in a 5% CO_2 atmosphere at 37 °C. LNCaP cells were also cultured in 10% charcoal stripped FCS (cFCS) and 1% (P-S). The cells were treated with polyphenols at designated concentrations with or without T at physiological concentrations (25 nM) in phenol red-free RPMI 1640 medium.

Western Blot Analysis

The Western blot analysis was conducted as described by our lab in a prior paper (*24*).

Cell Viability and PSA Protein Expression

LNCaP cells were seeded at 2×10^4 cells/well in 24 well plates. After 48 h the medium was changed to serum-free RPMI 1640 medium, and the cells were incubated for an additional 24 h to deplete endogenous steroid hormones prior to experiments. Cells were then treated with different concentrations of polyphenols with or without 25 nM T. After 4 days incubation, spent media were harvested, and the level of PSA in spent media were quantified by an enzyme linked immunoabsorbent assay (ELISA). The human PSA ELISA kit was obtained from Cytimmune, MA. Cell density was quantified by MTT assay. The protein levels of PSA were normalized by cell density measurements and expressed as a percentage over control (T alone).

Results

The Reaction Kinetics of 5α-Reductase

As described previously (*24*), the Km and Vmax values for 5α-reductase were determined in rat liver microsomes from Lineweaver-Burk plots using different concentrations of T. The mean value for Km in our experimental data was 2.45 μM and that for Vmax 3.11 nmol of DHT/min/mg protein.

Inhibition of 5α-reductase by Tea Polyphenols

TF-3, TF-2A, TF-2B, TF-1 and EGCG can inhibit 5α-reductase activity (Table I). TF-3 was found to be the most effective inhibitor of 5α-reductase. The maximum inhibition found in TF-3 was 89% at 20 μM. The IC_{50} values of these compounds are listed in Table I. The inhibitory effect was in the order: TF-3 > TF-2B > TF-2A > EGCG > TF-1. Among these compounds, TF-1 the only one without a gallate group, exhibited no appreciable inhibitory activity up to 40 μM.

Theaflavins Inhibit Testosterone-induced Cell Growth in LNCaP Cells

The LNCaP human prostate cancer cell line is a well-established and androgen-dependent cell line. LNCaP cells retain most of the characteristics of human prostate carcinoma, like the dependence on androgens, the presence of ARs, the production of acid phosphatase and PSA. Most importantly, LNCaP cells express only 5α-reductase-1 but not 5α-reductase-2 (*25*). For this reason, the LNCaP cell line becomes an attractive model for the *in vitro* studies on the biology of human prostate cancer. In this study, we first examined the effects of theaflavins on androgen- stimulated growth of LNCaP cells.

Table I. The Inhibitory Effects of Tea Polyphenols on 5α-reductase in Rat Liver Microsomes

Tea polyphenols	IC_{50} (μM)[a]
Theaflavin-3,3′-digallate (TF-3)	4.9
Theaflavin-3-gallate (TF-2A)	20.2
Theaflavin-3′-gallate (TF-2B)	7.9
Theaflavin (TF-1)	> 100
(-)-Epigallocatechin-3-gallate (EGCG)	92.6
Gallic acid	> 100
n-Propyl gallate	> 100

[a] IC_{50}, concentration (μM) of compounds producing 50% inhibition of 5α-reductase activity. Methods of enzyme assay were described in the text.

LNCaP cells were incubated with varying concentrations of theaflavins with T for 4 days. The MTT assay was performed to measure cell viability. In the absence of theaflavins, T alone apparently stimulates LNCaP cell number 50% on average above untreated control. TF-3, TF-2B or EGCG (10-40 μM) treatment resulted in a dose-dependent inhibition of LNCaP cell growth (Table II). TF-1 and TF-2A were less effective. TF-3 seems to exhibit higher inhibitory activity on LNCaP cell growth than TF-2A, TF-2B, TF-1 and EGCG. In contrast, TF-1 and TF-2A showed little inhibitory activity.

Inhibitory Effects of Tea Polyphenols on the Expression of AR Protein

After T is transformed into DHT by the enzyme 5α-reductase, DHT will bind to androgen receptor (AR) and result in a series of androgen actions. Since AR is the essential mediator for androgen action, we sought to determine the effects of theaflavins and EGCG on the AR protein expression. Therefore, a western blot analysis was performed to detect whether AR altered protein levels were induced by treatment with these compounds. Among the 5 tea polyphenols assessed, TF-3 significantly reduced the amount of AR protein at a dose of 40 μM. TF-2B, TF-2A, TF-1 and EGCG had lower effects at the same concentration (Table III).

Table II. Effects of Tea Polyphenols on the Growth of Prostate Carcinoma LNCaP Cells in the Presence of Testosterone

Tea polyphenols	*Concentration (μM)*	*Growth of LNCaP cells[a] as*	
		% of Control	*% of inhibition*
None (control)	·0	100	0
TF-3	10	45	55
	20	19	81
	40	18	82
TF-2A	10	85	15
	20	80	20
	40	65	35
TF-2B	10	60	40
	20	42	58
	40	38	62
TF-1	10	80	20
	20	82	18
	40	78	22
EGCG	10	60	40
	20	40	60
	40	42	58

[a] Cell growth was measured by MTT assay as described in the text.

Table III. Effects of Tea Polyphenols on the Expression of AR Protein in LNCaP Cells

Tea polyphenols	*Levels of AR protein[a]*	
	% of control	*% of inhibition*
Control (T alone)	100	0
TF-3	20	80
TF-2A	30	70
TF-2B	35	65
TF-1	40	60
EGCG	95	5

[a] The LNCaP cells were grown in RPMI medium with 10% cFCS for 48 h. After 48 h incubation, whole cell lysates were prepared from cell treated with specified polyphenols (40 μM) and testosterone (25 nM) for 24 h. The levels of androgen receptor were analyzed by western blot analysis as described in the text..

Inhibitory Effect of TF-3 on Expression of PSA

Evidence shows that alteration of 5α-reductase or AR levels modulates androgen responsive genes. Prostate specific antigen (PSA) is one of the androgen-responsive genes because the promoter of PSA gene contains functional androgen responsive element (ARE). PSA is specially produced by both prostate epithelial cells and prostate cancer and is the most commonly used serum marker for diagnosing cancers. In patients with prostate carcinomas, an increase of serum PSA level is observed (*26*). T strongly up-regulates PSA production in LNCaP cells. In the presence of T, treatment of cells varying concentrations of TF-3, significantly decreases the secretion levels of PSA in a dose-dependent manner (*24*).

Discussion

Recent epidemiological evidence has linked elevated serum levels of insulin-like growth factor-1 (IGF-1) with the development of prostate cancers (*27*). IGF-1 is capable of both enhancing proliferation and inhibiting apoptosis in normal and malignant prostate epithelial cells in culture. IGF-1-related signal transduction might, therefore, be an important factor in the development of prostate cancer. In many cell types, the binding of IGF-1 to the IGF-1 receptor causes rapid phosphatidylinositol-3-kinase-dependent activation of Akt through phosphorylation of specific threonine (Thr-308) and serine (Ser-473) residues. The down-stream effects of Akt activation, including increased cellular proliferation and protection from apoptotic stimuli, have been attributed in part to enhanced phosphorylation of Bad, glycerol synthase kinase-3 (GSK-3), caspase-9 and forkhead proteins and increased cyclin D protein levels (*28*).

Pretreatment of human normal prostate epithelial cells and DU145 prostate carcinoma cells with doses as low as 20 μg/mL of a mixture of black tea polyphenols (BTP) substantially reduced IGF-1-mediated Akt phosphorylation. This effect of BTP appears to be due partially to the reduced autophosphorylation of IGF-1 receptor in BTP- treated cells. BTP pretreatment also decreased downstream effects of Akt activation including phosphorylation of GSK-3, increased cyclin D1 protein levels and increased DNA synthesis (*29*). These results indicate that black tea polyphenols inhibit the IGF-1 signal transduction pathway, which has been linked to increased prostate cancer incidence in human populations and therefore provide further support for the potential of BTP to prevent prostate cancer.

Very high intake of soy products is associated with reduced prevalence of aggressive prostate cancer among Asian men (*30*). Black tea polyphenols have been proposed as potential chemopreventive agents primarily because of their high intake by this population with reduced cancer incidence and their reported ability to inhibit proliferation and increase apoptosis in prostate cancer cells in culture (*29*). Considerable evidence from animal studies suggests that combinations of agents can be more effective for the prevention of cancer than any single constituent (*31*). A recent study was carried out to identify possible synergistic effects between soy and tea components on prostate cancer progression (*32*). A soy phytochemical concentrate (SPC), black tea and green tea were compared with respect to tumorigenicity rate, primary tumor growth, tumor proliferation index, serum androgen level and metastasis to lymph node. SPC, black tea and green tea significantly reduced tumorigenicity. SPC and black tea also significantly reduced final tumor weights. Green tea did not reduce final tumor weight, although it tended to elevate the serum DHT concentration. Furthermore, the combination of SPC and black tea synergistically inhibited prostate tumorigenicity (from 43.8% to 18.8%), final tumor weight, and metastasis to lymph node in vivo (from 31.3% to 6.3%) (*32*). Meanwhile, the combination of SPC and green tea also synergistically, but in less extent, inhibited prostate tumorigenicity (from 43.8% to 35.7%), final tumor weight, and metastasis to lymph node (from 31.3% to 14.3%) (*32*). It is appeared that the synergistic effect of black tea and SPC is more prominent than that of green tea and SPC. The mechanisms of this different is worthy for further investigation.

Environmental factors such as diet and dietary supplements have an important role in modulating cancer incidence and mortality, and differences in diet may explain geographical differences in prostate cancer mortality. Since androgens regulate the growth and function of the normal prostate and prostate cancer, dietary components capable of altering this growth signaling pathway in the prostate may affect prostate cancer development and progression. Flavonoids and flavanoids are naturally-occurring polyphenolic compounds widely distributed in fruits, vegetables and beverages. In recent years, many of these flavonoids and flavanoids have been shown to exhibit cancer chemopreventive effects. Green tea polyphenols have been found to inhibit prostate carcinogenesis *in vitro* and *in vivo* (*33,34*). Relatively, black tea polyphenols can inhibit IGF-1 induced signaling in DU145 prostate cancer cells (*29*). In this study, we discovered that TF-2A, TF-2B and TF-3 which contain several galloyl groups inhibit prostate cancer by attenuating the function of androgen and AR. Both theaflavins and EGCG can inhibit 4-α-reductase activity. Interestingly, we found that the IC_{50} value of TF-3 was 18.9 fold less than EGCG (Table I). These data suggest that black tea polyphenols are stronger 5α-reductase inhibitors than green tea polyphenols. Compared with the structures of the structures of these compounds, we found that gallate group containing polyphenols are better inhibitors of 5α-reductase.

Acknowledgements

This study was supported by the National Science Council under the research projects, NSC93-2311-B-002-001; NSC93-2320-B-002-111 and NSC93-2320-B-002-127.

References

1. Carter, H.B.; Coffey, D.S. *Prostate* **1990**, *16*, 39-48.
2. Hovenian, M.S.; Deming, C. L. *Surg. Gynecol. Obster.* **1948**, *86*, 29-35.
3. Ross, R.; Berstein, L.; Judd, H.; Hanisch, R.P.; Ke, M,; Henderson, B. *J. Natl. Cancer Inst.* **1986**, *76*, 45-48.
4. Brawley, O.W.; Knopf, R.; Thompson, I. *Semin. Urol. Oncol.* **1998**, *16*, 193-201.
5. Gupta, S.; Hastak, K.; Ahmad, N.; Lewin, J.S.; Mukhtar, H. *Proc. Natl. Acad. Sci. USA*, **2001**, *98*, 10350-10355.
6. Gelmann, E.P. *J. Clin. Oncol.* **2002**, *20*, 3001-3015.
7. Balk, S. P. *Urology* **2002**, *60*, 132-139.
8. Fang, W.Z.; Wang, Y.; Ai, N.; Hou, Z.; Sun, Y.; Lu, H.; Welsh, W.; Yang, C. S. *Cancer Res.* **2003**, *63*, 7563-7570.
9. Adhami, V.M.; Ahmad, N.; Mukhtar, H. *J. Nutr.* **2003**, *133*, 2417s-2424s.
10. Caporali, A.; Davalli, P., Astancolle, S.; D'Arca, D.; Brausi, M.; Bettuzzi, S.; Corti, A. *Carcinogenesis* **2004**, *25(11)*, 2217-2224.
11. Pezzato, E.; Sartor, L.; Dell'Auca, I.; Dittadi, R.; Gion, M.; Belluco, C.; Lise, M.; Garbisa, S. *Int. J. Cancer* **2004**, *112(5)*, 789-792.
12. Chung, L.Y.; Cheung, T.C.; Kong, S.K.; Fung, K.P.; Choy, Y.M.; Chan, Z.Y.; Kwok, T.T. *Life Sci.* **2001**, *68*, 1207-1214.
13. Liang, Y.C.; Lin-Shiau, S.Y.; Chen, C.F.; Lin, J.K. *J. Cell. Biochem.* **1997**, *67*, 55-65.
14. Liang, Y.C.; Lin-Shiau, S.Y.; Chen, C.F., Lin, J.K. *J. Cell. Biochem.* **1999**, *75*, 1-12.
15. Lin, J.K. *Arch. Pharm. Res.* **2002**, *25(5)*, 561-571.
16. Haller, J.; Bhaduri, A.N. *Biochem. Biophys, Res. Commun.* **1988**, *244*, 903-907.
17. Lu, Y.P.; Lou, Y.R.; Xie, J.G.; Yen, P.; Huang, M.T.; Conney, A.H. *Carcinogenesis* **1997**, *18*, 2163-2169.
18. Liang, Y.C.; Chen, Y.C.; Lin, Y.L.; Lin-Shiau, S.Y.; Ho, C.-T.; Lin, J.K. *Carcinogenesis* **1999**, *20*, 733-736.
19. Pan, M.H.; Lin-Shiau, S.Y.; Ho, C.-T.; Lin, J.H.; Lin, J.K. *Biochem. Pharmacol.* **2000**, *59*, 357-367.

20. Chen, Y.C.; Liang, Y.C.; Lin-Shiau, S.Y.; Ho, C.-T.; Lin, J.K. *J. Agric. Food Chem.* **1999**, *47*, 1416-1421.
21. Chen, C.W.; Ho, C.-T. *J. Food Lipids* **1995**, *2*, 35-46.
22. Liang, T.; Heiss, C.E.; Ostrove, S.; Rasmusson, G.H.; Cheung, A. *Endocrinology* **1983**, *112*, 1460-1468.
23. Liang, T.; Heiss, C.E. *J. Biol. Chem.* **1981**, *256*, 7998-8005.
24. Lee, H.H.; Ho, C.-T.; Lin, J.K. *Carcinogenesis* **2004**, *25(7)*, 1109-1118.
25. Negli-Cesi, P.; Poletti, A.; Colciago, A.; Magri, P.; Martini, P.; Motta, M. *Prostate* **1998**, *34*, 283-291.
26. Stamey, T.A.; Yang, N.; Hay, A.R.; McNeal, J.E.; Freiha, F.S.; Redwine, E. *N. Engl. J. Med.* **1987**, *317*, 909-916.
27. DiGiovanni, J.; Kiguchi, K.; Frijhoff, A.; Wilker, E.; Bol, D. K.; Beltran, L.; Moats, S.; Ramirez, A.; Jorcano, J.; Conti, C. *Proc. Natl. Acad. Sci. USA*, **2000**, *97*, 3455-3460.
28. Kandel, E.S.; Hay, N. *Exp. Cell Res.* **1999**, *253*, 210-229.
29. Klein, R.D.; Fischer, S.M. *Carcinogenesis* **2002**, *23(1)*, 217-221.
30. Parkin, D.M.; Pisani, P.; Ferley, J. *CA Cancer J. Clin.* **1999**, *49*, 33-64.
31. Chemoprevention Working Group, *Cancer Res.* **1999**, *59*, 4743-4758.
32. Zhou, J.R.; Yu, L.; Zhong, Y.; Blackburn, G.L. *J. Nutr.* **2003**, *133*, 516-521.
33. Hiipakka, R.A.; Zhong, H.Z.; Dai, W.; Dai, Q.; Liao, S. *Biochem. Pharmacol.* **2002**, *63*, 1165-1176.
34. Gupta, S.; Hussain, T.; Mukhtar, H. *Arch. Biochem. Biophys.* **2003**, *410*, 177-183.

Fruits: Their Bioactives and Health Effects

Chapter 11

Morinda Citrifolia (Noni): Its Effect on Insulin Secretion by G-Protein-Coupled Receptor Systems

Chen X. Su, Jarakae C. Jensen, and Bing N. Zhou

R&D, Tahitian Noni International, Inc., 737 E 1180 S, American Fork, UT 84003

The American Diabetes Association estimates that 14 millions people in the USA are known to have diabetes, with 9% being type 1 diabetes and 91% being type 2 diabetes. Type 2 diabetes occurs when insulin secretion is no longer sufficient from the degenerated or diseased beta cells in the islet of langerhans. At the molecular level, remedies which may influence insulin secretion include 1) functioning through the ion channels of beta cell, 2) functioning through the second messenger pathways, in which G-Protein-Coupled Receptor Systems are the most important. Increased content of cAMP inside the beta cell, adenosine 3,5-cyclic monophosphate and a second messenger generated in the G-Protein-Coupled Receptor Systems, has been found by much research to be positively related to the increased insulin secretion. *Morinda citrifolia* (L) (Noni), a natural plant growing in tropical areas such as Tahiti and Hawaii has historical uses and anecdotal reputations in treating type 2 diabetes. In this study, we found that the dried form of Noni fruit was an inhibitor of PDE3, phosphodiesterase 3 which hydrolyzes cAMP, and an agonsit of P2Y receptor which was a G-Protein-Coupled receptor on the beta cell. The agonists of P2Y might increase the secretion of insulin by stimulating the cAMP production, as indicated by other previous research results. Therefore, our findings suggested that one of the mechanisms of Noni fruit anti-diabetic effects might be through the G-Protein-Coupled Receptor Systems.

Morinda citrifolia (L) (Noni) is an edible plant which grows in tropical areas such as Tahiti and Hawaii. Its historical uses as herb remedies can be traced back to the 1700's (*1*). Native has used the plant to treat a variety of illnesses including pain, tumors, diabetes, high blood pressure, gastric ulcers, etc. In recent years, much scientific research has shown evidences to support the medical use of Noni plant. For examples, Hirazumi found that the polysaccharides derived from Noni fruit inhibited the lung tumor in C57 B1/6 mice and suggested an immune enhancing mechanism (*2*). Su et al. later found that the Noni polysaccharides showed an anti-metastatic effect on C57/B1/6 mice bearing the lung tumor (*3*). The analgesic effect of Noni root and fruit has been investigated by Younos et al. and Wang et al. (*4,5*) Many other scientific studies about Noni included its anti-viral activity (*6*), anti-tubercular effects (*7*), hypotensive activity (*8*), immunological activity (*9*), etc. However, research about Noni's effect on diabetes is scarce.

Noni fruit has historically and anecdotally been known to treat type 2 diabetes. In general, it is believed that type 2 diabetes occurs when the balance of insulin secretion and insulin resistance is disrupted (*10*). Increase of insulin synthesis and secretion and decrease of insulin resistance to restore the balance is therefore being the medical efforts to treat type 2 diabetes.

In this study, we investigated the impact of Noni fruit on insulin secretion. Ion channels and second messages are the two components of pathways of insulin secretion (*11*). Ion channels include the potassium and calcium channels, while the second messages are G-Protein-Coupled Receptor Systems. We did a series of in vitro assays to screen the bioactivities of Noni fruit involving the two components, the blockage of potassium channel K_{ATP}, enhancement of Glucose-Like Peptide-1, inhibition of cyclic nucleotide phosphodiesterase 3, PDE3, inhibition of Peptidase, dipeptidyl Peptidase IV (DDP IV), and activation of P2Y receptor in an effort to reveal the Noni fruit's anti-diabetic mechanism.

Sample Preparation

Yellow-greenish noni fruits from Tahiti were picked, covered, and placed on a screened table under the sun for one week or until completed harvested. The harvested fruits were processed into puree by a fruit processor, where the coarse residues and seeds were screened off. The derived puree was centrifuged to obtain the clear juice. Concentrated syrup was first obtained from the clear juice by a rotary evaporator and it was further dried into a powder using freeze drying technique. The dried form of Noni fruit (DFNF) was 7% of the clear juice. DFNF was used throughout the experiments.

Potassium Channel K_{ATP}

The gating of K_{ATP} on the β-cell plays an important role in insulin secretion. The K_{ATP} consists of two subunits: a channel pore-forming subunit (Kir6.2) and a sulfonylurea binding site, SUR1 (*12*). Binding to SUR1 closes the K_{ATP} channels, reducing potassium efflux, depolarizing the cell and opening voltage-dependent calcium influx channels. The raised level of intracellular calcium led to the increase in insulin secretion.

Experiment

This assay measures binding of [^{3}H]glyburide to voltae insensitive ATP-sensitive potassium channel sites [K_{ATP}]. HIT-T15 Syrian hamster pancreatic β cells are used to prepare membranes in modified MOPS pH 7.4 buffer using standard techniques. A 100 μg aliquot of membrane is incubated with 5 nM [^{3}H]glyburide for 120 minutes at 25 °C. Noni-specific binding is estimated in the presence of 1 μM glyburide. Membranes are filtered and washed 3 times and the filters are counted to determine [^{3}H]glyburide specifically bound. Glyburide was used as the reference compound. DFNF was assayed at 100 μg/mL.

Glucose-Like Peptide-1

GLP-1 is an incretin hormone. One of the important biological activities of GLP-1 is to stimulate the glucose-dependent insulin secretion and insulin biosynthesis. GLP-1 has been shown to activate phosphoinositide 3-kinase in β-cell[13], which enhanced the voltage-dependent calcium channel current, and to increase β-cell mass by stimulating islet cell neogenesis (*14*) and by inhibiting apoptosis of islets (*15*). GLP-1 receptor agonism may stimulate insulin release.

Experiment

This assay measures binding of [^{125}I]GLP (7-36) amide to human glucagons-like peptide (GLP-1) receptors. CHO-K_1 cells stably transfected with a plasmid encoding the human GLP-1 receptor are used to prepare membranes in modified Tris-HCl pH 7.4 buffer using standard techniques. A 14 μg aliquot of membrane is incubated with 0.03 nM [^{125}I]GLP (7-36) amide for 90 minutes at 37 °C. Non-specific binding is estimated in the presence of 0.1 μM GLP (7-36) amide. Membranes are filtered and washed 3 times and the filters are counted to determine [^{125}I]GLP (7-36) amide specifically bound. GLP-1 (7-36) amide was used as the reference compound. DFNF was assayed at 100 µg/mL.

Result

No binding was observed.

DDP IV

DDP IV is a serine exopeptidase that cleaves X-proline dipeptides from the N terminus of polypeptide and has been suggested to degrade physiological levels of GLP-1. The inhibitors of DDP IV are therefore useful in the management of type 2 diabetes.

Experiment

Human recombinant dipeptidyl peptidase IV expressed in insect Sf9 cells is used. The dried form of Noni fruit and vehicle is preincubated with 0.02 µg/mL enzyme in Tris-HCl buffer pH 8.0 for 15 minutes at 37 °C. The reaction is initiated by addition of 20 µM Ala-Pro-AFC for another 30 minute incubation period.n Determination of the amount of AFC formed is read spectrofluorimetrically with excitation at 400 nm and emission at 510 nm. Dipepidylpeptidase IV inhibitor I was used as the reference compound. DFNF was assayed at 100 µg/mL.

Result

No significant inhibition was observed.

PDE3

PDEs, cyclic nucleotide phosphodiesterases are a family of enzymes (PDE1-PDE11) to degrade cAMP. There are a host of studies showing that the increase of cAMP is positively correlated with the increase of insulin release (*16*). The action of cAMP, a second message generated in G-protein-coupled receptor systems, in augmentation of insulin may involve in calcium channels and insulin gene promoter activities. Studies have shown that beta cells contain PDE1, PDE3, and PDE4 with PDE3 being the most important enzyme related to insulin release. Inhibitors of PDE3 increase the level of cAMP, therefore, can potentially increment the insulin secretion.

Experiment

PDE3 partially purified from human platelets is used. The dried form of Noni fruit and vehicle is incubated with 1 μM [^{3}H]cAMP in Tris buffer pH 7.5 for 20 minutes at 30 °C. The reaction is terminated by boiling for 2 minutes. The resulting AMP is converted to adenosine by addition of 10 mg/ml snake venom nucleotidase and further incubation at 30 °C for 10 minutes. Unhydrolyzed cAMP is bound to AG1-X2 resin, and remaining [^{3}H]Adenosine in the aqueous phase is quantitated by scintillation counting. IBMX was used as the reference compound. DNFN was assayed at 0.1, 1, 10, 100, 1000 μg/mL.

Result

DNFN showed inhibiting effect in a dose response manner. IC50 was extracted at 45 μg/mL.

Purinergic P2Y

P2Y belongs to the superfamily of G protein-coupled receptors' systems and presents on the pancreatic β cells. P2Y receptor agonists were found effective in stimulating insulin release in human pancreatic islets (*17*). Different mechanisms of the P2Y agonists' effectiveness have been investigated. It has been demonstrated that activation of P2Y receptor promoted Ca^{2+} releases through voltage-independent channels (*18*) and mediated accumulation of cAMP in bovine vascular smooth muscle cells (*19*).

Experiment

Binding Assay

Whole brains of male Wistar rats weighing 175 ± 25 g are used to prepare purinergic P2Y receptors in Tris-HCl buffer pH 7.4. A 200 μg aliquot is incubated with 0.1 nM [^{35}S]ATP-αS for 60 minutes at 25 °C. Non-specific binding is estimated in the presence of 10 μM ADP-βS. Membranes are filtered and washed, the filters are then counted to determine [^{35}S]ATP- αS specifically bound. ATP was used as the reference compound. DNFN was assayed at 30, 100, 300, 1000 μg/mL.

Result

DNFN showed binding effect in a dose response manner. IC50 was extracted at 135 μg/mL.

Agonist Activity Assay

A segment of taenia coli obtained from Duncan Hartley derived male or female guinea pigs weighing 325 ± 25 g and sacrificed by CO_2 overexposure is used. The tissue is placed under 1.75 g tension in a 10 ml bath containing Krebs solution pH 7.4 at 28 °C. Test substance (30 μM) reduction of submaximal carbachol (30 nM)-induced isotonically recorded contractions by 50 percent or more (≥50%) within 5 minutes, relative to control 0.30 μM 2-methylthio-ATP response, indicates possible P2Y agonist activity. 2-Methythio-ATP is used as the reference compound. DFNF was tested at 1, 3, 5 mg/mL.

Result

DFNF showed P2Y agonist activity in a dose response manner.

Discussion

At the tested level, 100 μg/mL, DFNF didn't show any activities in the pathways of potassium K_{ATP} channel, GLP-1, and DDP-IV. DFNF demonstrated activities as a PDE inhibitor and P2Y agonist in a dose response manner.

This is the first scientific evidences of Noni fruit having anti-diabetic effects.

Working as a PDE inhibitor, Noni fruit exerted its anti-diabetic effect by raising the level of cAMP to increase insulin secretion. Functioning as a P2Y agonist, it implied that Noni fruit may not only act to increase the level of cAMP, but also to promote Ca^{2+} releases. These data suggested that Noni fruit may improve insulin secretion and one of the major pathways was to increment the level of cAMP.

Further research is needed. For example, the direct effect on calcium channel is not clear. Meanwhile, this study focused on insulin secretion and no research has been done to see if Noni fruit works to improve the insulin resistant, another side the balance to maintain glucose homeostasis.

References

1. Allen, W.H.; London, C. In *The useful plants of India.* 1873.
2. Hirazumi, A., Furusawa, E., Chou, S.C., Hokama, Y. *Proc. West Pharmacol. Soc.* **1994**, *37*, 145-146.
3. Su, C.X.; Wang, M.Y.; Jensen, J. C.; Kim, H.A.; West, B.; Jensen, S.; Palu, A.; Zhou, B.N.; Friz, J.W. *Proc. Amer. Asso. Cancer Res.* **2004**, *45*, 29.
4. Younos, C.; Rolland, A.; Fleurentin, J.; Lanhers, M.C.; Misslin, R. *Planta Med.* **1990**, *56*, 430-434.
5. Wang, M.Y.; West, B.J.; Jensen, C.J.; Nowicki, D.; Su, C.; Palu, A.K.; Anderson, G. *Acta Pharmacol. Sin.* **2002**, *23,* 1127-1141.
6. Umezawa, K.; Japan, Kokai Tokyo Koho JP 0687, 1992, *736,* 94-87, 736.
7. Saludes, J.P.; Garson, M.J.; Franzblau, S.G.; Aguinaldo, A.M. *Phytother. Res.* **2002**, *16*, 683-685.
8. Moorth, N.K.; Reddy, G.S. *Antiseptic* **1970**, *67*, 167-171.
9. Hokama, Y. *FASEB J.* **1993**, *7*, A866.
10. Maire, E.D. ; Josephine, M.E. *Pharmacol. Rev.* **2003**, *55*, 105-131.
11. Wang, S. ; Ferguson, K.C. ; Buitis, T.P. ; Dhurandhar, N.V. *Expert Opin. Investig, Drugs* **1999**, *8*, 1117-1125.
12. Proks, P. ; Ashcroft, F.M. *Proc. Natl. Acad. Sci.* **1997**, *94*, 11716-11720.
13. Viard, P.; Butcher, A.J.; Halet, G.; Davies, A.; Nurnberg, B.; Heblich, F.; Dolphin, A.C. *Nat. Neurosci.* **2004**, *7*, 939-946.
14. Paris, M.; Tourrel-Cuzin, C.; Plachot, C.; Ktoza, A. *Exp. Diabesity Res.* **2004**, *5*, 111-121.
15. Farilla, L.; Bulotta, A.; Hirshberg, B.; Li Caliz, S.; Khoury, N.; Noushmehr, H.; Bertolotto, C.; Di Mario, U.; Harlan, D.M.; Perfetti, R. *Endocrinology* **2003**, *144*, 5149-5158.
16. Sharp, G.W.G. *Diabetologia* **1979**, *16*, 287-296.
17. Fernandez-Alvarez, J.; Hillaire-Buys, D.; Loubatieres-Mariani, M.M.; Gomis, R.; Petit, P. *Pancreas* **2001**, *22*, 69-71.
18. Tang, J.; Pugh, W.; Polonsky, K.S.; Zhang, H. *Am. J. Physiol.* **1996**, *270 (3 Pt 1)*, E504-512.
19. Tada, S.; Okajima, F.; Mitsui, Y.; Kondo, Y.; Ui, M. *Eur. J. Pharmacol.* **1992**, *227*, 25-31.

Chapter 12

Antiinflammatory Constituents in Noni (*Morinda citrifolia*) Fruits

Haiqing Yu[1], Shiming Li[1], Mou-Tuan Huang[2], and Chi-Tang Ho[1]

[1]Department of Food Science, Rutgers, The State University of New Jersey, 65 Dudley Road, New Brunswick, NJ 08901–8520
[2]Susan Lehman Cullman Laboratory of Cancer Research, Ernest Mario School of Pharmacy, Rutgers, The State University of New Jersey, 164 Frelinghuysen Road, Piscataway, NJ 08854–8020

Morinda citrifolia (Noni) is a tropical plant whose leaves, barks, roots and fruits have been used as traditional remedy for various diseases. Although inflammation was considered as one of the indications for noni fruits, and fruit extract was demonstrated to have immunomodulatory activity *in vitro*, no principle has been identified. In our study, we attempted to identify constituents that may be responsible for anti-inflammatory activity, and to elucidate the possible roles on anti-inflammatory mediators for interested compounds. Solvent extraction and column chromatography were major techniques used for isolation of compounds, while structures were elucidated by integration of data from UV, IR, MS and NMR analysis. Anti-inflammatory activity was assessed by mouse ear edema model. Subsequent Elisa analysis was carried out for their effects on anti-inflammatory mediators such as IL-1β, IL-6, PGE_2 and myeloperoxidase. Scopoletin, quercetin, ursolic acid were identified as major anti-inflammatory constituents. Since ursolic acid was known to have anti-inflammatory activity, we characterized the mode of action of scopoletin and quercetin in mouse ear inflammation model. The most potent inhibition occurred at 0.5 μmol level for both quercetin (62.3%) and scopoletin (50.4%). Scopoletin mainly inhibited production of myeloperoxidase and PGE_2 while quercetin had significant suppressing effect on IL-6 production. Both compounds showed moderate inhibition on IL-1β production.

Morinda citrifolia L. (Rubiaceae), common name noni, is a tropical evergreen shrub mainly distributed in French Polynesia, Hawaii, India and Southeast Asia. The leaves, fruits, barks and roots have been used traditionally for treatment of cancer, hypertension, diabetes, etc. (*1*). Exploration of chemical constituents of noni flower started as early as in 1960s, however, information on other plant tissues continues to accumulate.

Noni leave extract significantly upregulated LDL receptor *in vitro*, although it doesn't have a significant antioxidant activity (*2*). This result suggests that the cardiovascular protective effect was not due to prevention of LDL oxidation, but due to enhanced activity of LDL receptor. Noni leaves were also reported as containing antitubercular constituents (E-phytol, cycloartenol, stigmasterol, beta-sitosterol, campesta-5,7,22-trien-3β-ol and the ketosteroids stigmasta-4-en-3-one and stigmasta-4,22-dien-3-one) in the nonpolar fraction (*3*). Sang *et al.* isolated citrifolinin B, citrifolinoside A, citrifolinoside 1 and five flavonol (quercetin or kaempferol) glycosides. Among those compounds, citrifolinin B is a new iridoid glycoside with weak DPPH scavenging activity, while citrifolinoside 1, significantly inhibited UVB induced Activator Protein-1 (AP-1) activity *in vitro* (*4-6*).

Anthraquinones such as damnacanthal, 7-hydroxy-8-methoxy-2-ethylanthraquinone, morenone 1, morenone 2 (*7,8*), and anthraquinone glycosides (Physcion, morindone, Physcion-8-*O*-alpha-L-arabinopyranosyl (1→3) β-D-galactopyranosyl (1→6)-β-D-galactopyranoside) were isolated from the Noni root (*9*). Among those, damnacanthal was proved to have induction effect on normal phenotypes in ras-transformed cells (*10*). In the mean time, damnacanthal has potent inhibitory activity towards tyrosine kinases, phosphorylation was enhanced for both extracellular signal regulated kinases (ERK) and stress-activated protein kinases (SAPK) pretreated with damnacanthal followed by UV treatment *in vitro* (*11*).

Short Chain fatty acids such as hexanoic acid and octanoic acid were found to be abundant in ripe fruit extract. Specifically, octanoic acid was tested as toxic to Drosophila species (*12*). These short chain fatty acids are the major source of the soapy foul flavor of noni fruit. Interestingly, hexanoic acid and octanoic acid are mainly present as esters of di- or tri-saccharides in the fruit extract (*13,14*). Among them, 6-*O*-(β-D-glucopyranosyl)-1-*O*-octanoyl-β-D-glucopyranose was found to inhibit AP-1 transduction *in vitro*, thus affecting cell transformation in the cancer proliferation stage (*15*). Along with the above compounds, 3 glycoside (3-methylbut-3-enyl 6-*O*-β-D-glucopyranosyl, asperulosidic acid, rutin) were also isolated (*13,14*).

It has been known that elevated level of mediators including pro-inflammatory cytokines (IL-1β, TNFα, IL6) and arachidonic acid metabolites (prostaglandins and leukotriens) are correlated with activated inflammation process (*16-22*). Although traditionally, inflammation was considered at one of the indications for noni fruits, evidence has been discrete. Li *et al* reported only moderate anti-inflammatory activity of Noni fruit extract compared to 23 other herbs, with IC_{50} of 163 μg/ml on inhibition of COX-1 (*23*). While Hirazumi reported its effectiveness in stimulating the release of several anti-inflammatory mediators, including TNF-α, IL-1β, IL-10, IL-12p70, IFN-γ, and nitric oxide (NO). It has been suggested that the polysaccharide constituents in noni are responsible for its immunomodulatory activity (*24*). However no principle has been isolated and elucidated. More interestingly, immunomodulation was suggested to be responsible for anticancer activity of Noni (*25,26*). So it would be of interest to confirm and explore the anti-inflammatory principles in Noni. The present work attempts to identify new constituents that may be responsible for anti-inflammatory effects. Further studies will be carried out on preliminary elucidation of anti-inflammatory mechanism of interested compounds.

Materials and Method

Plant Material

Tahiti noni fruits were collected and water was partially removed by drum drying to yield a paste.

Solvents and Reagents

Solvents including methanol, water, acetonitrile, chloroform, ethyl acetate, hexane, n-butanol and acetone were of HPLC grade, and purchased from Fisher Scientific (Springfield, NJ). TPA (12-*O*-tetradecanoyl phorbol 13-acetate), deuterated solvents including $(CD_3)_2SO$, $CDCl_3$, and CD_3OD were obtained from Sigma (St. Louis, MO). Silica gel (60Å, 32-63 μm) for normal phase chromatography was purchased from Sorbent Technology Inc. (Atlanta, GA). Octadecyl (C_{18}) derivatized silica gel (60Å) for reversed phase chromatography, and Sephadex LH-20 were purchased from Sigma (St. Louis, MO). Thin Layer Chromatography (TLC) plates were purchased from Fisher Scientific (Springfield, NJ). ELISA assay kits were purchased from BioSource International, Inc (Camarillo, CA).

Instruments

NMR spectra were recorded on a Varian 300 Spectrometer (Varian Inc., Palo Alto, CA). ^{1}H-NMR and ^{13}C-NMR was recorded at 300MHz with TMS served as internal standard. APCI-MS spectra were obtained on a VG-platform II (Micromass, Beverly, MA). ESI-MS spectra were obtained on a Quattro II triple quadrupole mass analyzer (Micromass, Beverly, MA). HPLC system from LC-MS was composed of a Shimadzu (Piscataway, NJ) SCL-10A system controller, 2 LC-10AS liquid chromatograph pumps, a Supelco Discovery C18 4.6mm X 25cm column, a variable wavelength Varian 2050 UV detector and a Waters 717 autosampler. Analytical HPLC system was composed of a Varian Vista 5500 system controller, with a variable UV wavelength Varian 9065 polychrom detector (Walnut Creek, CA). Preparative HPLC system was composed of a Water Model 590 pump, a Zorbax Rx-C18 9.4mm x 25cm column, a Milton Roy spectromonitor 3100 UV detector, and Peaksimple data system.

Extraction and Isolation

The noni paste was extracted with 95% ethanol at room temperature for one week. The extract was filtered and the residue was re-extracted for anther 2 times. The filtrates were combined and evaporated under reduced pressure to yield brown syrup-like residue. The residue was then suspended in 500 mL water and transferred to a separatory funnel, where 500 mL of hexane (3 times), 500 mL of ethyl acetate (3 times), and 500 mL of *n*-butanol (3 times) were added successively for liquid-liquid partition. Each fraction was collected and dried under reduced pressure.

The ethyl acetate fraction was loaded on silica gel column, eluted with a gradient of ethyl acetate: methanol: acetic acid 50:1:0.01, 20:1:0.01, 10:1:0.01, 5:1:0.01, 1:1:0.01 consecutively. Ten fractions were collected. Each fraction were dried under reduced pressure, and further purified on repeated silica gel column, RP-18 column and Sephadex LH-20 column to yield pure compounds. Scopoletin and quercetin was purified from 50:1:0.01 fraction, while ursolic acid, β-sitosterol and campesterol were isolated from 20:1:0.01 fraction.

Anti-inflammatory Assay

Eight Female CD-1 mice aged 5-6 weeks were grouped into acetone + acetone (negative control) group. 6 mice were grouped into actone + TPA (0.8 nmol) (positive control) group. All other test groups were composed of 5 mice. Both ears of test mouse were treated topically with 20 μL acetone or test compound (contained 0.5 μmol, 1 μmol and 2 μmol respectively) in acetone, at 20 minutes before acetone or TPA (0.8 nmol) application once a day for 4 days. The mice were killed by cervical dislocation at 7.5 hours within the last dose of TPA treatment. Ear punch (6 mm in diameter) were taken and weighed. The ear sample were homogenized and kept at –80 °C freezer until ELISA analysis.

$$\text{Inhibition \%} = \frac{\text{(ear weight of test compound)} - \text{(ear weight of negative control)}}{\text{(ear weight of positive control)} - \text{(ear weight of negative control)}} \times 100\%$$

ELISA analysis

Samples and standards, as well as analysis methods for PGE_2, IL-1β and IL-6 assays were prepared as directed from ELISA reagent kits. Five standards were used to establish a standard curve for quantification of each protein.

Results and Discussion

Chemistry of Potential Anti-inflammatory Constituents

From the ethyl acetate fraction of noni extract, a total of 5 compounds were identified (Figure 1), which by nature are flavone, coumarins, steroid and terpenoids respectively.

Identification of Scopoletin

Scopoletin was obtained as a white power. APCI MS showed a pseudo-molecular ion $[M-H]^-$ at m/z 191 and $[M+H]^+$at m/z 193. Additionally, a daughter ion at m/z 176 on $APCI^-$ spectrum indicated loss of a methyl group. IR spectrum showed an intense absorption at wave number 3341.38cm^{-1}, indicating stretching vibration of O-H bond, while an strong sharp absorption at 1708.51cm^{-1}, indicating a carbonyl group. ^{1}H-NMR (DMSO-d_6, 300 MHz): 10.29 (1H, s, OH), 7.89 (1H, d, J= 9.3 Hz, H-3), 7.20 (1H, s, H-8), 6.76 (1H, s, H-5), 6.19 (1H, d, J = 9.3Hz, H-4), 3.80 (3H, s, CH_3). ^{13}C-NMR (DMSO-d_6, 300 MHz): 160.63 (s, C-2), 151.08 (s, C-7), 149.45 (s, C-6), 145.20 (s, C-8), 144.44 (s, C-9), 111.66 (s, C-5), 110.50 (s, C-3), 109.54 (s, C-4), 102.73 (d, C-10), 55.96 (s, C-11).

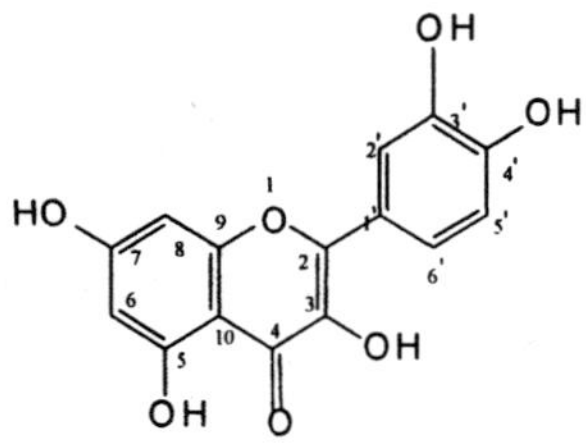

Scopoletin

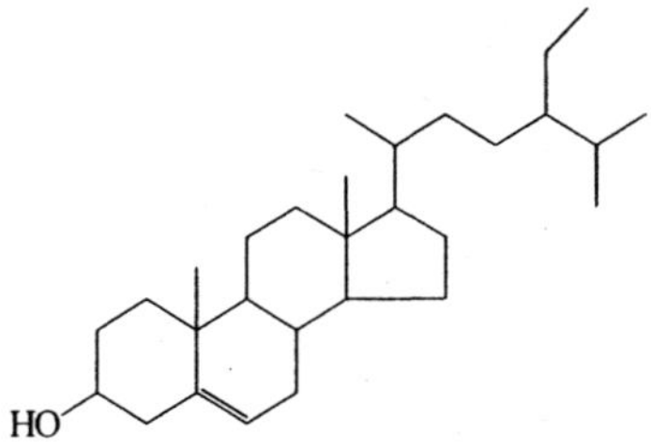

β-Sitosterol

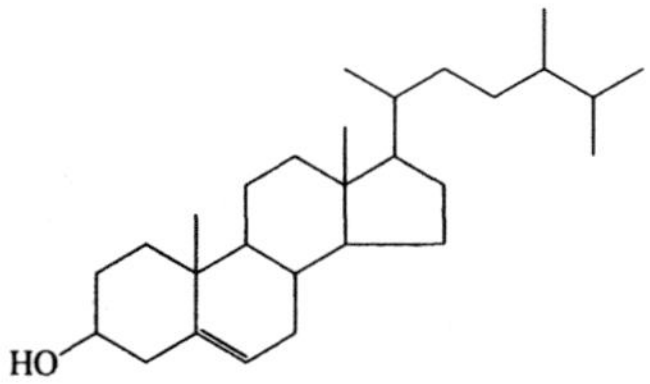

Campesterol

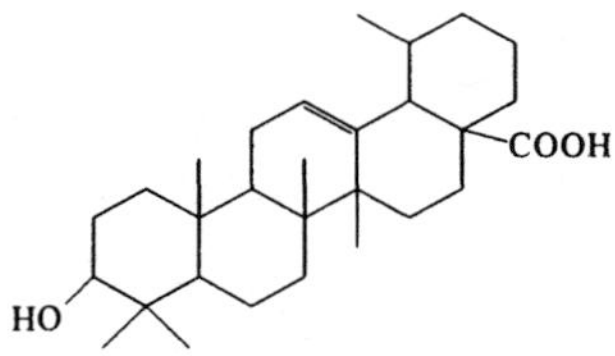

Ursolic Acid

Figure 1. Compounds Isolated from Ethyl Acetate Fraction

Identification of Quercetin

Quercetin was obtained as a yellow power. ESI MS spectrum showed pseudo-molecular ion $[M-H]^-$ at m/z 301 and $[M+H]^+$ at m/z 303. ^{1}H-NMR (DMSO-d_6, 300 MHz): 12.48 (1H, s, OH-3), 10.79 (1H, s, OH-5), 9.60 (1H, s, OH-7), 9.37 (1H, s, OH-4′), 9.30 (1H, s, OH-3′), 7.66 (1H, d, J= 2.4 Hz, H-2′), 7.52 (1H, dd, J=8.4 Hz, 2.4 Hz, H-6′), 6.86 (1H, d, J=8.4 Hz, H-5′), 6.39 (1H, d, J= 1.8 Hz, H-8), 6.17 (1H, d, J= 1.8Hz, H-6). ^{13}C-NMR (DMSO-d_6, 300MHz, Appendix 9): 175.80, 163.85, 160.67, 156.08, 147.66, 146.74, 145.02, 135.69, 121,9, 119.94, 115.57, 115.02, 102.97, 98.16, 93.31.

It has been reported that in the n-butanol fraction of noni fruits extract, rutin (quercetin glycoside) was identified. Here we confirmed the existence of quercetin aglycone in less polar fraction.

Identification of Ursolic Acid

Ursolic acid was obtained as a white powder. The EI MS showed molecular ion at m/z 456 (19%), with daughter ions at m/z 438 (15%), 410 (32%), 395

(10%), 322 (9%), 300 (27%), 287 (11%), 273 (8%), 248 (100%), 207 (32%), 203 (50%), 133 (46%). Among those, the base peak m/z 248 and the peak at 207 were products of Retro-Diels-Alder reaction. ^{1}H-NMR (DMSO-d_6, 300 MHz, Appendix 5) and ^{13}C-NMR (DMSO-d_6, 300 MHz) obtained agreed with literature data reported earlier (*27*).

Ursolic acid was reported previously for its activity of stimulation of splenocytes proliferation in mice (*28*). Huang *et al* reported that ursolic acid is able to inhibit TPA-induced mouse ear inflammation (*29*). It was also identified from Plantago major, and demonstrated to be a selective inhibitor of COX-2 enzyme (*30*). Meanwhile, Suh *et al* reported it as a suppressing agent on iNOS enzyme in mouse macrophages (*31*). Most recently, Chiang *et al* reported ursolic acid strongly enhanced the secretion of IFN-γ in human peripheral blood mononuclear cells (*32*). All these evidence suggest that ursolic acid is a potential anti-inflammatory agent function by modulating pro-inflammatory cytokines and inflammatory enzymes.

Anti-inflammatory Activity of Scopoletin and Quercetin

Among the isolated compounds, scopoletin and quercetin was selected for further bioactivity test. Scopoletin was previously reported as dose dependently inhibited iNOS enzyme (*33*), indicating its possible role as having effects on anti-inflammatory mediators, as iNOS induction can be mediated by different pro-inflammatory cytokines and other stimulants. Quercetin is another abundant constituent. It is well known as a strong antioxidant and modifies eicosanoid biosynthesis (*34*), indicating that quercetin may possess anti-inflammatory activity.

Anti-inflammatory Effects on TPA-induced Mouse Ear Edema Model

Quercetin and scopoletin were tested for their *in vivo* anti-inflammatory using TPA-induced mouse ear edema model. TPA (12-*O*-tetradecanoylphorbol-13-acetate), a phorbol ester, is a natural product exists in seed of *Croton tiglium L.* It is a potent tumor promoter and inflammation inducer that can be used to stimulate production of pro-inflammatory cytokines (*35-37*). It was reported that multiple topical applications of TPA to mouse ears will induce a prolonged inflammatory reaction, which results in increases in ear weight, inflammatory cell infiltration and epidermal hyperplasia (*35*). Therefore, in our model, the weight increase of mouse ears will be an indicator of inflammation.

Three levels of each compound and a combination of them both at 0.5 μM level were evaluated. Results were shown in Figure 2. Both quercetin and scopoletin showed anti-inflammatory effects on TPA induced mouse ear edema at the test levels. The most potent inhibition occurred at 0.5 μmol for both quercetin (62.3%) and scopoletin (50.4%).

For quercetin, we speculate that this phenomenon may be due to its antioxidant properties, who at lower concentrations, serve as antioxidant while at higher concentrations, pro-oxidant. However, for scopoletin, it may be due to a different mechanism since it was shown to have minimal antioxidant activity (*38*).

Combination of two compounds did not enhance each other in their anti-inflammatory activities. The mixture showed an inhibition rate of 50.1%, which is weaker than quercetin alone and not significantly different from scopoletin alone according to student's T test.

Inhibitory Effects on Pro-inflammatory Mediators

PGE_2

In the test against prostaglandin E_2 (PGE_2) expression, both quercetin and scopoletin showed stronger inhibitory effect on production of PGE_2 release at lower concentrations. At a concentration of 0.25 μM, scopoletin seemed to be a stronger inhibitor than quercetin, which suppressed production of PGE_2 by 57.7%.

The combination of compounds showed no synergistic effect on their activity, decreasing from 57.7% in the case of scopoletin and 55.4% in the case of quercetin to 44.6% (Figure 3).

The inhibition effects on PGE_2 indicate that both compounds involves in the arachidonic acid metabolism pathway. Also, as PGE_2 is able to induce production of other pro-inflammatory cytokines, we can foresee that a wide array of inflammation cascade would be suppressed following the reduction of PGE_2.

IL-1β

When tested against inhibition of IL-1β, quercetin has a significant inhibitory effect at 0.5 μmol level. As for scopoletin, the same trend seems to exist. At a low level of 0.25 μmol level, IL-1β release is reduced by 26.1 %. This trend is contrary to the case of PGE_2 assay. This is surprising since IL-1β is reported to be able to enhance phospholipase (PLA_2) activation and production of prostaglandins. In this case, other factors that act as better PGE2 enhancer than IL-1β may play a more important role. Further study on other mediators would help to explain this phenomenon.

In the meantime, an examination of the effect of combination of both compounds showed no synergistic effect on the inhibition of IL-1β at tested condition (Figure 4).

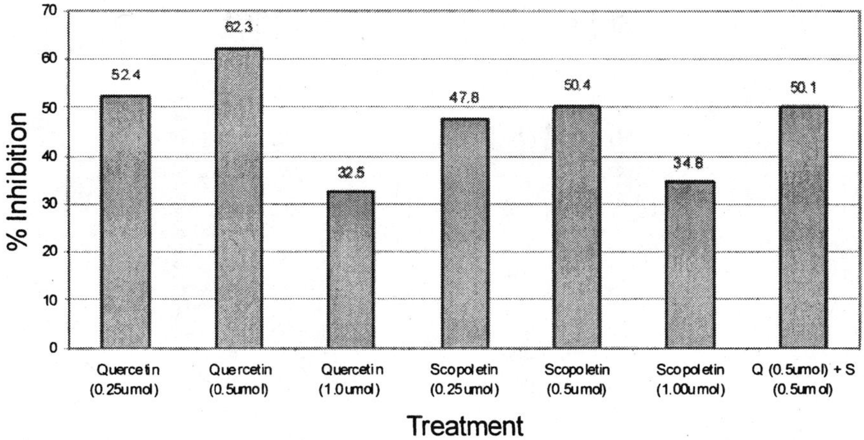

Note: Q: Quercetin; S: Scopoletin; * indicates statistical significance.

Figure 2. Effects of quercetin and scopoletin on TPA-induced inflammation.

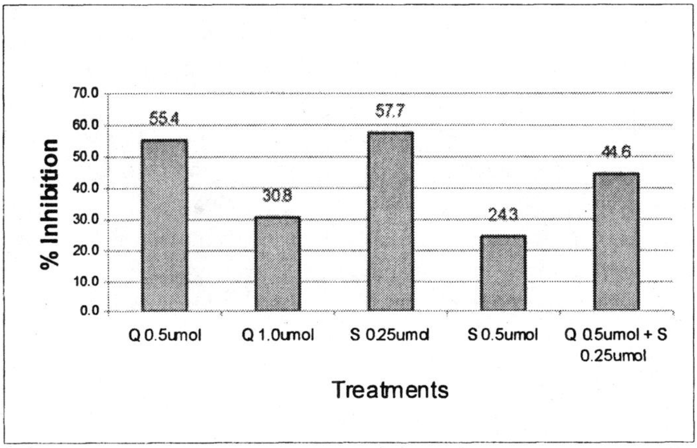

Q: Quercetin; S: Scopoletin; * indicate inhibition is statistically significant

Figure 3. Inhibition effects of quercetin and scopoletin on PGE_2

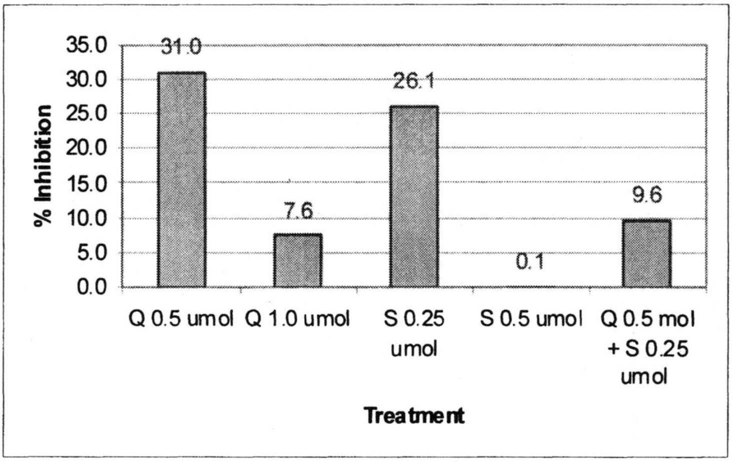

Q: Quercetin; S: Scopoletin; * indicates inhibition is statistically significant

Figure 4. Inhibition effects of quercetin and scopoletin on IL-1β.

IL-6

Figure 5 shows the effect of both compounds against IL-6. Quercetin significantly inhibited IL-6 release at both test levels, with higher inhibition at 0.5 μmol, with an inhibition rate of 51.4%. Similarly, scopoletin inhibited IL-6 protein synthesis at both test levels too, and the maximun activity occurred at 0.25 μmol with a reduction rate of 28.0%. The fact that both compounds showed higher suppression effect at lower concentration seems to agree with that of IL-1β. This is understandable since IL-1β is responsible for stimulating the production of other pro-inflammatory cytokines, including IL-6.

Additionally, quercetin was shown to be a stronger inhibitor of IL-6 when applied at the same concentration with scopoletin.

Again, the combination of scopoletin with quercetin did not bring any additional enhancing benefit at the test condition.

References

1. Wang, M.Y.; West, B.J.; Jensen, C.J.; Nowicki, D.; Su, C.; Palu, A.K.; Anderson, G. *Acta Pharmacol. Sin.* **2002**, 1127-1141.
2. Salleh, M.N.; Runnie, I.; Roach, P.D.; Mohamed, S.; Abeywardena, M.Y. *J. Agric. Food Chem.* **2002**, *50*, 3693-36973.

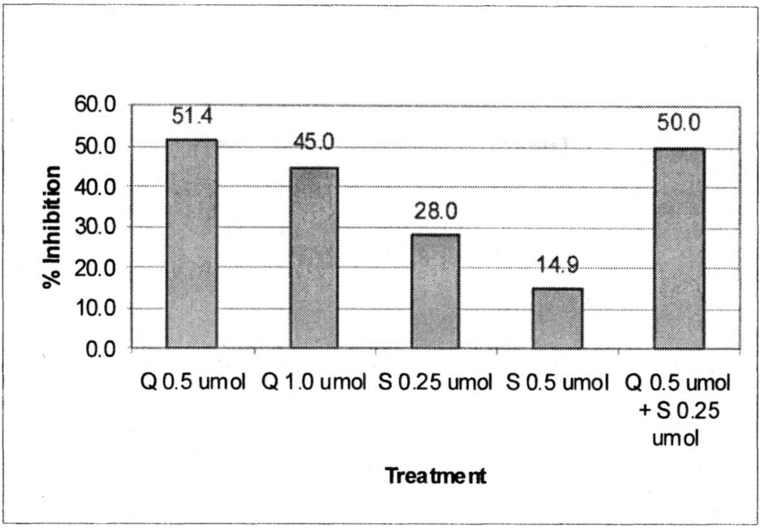

Q: Quercetin; S: Scopoletin; * indicate inhibition is statistically significant.

Figure 5. Effects of quercetin and scopoletin on IL-6

3. Saludes, J.P., Garson, M.J.; Franzblau, S.G.; Aguinaldo, A.M.. *Phytother. Res.* **2002**, *16*, 683-685.
4. Sang, S.; Cheng, X.; Zhu, N.; Stark, R.E.; Badmaev, V.; Ghai, G.; Rosen, R.T.; Ho, C.-T. *J. Agric. Food Chem.* **2001**, *49*, 4478-4481.
5. Sang, S.; Cheng, X.; Zhu, N.; Wang, M.; Jhoo, J.W.; Stark, R.E.; Badmaev, V.; Ghai, G.; Rosen, R.T.; Ho, C.-T. *J. Nat. Prod.* **2001**, *64*, 799-800.
6. Sang, S.; He, K.; Liu, G.; Zhu, N.; Cheng, X.; Wang, M.; Zheng, Q.; Dong, Z.; Ghai, G.; Rosen, R.T.; Ho, C.-T. *Org. Lett.* **2001**, *3*, 1307-1309.
7. Rusia, K.; Srivastava, S. *Surr. Sci.* **1989**, 58, 249.
8. Jain, R.K.; Ravindra, K. *Proc. Natl. Acad. Sci. India, Sect. A,* **1992**, *62*, 11-13.
9. Srivastava, M.; Singh, J.A. *International J. Pharmacognosy* **1993**, *31*, 182-184.
10. Hiramatsu, T.; Imoto, M.; Koyano, T.; Umezawa, K. *Cancer Lett.* **1993**, 161-166.
11. Hiwasa, T.; Arase, Y.; Chen, Z.; Kita, K.; Umezawa, K.; Ito, H.; Suzuki, N. *FEBS Lett.*, **1999**, *444*, 173-176.
12. Legal L.; Chappe, B.; Jallon, J.M. *J. Chem. Ecology*, 1994, *20*, 1931-1943.
13. Wang, M.; Kikuzaki, H.; Csiszar, K.; Boyd, C.D.; Maunakea, A.; Fong, S.F.; Ghai, G.; Rosen, R.T.; Nakatani, N.; Ho, C.-T. *J. Agric. Food Chem.* **1999**, *47*, 4880-4882.

14. Wang, M.; Kikuzaki, H.; Jin, Y.; Nakatani, N.; Zhu, N.; Csiszar, K.; Boyd, C.; Rosen, R.T.; Ghai, G.; Ho, C.-T. *J. Nat. Prod.* **2000**, *63*, 1182-1183.
15. Liu, G.; Bode, A.; Ma, W.Y.; Sang, S.; Ho, C.-T.; Dong, Z. *Cancer Res.* **2001**, *61*, 5749-5756.
16. Bennett, A. *Prog. Lipid Res.* **1986**, *25*, 539-542.
17. Le, J.; Vilcek, J. *Lab. Invest.* **1987**, *56*, 234-248.
18. Weitzman, S.A.; Gordon, L.I. *Blood* **1990**, *76*, 655-663.
19. Duff, G.W. *Scand. J. Rheumatol. Suppl.* **1994**, *100*, 9-19.
20. Ward, A.J.; Olive, P.L.; Burr, A.H.; Rosin, M.P. *Environ. Mol. Mutagen.* **1994**, *24*, 103-111.
21. Balsinde, J.; Winstead, M.V.; Dennis, E.A. *FEBS Lett.* **2002**, *531*, 2-6.
22. Bazan, N.G.; Colangelo, V.; Lukiw, W.J. *Prostaglandins Other Lipid Mediat.* **2002**, *68-69*, 197-210.
23. Li, R.W.; Myers, S.P.; Leach, D.N.; Lin, G.D.; Leach, G. *J. Ethnopharmacol.*, **2003**, *85*, 25-32.
24. Hirazumi, A.; Furusawa, E. *Phytother. Res.* **1999**, *13*, 380-387.
25. Hirazumi, A.; Furusawa, E.; Chou, S.C.; Hokama, Y. *Proc. West Pharmacol. Soc.* **1994**, *37*, 145-146.
26. Hirazumi, A.; Furusawa, E.; Chou, S.C.; Hokama, Y. *Proc. West Pharmacol. Soc.*, **1996**, *39*, 7-9.
27. Krisraski, R.W.; Pitner, T.H. *Pharmaceutical Res.* **1989**, *6*, 531-554.
28. Hsu, H.Y.; Yang J.J.; Lin C.C. *Cancer Letter* 1997, *111*, 7-13.
29. Huang, M.T.; Ho, C.-T.; Wang, Z.Y.; Ferraro, T.; Lou, Y.R.; Stauber, K.; Ma, W.; Georgiadis, C.; Laskin, J.D.; Conney, A.H. *Cancer Res.* **1994**, *54*, 701-708.
30. Ringbom, T.; Segura, T.; Noreen, Y.; Perera, P.; Bohlin, L. *J. Natural Prod.* **1998**, *61*, 1212-1215.
31. Suh, N.; Honda T.; Finlay, HJ.; Barchowsky, A.; Williams, C.; Benoit, N. E.; Xie, Q.W.; Nanthan, C.; Bribble, B.W.; Sporn, M.B. *Cancer Res.* **1998**, *58*, 717-723.
32. Chiang, L.C.; Ng, L.T.; Chiang, W.; Chang, M.Y.; Lin, C.C. *Planta Med.* **2003**, *69*, 600-604.
33. Kang, T.; Pae, H.; Jeong, S.; Yoo, J.; Choi, B.; Jun, C.; Chung, H.; Miyamoto, T.; Higuchi, R.; Kim, Y. Planta Med. **1999**, *65*, 400-403.
34. Formica, J.V.; Regelson, W. *Food Chem. Toxicol.* **1995**, *33*, 1061-1080.
35. Stanley, P.L.; Steiner, S.; Havens, M.; Tramposch, K.M. *Skin Pharmacol.* **1991**, *4*, 262-271.
36. Giacosa, A.; Filiberti, R. *Eur. J. Cancer Prev.* **1996**, *5*, 307-312.
37. Nomura, E.; Hosoda, A.; Morishita, H.; Murakami, A.; Koshimizu, K.; Ohigashi, H.; Taniguchi, H. *Bioorg. Med. Chem.* **2002**, *10*, 1069-1075.
38. Foti, M.; Piattelli, M.; Baratta, M.T.; Ruberto, G. *J. Agric. Food Chem.* **1996**, *44*, 497-501.

Chapter 13

Polymethoxyflavones: Chemistry, Biological Activity, and Occurrence in Orange Peel

Shiming Li[1], Chih-Yu Lo[1], Slavik Dushenkov[2], and Chi-Tang Ho[1]

[1]Department of Food Science, Rutgers, The State University of New Jersey, 65 Dudley Road, New Brunswick, NJ 08901–8520
[2]WellGen Inc., 63 Dudley Road, New Brunswick, NJ 08901

Polymethoxyflavones (PMFs) are of particular of interest, due to their broad spectrum of biological activities, including anti-inflammatory, anti-carcinogenic, anti-viral, anti-oxidant, anti-thrombogenic and anti-atherogenic properties. The unique feature of the chemical structure of PMFs is the polymethylation of polyhydroxylated flavonoids, one of the major naturally occurring polyphenolic compounds. PMFs are exclusively found in citrus plants. There has been increasing interest in the exploration of beneficial health properties of PMFs. Isolation and characterization of PMFs from different *Citrus* species, such as sweet orange (*Citrus sinensis*) peel, has provided considerable material for research into the biological activity of different PMFs against inflammatory, cancer, and cardiovascular disease.

Flavonoids, a class of chemically related polyphenols of plant origin, exist ubiquitously in nature and also exhibit a broad spectrum of pharmacological properties (*1-10*). These polyphenolic compounds have a basic 15-carbon skeleton and can be represented as C6-C3-C6, consisting of two benzene rings (C6) joined by a linear three carbon chain (C3) (Figure 1). Flavonoids from citrus fruits have a benzo-γ-pyrone skeleton with a carbonyl group at the C3 position (Figure 1). These particular flavonoids consist of two classes of compounds named flavanones and flavones. Flavanones are found in tomatoes and some aromatic plants such as mint, but are generally present in high concentrations only in the citrus fruits. The most prevalent flavanones are hesperetin from oranges, naringenin from grapefruit, and eriodictyol from lemons (*11*). These flavanones are generally glycosylated by a disaccharide at position 7. Hesperetin and naringenin are found in the fruit tissue and peel, largely as their glycosides, hesperidin and naringin. Flavones are not common in either fruits and vegetables. The major flavones are glycosides of luteolin and apigenin. It is reported that the edible source of flavones are limited in parsley and celery (*11*). Two polymethoxylated flavones relatively common in citrus are tangeretin and nobiletin, both present in sweet orange peel (*Citrus sinensis* (L.) Osbeck) and in the peel of bitter orange (*Citrus aurantium* L.) (*12*). Sinensetin is the third common polymethoxyflavone found in the skin of citrus fruits (*11*).

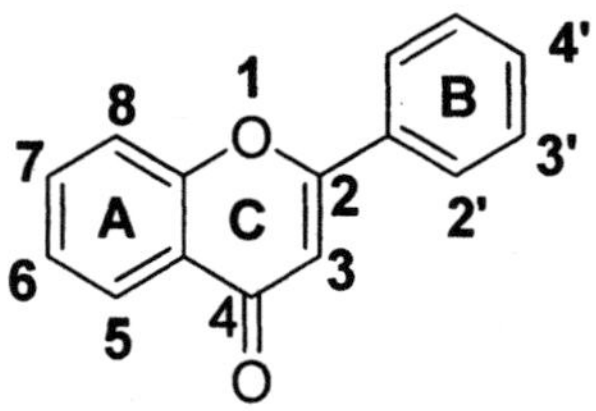

Figure 1. Basic skeleton of flavonoids

More than 4000 naturally occurring flavonoids have been identified and characterized. Significant scientific evidence has shown that flavonoids have many beneficial health effects for human beings (*1-4*). Citrus flavonoids are of particular interest because many of these flavonoids exhibit a broad spectrum of biological activity, including anti-inflammatory, anti-carcinogenic, anti-viral, anti-oxidant, anti-thrombogenic and anti-atherogenic properties (*1-10*). It is suggested that cancer induction can be prevented by ingestion of certain food ingredients: For example, Citrus flavonoids in citrus fruits and juices are among the prominent cancer preventing agents (*13*).

Natural Occurrence of Polymethoxyflavones

Citrus production worldwide in major citrus producing countries in 2003/2004 was 73.1 million metric tons (161 billion lbs). Total citrus production in the United States was 14.85 million metric tons (33 billion lbs) (National Agricultural Statistics Service). Around 34% of these products were used for juice production (*14*), yielding approximately 44% (4-5 billion lbs in the U.S.A.) of peels as by-products.

Orange peel is an excellent natural protective coating and has innate resistance against UV light, Fungi, insects, mechanical abrasion, *etc.* Medicinally, orange peel has been used in traditional medicine in some Asian countries for relieving stomach upset, skin inflammation, muscle pain, and ringworm infections. Flavonoids, consisting of mainly polymethoxylated flavonoids, terpenoids, such as limonene and linalool, and other volatile oils are the major ingredients of orange peel. Traditionally, orange peels were processed to obtain the volatile and nonvolatile fraction for various applications in the food, drug, and cosmetic industries. However, application for these extracted chemicals is limited because the overall demand for these compounds relative to the quantity produced is of little significance.

Polymethoxylated flavonoids (PMFs) are found in abundance in citrus fruits, especially in the peel. Therefore, the isolation, characterization, and exploration of biological effects of polymethoxylated flavonoids (PMFs) from orange peel may lead to new pharmacological applications.

PMFs exist almost exclusively in *Citrus* genus, particularly in the peel of two citrus species, sweet orange (*C. sinensis*) and mandarin (*Citrus reticulata* Blanco). So far, there are more than 20 polymethoxylated flavonoids that have been isolated and identified from different tissues of the citrus plant (*14-18*). The types and content of PMFs vary between different varieties of citrus species. Thus the types and concentrations of PMFs may also serve for taxonomic purposes in the botanical and agricultural sciences. As an example, PMF concentration in Dancy tangerine peels are almost 10 fold higher than that of other varieties of tangerine and oranges (*19*). Relative to the types of PMFs, the only PMF detected in the Ambersweet orange peel was tangeretin, Temple orange peel and Valencia orange peel were similar except that no 3,5,6,7,8,3′,4′-heptamethoxyflavone was found in Temple orange peel (*20*).

Chemistry of PMFs

The structures of PMFs usually differ in the numbers, types or positions of substitution on the 2-phenyl-γ-pyrone skeleton. The general structures of PMFs are illustrated in Figure 2. Interestingly, the majority of the identified compounds are derivatives from the base structure of flavones, though two flavanones were identified from Dancy tangerine leaves (*15*). Among those identified PMFs, tangeretin and nobiletin have been intensively studied recently because of their relatively ready availability and pharmacological properties.

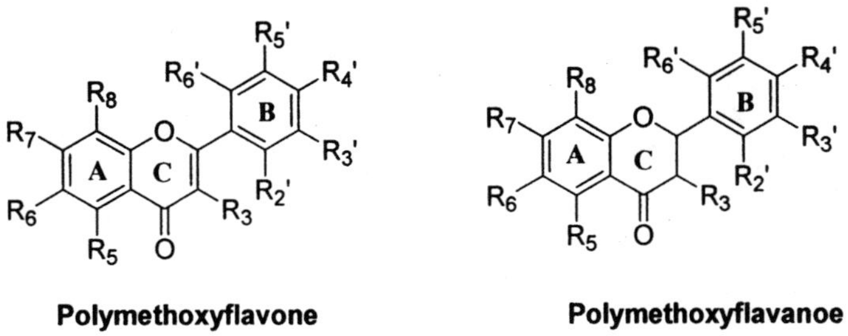

(R_3 to R_8 & R_2' to R_6' = H, OH, or OMe, two or more total methoxy groups)

Figure 2. General structure of polymethoxyflavones (PMFs).

Biological Activities of Polymethoxyflavones

Citrus fruit-derived polymethoxyflavones and their metabolites exhibit a wide variety of biological activities, such as anti-inflammatory, anti-cancer, and anti-atherogenic activities. (*1-10*). For example, polymethoxyflavones have been shown to block adhesion molecule biosynthesis by cytokine-induced endothelial cells; to block activation-induced deregulation of neutrophils and mast cells; to inhibit expression of tumor necrosis factor–α (TNFα); to reduce the invasiveness of tumors in animal models; to induce the differentiation of myeloid leukemic cells; to suppress cell proliferation while promoting apoptosis; to reduce lymphocyte proliferation and platelet aggregation; and to suppress ethanol-induced gastric hemorrhagic lesions while promoting chloride secretion by human colonic epithelial cells, *etc.* (*5-7*). Significant scientific data demonstrates that citrus flavonoids promote metal chelation, scavenging of free radicals, alteration of phase I cytochrome P-450 (CYP) CYP1A1 enzyme activation, and stimulation of phase II conjugation enzymes (*2*). Epidemiological studies indicate that flavonoid ingestion is associated with a reduced risk of certain forms of cancer and chronic disease (*7*). Due to the hydrophobic nature of methoxy groups comparing to hydroxyl groups, PMFs, such as nobiletin and tangeretin, are more lipophilic compared to polyhydroxylated flavonoids, such as quercetin, luteolin and narigenin. Consequently, PMFs may have higher permeability through the small intestine and more readily absorbed into the blood circulation system of the human body (*2*).

Anti-inflammatory Activities of PMFs

Inflammation and Human Health

Inflammation is an immune response to bacterial, viral, mechanical, or chemical damages on tissues. It involves activation and directed migration of leukocytes (neutrophils, monocytes and eosinophils) from the venous system to damage sites. Persistent accumulation and activation of leucocytes is an indication of chronic inflammation (*21*).

A variety of proteins act as mediators in the process of leucocyte migration and activation. Adhesion molecules (L- P-, and E-selectin), cytokines (TNF, interleukin-1 (IL-1), IL-6, *etc*) and leukocyte-activating molecules, endothelial vascular cell-adhesion molecule-1 (VCAM-1), MadCAM-1, and matrix metalloproteinases (MMPs) are involved in the process of recruitment and activation of neutrophils, which are the first to migrate to the injury site to begin the process of tissue repair. Monocytes, the second group of leucocytes to be recruited, are mediated by chemokines. After reaching damaged tissue, monocytes will differentiate into macrophages, which become the major source of growth factor and cytokines. During the inflammation process, chemical agents that have the ability to modulate the above-mentioned proteins will lead to either pro- or anti-inflammatory effects (*21*).

There are a few key mediators in modulation of inflammation. It has been well established that the production of cytokines (IL-1β, TNF-α, IL-6, IL-12, IFN-γ), nitric oxide (NO), prostaglandins, and leukotrienes are correlated with the inflammation process. The pro-inflammatory mediators, prostaglandins and NO, are produced by actions of cyclooxygenases (COX-2) and inducible nitric oxide synthase (iNOS), respectively. These inflammatory mediators are soluble and diffusible molecules that act locally at the site of tissue damage and infection, and also at more distant sites. COX-2 and iNOS are important enzymes that mediate most of the inflammatory processes. Improper up-regulation of COX-2 and/or iNOS has been associated with pathophysiology of certain types of human cancers as well as inflammatory disorders (*22*).

Several studies have shown that the eukaryotic transcription factor, nuclear factor-kappa B (NF-κB) is involved in regulation of COX-2 and iNOS expression. NF-κB is activated by a variety of pro-inflammatory stimuli such as cytokines, phorbol esters, bacterial or viral products, oxidants and UV radiation,. The most likely mechanism underlying inhibition of NFκB activation is via inhibition of Inhibitory Kappa B kinase (IKK), which prevents the degradation of inhibitory-kappa B (IκB) and thereby hampers subsequent nuclear translocation of the functionally active subunit of NF-κB (*23*).

Activator Protein-1 (AP-1) is also an important transcription factor that is involved in the transcriptional regulation of cytokines and mediators. Many chemical agents affect the formation and activation of AP-1 proteins. Following activation, AP-1 binds to specific recognition sequences in the promoter regions of target genes causing modulation of gene transcription (*24*). COX-2 gene promoter region has a binding site for AP-1, indicating that COX-2 gene transcription is not only regulated by NFκB, but also by AP-1. There are also reports that activation of both AP-1 and NFκB may result in greater inflammation than if either transcription factor alone is activated *(24, 25)*.

A detailed mechanism of action about how phytochemicals play an active role in the prevention of inflammation has been elucidated (*26*). Major phytochemicals that have anti-inflammatory properties include curcuminoids, flavonoids, coumarins, and terpenoids. There is abundant research in the anti-inflammatory properties of Curcumin and it has been studied in a phase I clinical trial (*27*). The anti-inflammatory activity of terpenoids has also recently been highlighted as research show the anti-inflammatory activity associated with linalool, oleanolic acid, and ursolic acid. appears to enhance the secretion of interferon-gamma (IFN-γ, *28*). Research has demonstrated that flavonoids have been shown to modulate key enzymes involved in the inflammation process; some flavonoids are potent inhibitors of key enzymes involved in prostaglandin biosynthesis, i.e., phospholipase, lipoxygenase, and cyclooxygenase (*29*).

Anti-inflammatory Activities of PMFs

One of the most significant biological properties of the PMFs is their anti-inflammatory activity. As early as 1936, the capillary protective effects of certain citrus flavonoids in combination with ascorbic acid were reported (*5*). Subsequent research data not only confirmed the earlier findings, but also provided additional supporting evidence demonstrating that citrus flavonoids, especially polymethoxyflavones, are directly associated with the inhibition of enzymes involved in the inflammation.

The anti-inflammatory activity of flavonoids is associated with their free radical scavenging property. Free radical species, such as superoxide radical are ubiquitous. These radicals may be transformed to become more reactive and damaging radicals; super oxide radical is readily converted to hydroxyl radical by reaction with adjacent water molecules or ferric ions in the Haber-Weiss and Fenton reactions (*30-32*). In view of the pharmacological interest in flavonoids, Fazilatiun, *et al.* determined the relative super oxide radical scavenging activities of some flavonoids. According to this research, the super oxide radical scavenging capacity of flavonoids at 100 μM shows a decreasing capacity for the following flavonoids: quercetin > luteolin > 5,7,3′,5′-tetrahydroxyflavone > blumeatin > rhamnetin > tamarixetin > luteolin-7-methyl ether > dihydroquercetin-4'methyl ether > dihydroquercetin-4′,7-dimethyl ether (*33*). there was also evidence that the ability of scavenging reactive oxygen species (ROS) of a flavonoid antioxidant from spinach leaves, 6-(3,4-dihydroxy-phenyl)-9-hydroxy-7-methoxy-[1,3]dioxolo-[4,5-g]chromen-8-one-4′-β-glucuronid (GF, Figure 3) was also shown to be more effective than EGCG and Vitamin E. Consequently, this natural antioxidant from spinach may be considered as an effective source for combating oxidative damage (*34*).

From these works, it was felt that flavonoids with a free hydroxyl group are more physiologically active than methylated compounds in their free radical scavenging activity. Hence, it is an observation that hydroxylated PMFs have the property of scavenging free radical species, whereas fully methoxylated flavonoids can effectively inhibit the enzymes like inducible nitric oxide synthase (iNOS) and NADPH oxidase that generate free radicals like NO and super oxide anion.

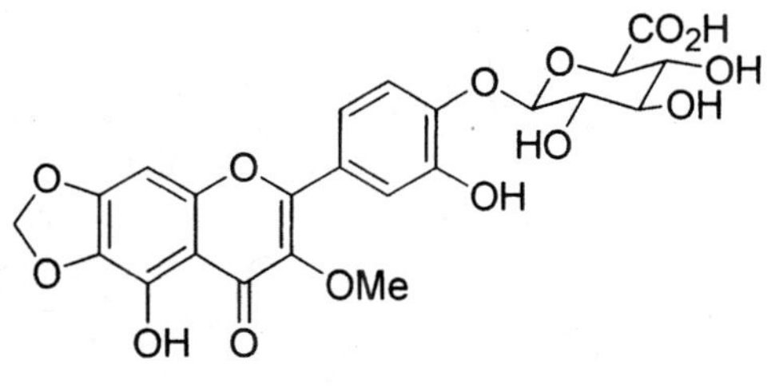

Figure 3. GF from spinach leaves

One example of a potential anti-inflammatory agent within the PMF family, nobiletin, has been shown to inhibit matrix degradation of the articular cartilage and pannus formation in osteoarthritis and rheumatoid arthritis by effectively inhibiting the production of promatrix metalloproteinase-9 (MMP-9) and prostaglandin E_2 (PGE_2) in human synovial fibroblasts by selectively downregulating cyclooxygenase-2 (COX-2) activity (*6*). Molecular biological evidence has shown that nobiletin suppresses gene expression and production of some matrix metallproteinases (MMP-1, MMP-3 and MMP-9) in rabbit articular chondrocytes and synovial fibroblasts (*35*). Additional recent work from Ito's group has demonstrated that proMMP-1 and proMMP-3 were also inhibited by nobiletin, while the endogenous MMP inhibitor TIMP-1 was upregulated. Gene expression of other pro-inflammatory cytokines, such as interleukin IL-1α, IL-1β, TNF-α, and IL-6 were found to be downregulated by nobiletin in mouse macrophages (*6*). It was shown in an LPS-induced, inflammatory response mouse macrophage model that nobiletin moderately attenuates COX-2 (gene or enzyme?) expression and significantly suppressed the activation of AP-1, NF-κB, and CREB (*36*). Ohigashi's research group reported that nobiletin significantly inhibited two distinct stages of skin inflammation induced by double TPA application (*8*). The important inflammatory mediator, PGE2, release was found to be inhibited by nobiletin. Of great interest was the finding that nobiletin selectively down regulates COX-2 mRNA expression, but not COX-1 (*6*). This feature resembles commercial anti-inflammatory steroidal drugs, but with fewer side effects. These findings indicate that nobiletin could be a novel anti-inflammatory and/or immunomodulatory potential drug. All of the above work is based mostly on *in vitro* evaluation, while these anti-inflammatory function have yet to be confirmed *in vivo*.

Anti-cancer Activity of PMFs

Inflammation and Cancer

There is numerous reports which show that inflammation and infection are closely associated with cancer initiation and proliferation. The proposed mechanisms includes: direct integration of viral DNA into the host genome; immunosuppression caused by viral infection leading to a failure to inhibit malignancy; production of ROS and RNS which cause damage to the host DNA. (*37*). Detailed studies have unveiled the relationship between inflammation and cancer. Chronic inflammation is induced by biological, chemical, and physical factors that have been found to be associated with the increased risk of human cancer at various sites. Inflammation causes the activation of a variety of inflammatory cells, which in turn induce and activate several oxidant-generating enzymes such as NADPH oxidase, iNOS, myeloperoxidase, and eosinophil peroxidase. These enzymes generate high concentrations of diverse free radicals and oxidants, including super oxide anion, nitric oxide, nitroxyl, nitrogen dioxide, hydrogen peroxide and hypochlorous acid. These free radicals and oxidants react with each other and produce more potent reactive oxygen and nitrogen species, such as peroxynitrite. The adverse effects of these reactive species include damage to DNA, RNA, lipids, and proteins by nitration, oxidation, chloronation, and bromination reactions, which lead to increased mutations and altered functions of enzymes and proteins, These effects contribute to the multistage carcinogenesis process (*37*). Because of the intimate relationship between inflammation and cancer, a thorough exploration of inflammation should be pursued, especially as it relates to early-stage cancer induction in human populations (*21,37,38*).

Anti-cancer Activity of PMFs

Many *in vitro* and *in vivo* studies indicate protective effects of polymethoxyflavones against the occurrence of cancer. PMFs inhibit carcinogenesis by mechanisms like blocking the metastasis cascade, inhibition of cancer cell mobility in circulatory systems, apoptosis, selective cytotoxicity, and antiproliferation. PMFs have been shown to reduce the invasion of tumors in animal models, to induce the differentiation of myeloid leukemic cells, and to suppress proliferation of cancer cells. PMFs have also been shown to reduce lymphocyte proliferation and platelet aggregation and can suppress ethanol-induced gastric hemorrhagic lesions, while promoting chloride secretion by human colonic epithelial cells (*39*).

Kandaswami, *et al.*, did comparative studies of the antiproliferative effects between PMFs, such as nobiletin and tangeretin and polyhydroxyflavonoids like quercetin and taxifolin on the *in vitro* growth of a human squamous-cell carcinoma cell line [HTB43]. They found that the two PMFs inhibited cell growth significantly, whereas the two polyhydroxyflavonoids showed no significant inhibition at any tested concentrations (*9*). A possible explanation for this observation was that the permeability of PMFs is much greater than that of the polyhydroxyflavonoids due to their non-polar nature. This suggests that the PMFs greater activity was likely due to this enhanced cellular absorption (*9*).

Because tangeretin and nobiletin are the most abundant PMFs in citrus peels, they have been the primary research focus in terms of their bioactivity. The anticancer activity of tangeretin is well accepted (*40,41*). The anti-cancer function of tangeretin has multiple mechanisms. It can inhibit the cancer initiation stage by modulating hepatic enzymes, thus affecting xenobiotic activation and detoxification in the liver. As an example, glutathione transferase activity is enhanced by tangeretin (*42*), while P450 1A2, 3A4, and 2B, 3A were reported to be inhibited respectively to different extents (*43,44*). Tangeretin also plays a multiple role in the cancer-cell proliferation stage. Although not a good free radical scavenger, it does inhibit 15-lipoxygenase activity (*45*), suggesting that it may exert a modulating effect on enzymatic lipid oxidation. Tangeretin is able to enhance gap-junction intercellular communication between normal cells and mutated cells and thus inhibit cancer cell proliferation (*46*). Tangeretin can arrest the cell cycle in the G1 phase by inhibiting cyclin-dependent kinases (Cdk) and enhancing Cdk inhibitor proteins (*47*).

Research on the anticancer activity of nobiletin originated from the anti-carcinogenic and anti-tumor activity of citrus flavonoids. Nobiletin can inhibit the proliferation of human prostate, skin, breast and colon carcinoma cell lines, inhibit the production of some MMPs, and inhibit the proliferation and migration of human umbilical endothelial cells (*8,48,49*). Tangeretin has shown an inhibition mechanism of human mammary cancer cells and cytolysis in which it inhibits extracellular-signaling-regulated kinase (ERK) phosphorylation (*50*).

Nobiletin was also shown to have anti-proliferative and apoptotic effects on a gastric cancer cell line and that are has a disruptive effect on cell-cycle progression. In an evaluation of 42 flavonoids, nobiletin showed the strongest antiproliferative activity against six human cancer cell lines (*7*). Nobiletin has been shown to suppress prostaglandin E2 (PGE2) production and COX-2 protein expression in vitro. COX-2, induced by several stimuli associated with inflammation, is involved in carcinogenesis including colon tumorigenesis of humans and rodents (*13*). During a recent investigation, it was discovered that nobiletin is a dual inhibitor of both NO and O_2^- generation in leukocytes, and inhibits tumor promotion in two-stage mouse skin tumorigenesis (*8*).

Anti-atherogenic Activity of PMFs

Strong evidence indicates that citrus flavonoids can reduce the hepatic production of cholesterol containing lipoproteins, thus reducing total cholesterol concentration in the plasma, and consequently reducing the occurrence of cardiovascular disease (*51-53*). A recent study suggests that nobiletin can reduce the circulating concentrations of very low density lipoproteins (VLDL) and low density lipoproteins (LDL) in the blood and can directly inhibit macrophage derived foam-cell formation at the site of lesion development within a vessel wall. It was recently reported that PMF food supplementation may ameliorate hypertriglyceridemia and its anti-diabetic effects in hamsters through adipocytokine regulation and peroxisome proliferators activated receptor-α (PPARα) and PPARγ activation (*54*).

Nobiletin anti-inflammatory activities are likely linked to the prevention of plaque formation during atherosclerosis. Nobiletin markedly reduced TPA-induced, lectin-like, ox-LDL receptor-1 (LOX-1) mRNA expression in THP-1 human monocyte-like cells in dose and time-dependent manners. It also suppressed the phosphorylation of extracellular signal-regulated protein kinase (ERK) 1/2, c-Jun NH2-terminal kinase (JNK) 1/2, and c-Jun (Ser-63), thereby inhibiting the transcriptional activity of activator protein-1. Furthermore, nobiletin attenuated the expression of SR-A, SR-PSOX, CD36, and CD68, but not CLA-1, mRNA, leading to the blockade of DiI-acLDL uptake (*55*). It seems likely that PMFs may play a role in the prevention of atherosclerosis because it has been demonstrated that there is an inflammatory basis for the genesis of atherosclerosis. (*4,53,56*).

Polymethoxyflavones from Orange Peel

Many polymethoxyflavones have been identified, isolated or synthesized (Table I) for a variety of reasons, such as natural products isolation and structural elucidation, screening and identification of bio-active molecules, or as intermediates or products of organic synthesis.

Using LC-APCI-MS and EI-MS spectroscopy methods, Ghai *et al.* identified a total of 14 PMFs from orange peel extracts, including polymethoxyflavones and hydroxylated polymethoxyflavones (*10*). The structures of these PMFs are shown in Table I. Although the structures of these PMFs were not confirmed by further instrumental analysis such as NMR, this study demonstrated the existence of an array of PMFs in orange. In a study of cytokine production inhibition by PMFs, Manthey, *et al.* performed a study using compositions of limocitrin derivatives and 5-demethylsinensetin and methods to inhibit neoplastic disease (*5*). Table I also illustrates the PMFs used in Manthey's study. Recently, in searching for anti-inflammatory agents, six polymethoxyflavones and 5-demethylnobiletin were isolated from sweet orange peel extract in Ho's laboratory. The structures of these seven PMFs (Table I) have been elucidated and confirmed by various analytical techniques including LC-MS and proton NMR. The identified compounds are tangeretin (#1), sinensetin (#2), nobiletin (#3), 5,6,7,4′-tetramethoxyflavone (#6), 3,5,6,7,8,3′,4′-heptamethoxyflavone (#8), 3,5,6,7,3′,4′-hexamethoxyflavone (#11), and number 29, 5-hydroxy-6,7,8,3′,4′-pentamethoxyflavone (*57*).

During the course of continuing isolation of PMFs from sweet orange peel extracts, we found that within the extract of sweet orange peel, there exists a number of hydroxylated polymethoxyflavones. There are no literature reports describing the systemic isolation of hydroxylated PMFs from sweet orange peel. Because of the unique bioactivity and immuno- modulatory effects of the PMFs, continued isolation and characterization of hydroxylated polymethoxyflavones should spur new interests for *in vitro* and *in vivo* studies of the PMFs, particularly the hydroxylated flavones. Furthermore, two previously unidentified polymethoxyflavanones have been isolated during the course of this study (Table I, Compounds #25,26).

Experimental

Materials

Sweet orange peel extract was obtained from Florida Flavors Company, Lakeland, Florida, USA. A pre-packed silica gel (60Å, 32-63 μm) column (330 g), for normal phase chromatography was purchased from Teledyne Isco, Inc. (Lincoln, NE). C-18 reversed phase high performance liquid chromatography analytical and preparative columns were, purchased from YMC, Inc. (Japan). Chiral analytical and preparative high performance liquid chromatography columns, Regis Welk-O 1 (R, R) were purchased from Regis Technologies, Inc. (Morton Grove, IL).

Flash Column System

An automated flash chromatography system (Model, Foxy 200, sg100, Teledyne Isco Inc., Lincoln, NE) equipped with a 330 g prepacked silica gel flash column was used. The mobile phase for normal phase flash column comprises ethyl acetate and hexanes and the flow rate was set at 90 mL/min. The eluent was monitored with a single channel UV detector at a wavelength of 254 nm.

HPLC Systems

A Preparative liquid chromatograph (Waters Corp., Milford, MA) equipped with two piston pumps, a UV-vis detector and an automatic injector was used. A Welk-O 1 (R,R) 450 gram column (Regis Technologies, Inc., Morton Grove, IL) was used for the preparative HPLC system. The mobile phase was 35% absolute ethanol and 65% hexanes with a flow rate set at 85 mL/min. The eluent was detected with a UV wavelength at 326 nm. Another automated HPLC system from Gilson Inc. was used for reverse phase preparative separations. The mobile phase was acetonitrile and water with a flow rate of 20 mL/min. The eluent was detected with dual wavelength at 254 nm and 326 nm.

NMR Instrument

NMR spectra were recorded on a Varian 500 Spectrometer (Varian Inc., Palo Alto, CA) with tetramethyl silane serving as internal standard, proton NMR spectra were recorded at 500 MHz; ^{13}C NMR spectra were obtained at 125 MHz.

Mass Spectrometer

ESI-MS spectra were obtained on a Micromass VG Platform II mass analyzer (Micromass, Beverly, MA). EI-MS spectra were obtained on a MicroMass AutoSpec HF (Micromass, Beverly, MA).

Liquid Chromatography-Electron Spray Ionization Mass Spectrometry (LC-MS)

An HPLC-MS system was composed of an auto-sampler injector, an HP1090 system controller, with a variable UV wavelength (190-500 nm) detector, an ELSD (Evaporizing Laser Scattered Deposition) detector and an ESI-MS detector. The mobile phase was comprised of acetonitrile and H_2O with 0.05% TFA using gradients ranging between 10 - 90 % acetonitrile.

General Separation Procedures from Crude Sweet Orange Peel Extract (OPE)

The crude OPE mixture (100 g) was dissolved in a mixture of methylene chloride (12 mL) and hexanes (8 mL) and loaded onto a 330 g of pre-conditioned silica gel flash column. The elution was started with 10% ethyl acetate and 90% hexanes and modified linearly to 40% ethyl acetate and 60% hexanes within 35 min. Then a isocratic mobile phase of 40% ethyl acetate and 60% hexanes was applied for another 15 min. The fractions showing a strong UV absorbance at 254 nm were analyzed by LC/MS and by thin layer chromatography (TLC) with a solvent system of 40% ethyl acetate and 60% hexane. The fractions were combined into several groups according to their molecular weight obtained from LC/MS analysis. Further separation of each group was done either on a (preparative) reverse phase HPLC system or on a chiral HPLC system.

Separation Procedures for Hydroxylated PMFs

The fractions that contain hydroxylated PMFs characterized by LC/MS were combined and concentrated. The residue was dissolved in acetonitrile and water. The dissolved solution was loaded onto a C-18 reverse phase Preparative HPLC system. A gradient method was used to increase between 25% acetonitrile vs. H_2O to 60% acetonitrile in 25 min. The fractions were analyzed by LC/MS. Both the pure compounds and mixtures were collected. The pure fractions, by LC-MS, were concentrated and lyophilized. The compounds were analyzed by EI-MS. ESI-MS and NMR. The mixtures were concentrated on rotary evaporator and dissolved using the minimal amount of methylene chloride. The solution was loaded onto the Preparative HPLC system equipped with the Regis column (Welk-O 1 R,R 450 gram). The monitoring UV absorbance was set at 326 nm. The fractions were collected and concentrated respectively. MS and NMR analysis was made for each of these fractions.

Summary

Using silica gel normal phase chromatography and C-18 reverse phase HPLC, as well as a chiral HPLC system, a total of seven hydroxylated polymethoxyflavones, two polymethoxyflavanones, and two polymethoxy-chalcones were newly isolated from the orange peel extract in our laboratory (*58*). These compounds were identified as the following (Table I): 5-hydroxy-6,7,4′-trimethoxyflavone, 5-hydroxy-6,7,8,4′-tetramethoxyflavone, 3-hydroxy-5,6,7,4′-tetramethoxyflavone, 3-hydroxy-5,6,7,8,4′-pentamethoxyflavone, 5-hydroxy-3,6,7,8,3′,4′-hexamethoxyflavone, 5-hydroxy-3,7,3′,4′-tetramethoxyflavone, 5-hydroxy-3,7,8,3′,4′-pentamethoxyflavone, 5,6,7,4′-tetramethoxy-flavanone, 5-hydroxy-6,7,8,3′,4′-pentamethoxyflavanone, 2′-hydroxy-3,4,4′,5′,6′-pentamethoxychalcone, and 2′-hydroxy-3,4,3′,4′,5′,6′-hexamethoxychalcone. All these newly isolated compounds were characterized by MS, NMR (^{1}H and ^{13}C) and 2D NMR. The analytical data of these compounds including ^{1}H NMR, ^{13}C NMR, UV, LRMS, EIMS and HRMS will be published in the near future.

The PMF profile of orange peel is illustrated in Table I. There were total of 12 polymethoxyflavones, ten hydroxylated polymethoxyflavones, one hydroxylated polymethoxyflavanone, one polymethoxyflavanone, two poly-methoxychalcones, and four hydroxylated PMFs identified.

Table I. PMFs from Orange Peel

NO	*Structure*	*Name*	*Isolation*	*Identi-Fication*	*Ref*
1	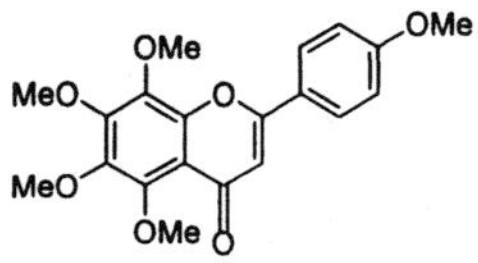	5,6,7,8, 4'-Pentamethoxy-flavone (Tangeretin)	Schneider Manthey Yu, Li	Ghai Swift	*5, 10 57-60*
2	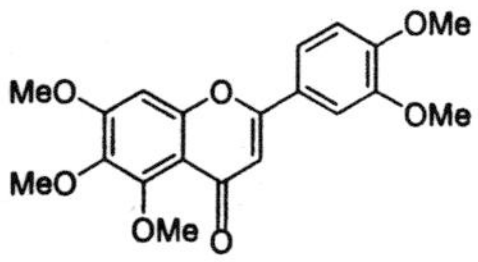	5,6,7,3',4'-Pentamethoxy-flavone (Sinensetin)	Tatum Schneider Yu, Li Manthey	Swift Ghai	*5, 10 57-61*
3	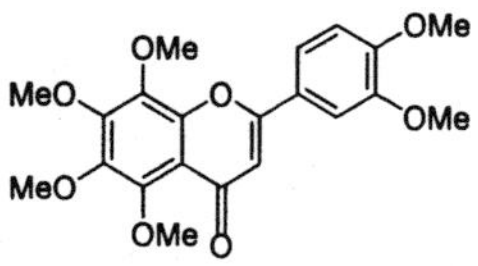	5,6,7,8,3',4'-Hexamethoxy-flavone (Nobiletin)	Tatum Schneider Yu, Li Manthey	Swift Ghai	*5, 10 57-61*
4	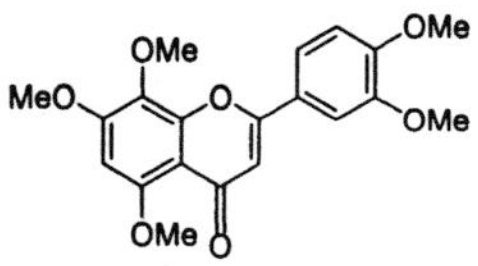	5,7,8,3',4'-Pentamethoxy-flavone	Schneider Manthey	Ghai	*5,10 60*
5	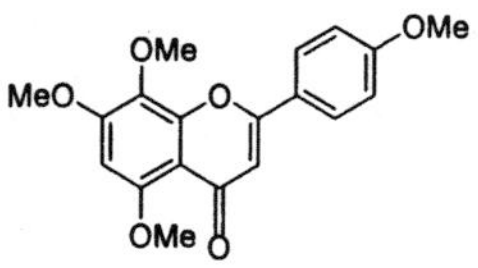	5,7,8,4'-Tetramethoxy-flavone	Tatum Schneider Manthey	Ghai	*5,10 59,61*
6	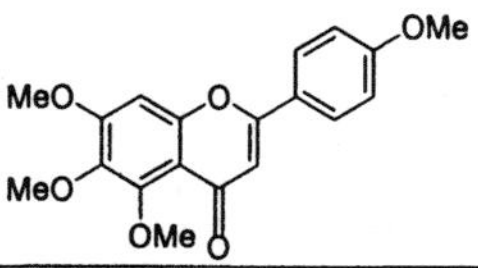	5,6,7,4'-Tetramethoxy-flavone	Tatum Yu, Li	Swift Ghai	*10,57 58,60 61*

Table I. *Continued.*

NO	*Structure*	*Name*	*Isolation*	*Identi-Fication*	*Ref*
7	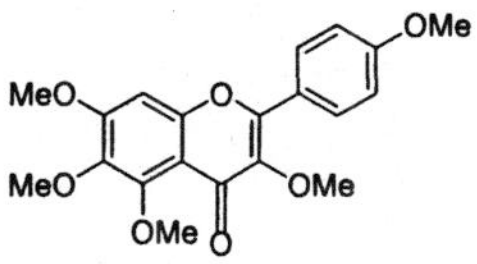	3,5,6,7,4'-Pentamethoxy-flavone	Boehm Yu, Li		*57,58* *62*
8	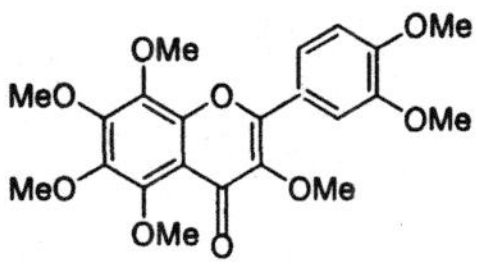	3,5,6,7,8,3',4'-Heptamethoxy-flavone	Tatum Yu, Li Manthey	Swift Ghai	*5,10* *58,59* *60,61*
9	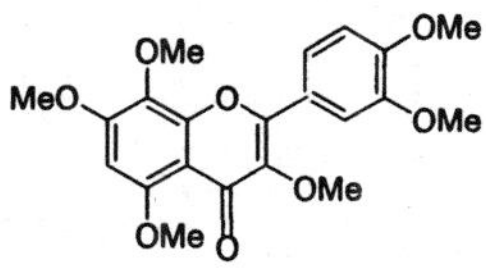	3,5,7,8,3',4'-Hexamethoxy-flavone	Tatum		*61*
10	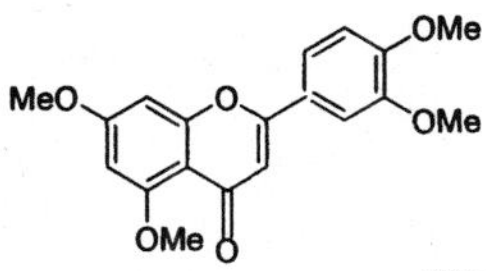	5,7,3',4'-Tetramethoxy-flavone	Prendergast Manthey	Mizuno	*5,63* *64,65*
11	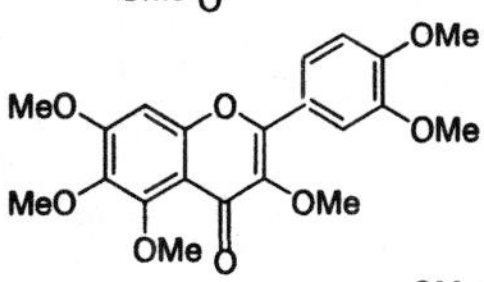	3,5,6,7,3',4'-Hexamethoxy-flavone	Tatum Yu, Li	Yu	*57,58,61*
12	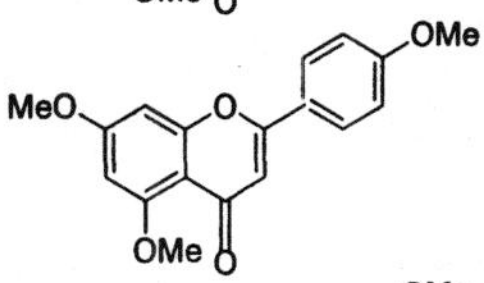	5,7,4'-Trimethoxy-flavone		Mizuno	*63*
13	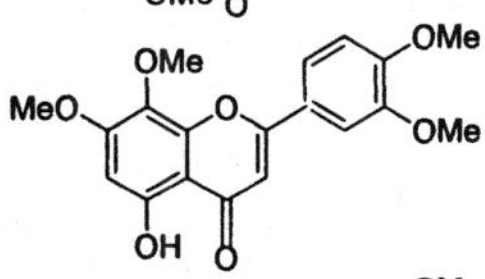	5-Hydroxy-7,8,3',4'-tetramethoxy-flavone		Ghai	*10*
14	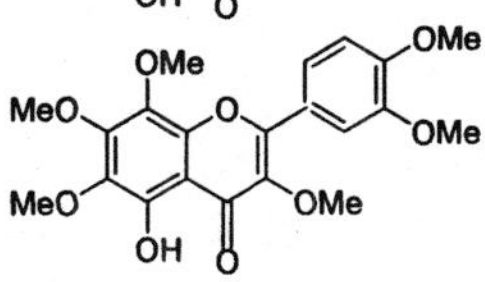	5-Hydroxy-3,6,7,8,3',4'-hexamethoxy-flavone	Li	Ghai Li	*10,58*
15	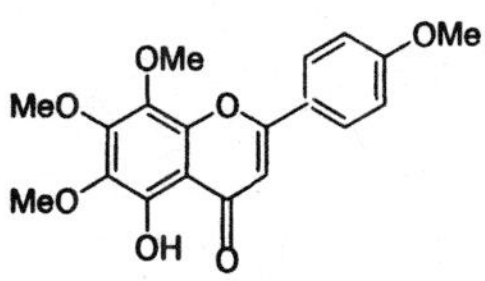	5-Hydroxy-6,7,8,4'-tetramethoxy-flavone (5-Demethyl-tangeretin)	Li	Ghai Li	*10,58*

Continued on next page.

Table I. *Continued.*

NO	Structure	Name	Isolation	Identi-Fication	Ref
16	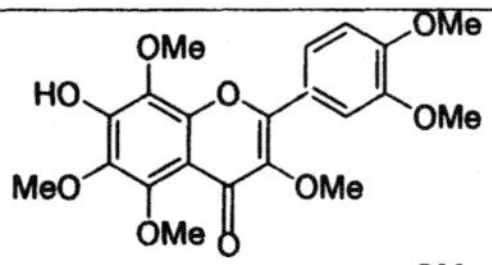	7-Hydroxy-3,5,6,8,3',4'-hexamethoxy-flavone		Ghai	*10*
17	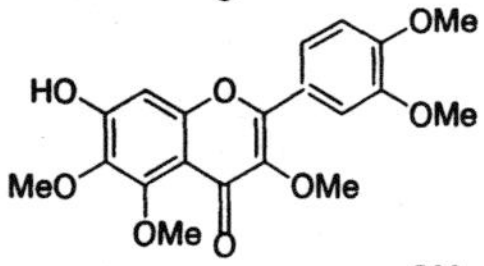	7-Hydroxy-3,5,6,3',4'-pentamethoxy-flavone		Ghai	*10*
18	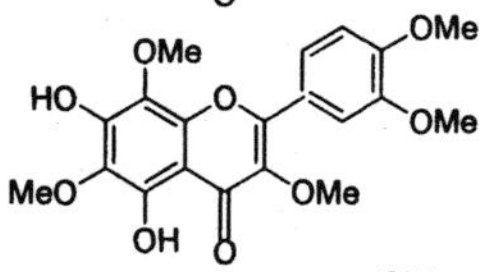	5,7-Dihydroxy-3,6,8,3',4'-pentamethoxy-flavone		Ghai	*10*
19	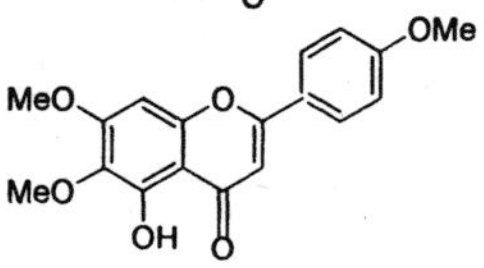	5-Hydroxy-6,7,4'-trimethoxy-flavone	Li	Li	*58*
20	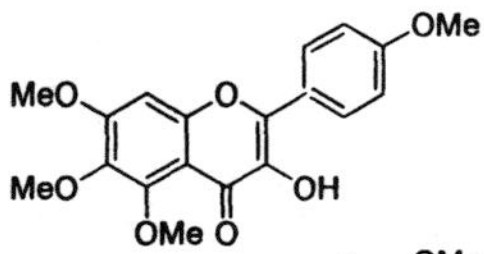	3-Hydroxy-5,6,7,4'-tetramethoxy-flavone	Li	Li	*58*
21	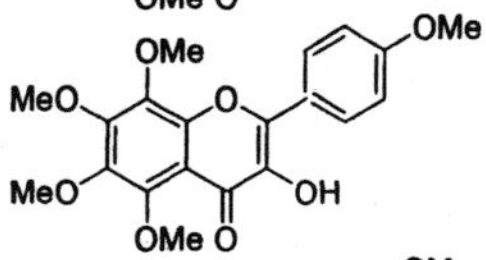	3-Hydroxy-5,6,7,8,4'-pentamethoxy-flavone	Li	Li	*58*
22	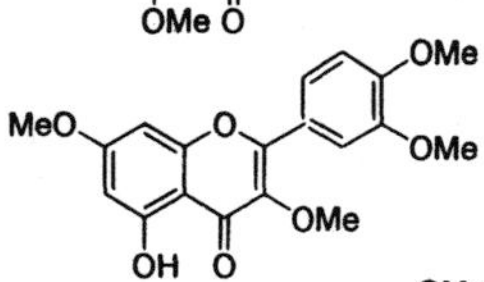	5-Hydroxy-3,7,3',4'-tetramethoxy-flavone	Li Manthey	Li	*5, 58*
23	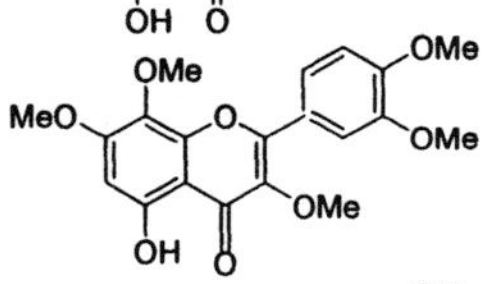	5-Hydroxy-3,7,8,3',4'-pentamethoxy-flavone	Li	Li	*58*
24	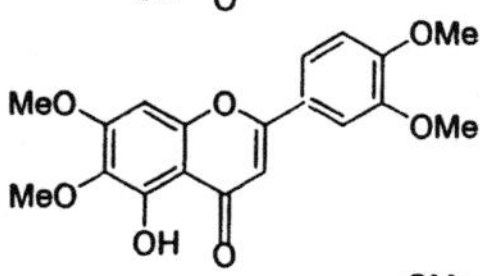	5-Hydroxy-6,7,3',4'-tetramethoxy-flavone	Manthey	Li	*5, 58*
25	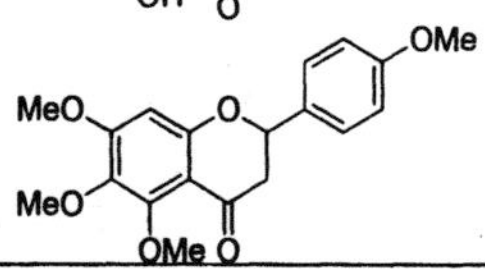	5,6,7,4'-Tetramethoxy-flavanone	Li	Li	*58*

Table I. *Continued.*

NO	*Structure*	*Name*	*Isolation*	*Identi-Fication*	*Ref*
26	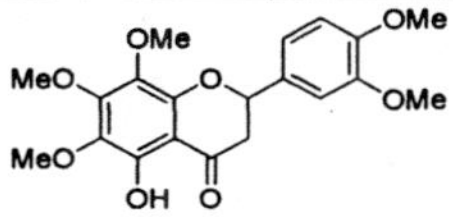	5-Hydroxy-6,7,8,3',4'-pentamethoxyflavanone	Li	Li	*58*
27	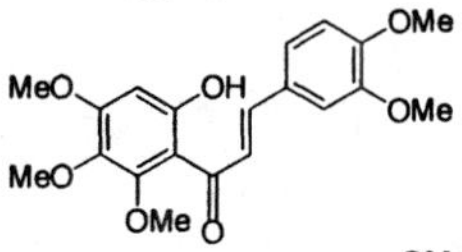	2'-Hydroxy-3,4,4',5',6'-pentamethoxy-chalcone	Li	Li	*58*
28	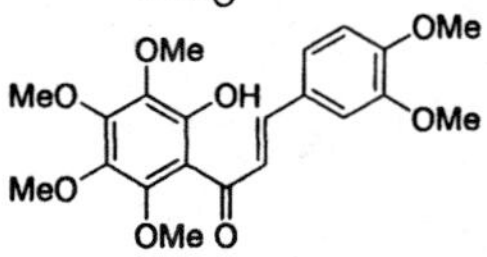	2'-Hydroxy-3,4,3',4',5',6'-hexamethoxy-chalcone	Li	Li	*58*
29	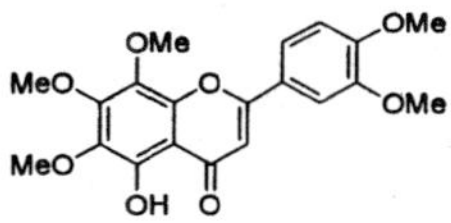	5-Hydroxy-6,7,8,3',4'-pentamethoxyflavone	Yu, Li	Yu, Li	*57,58*

References

1. Middleton, E.; Kandaswami, C.; Theoharides, T.C. *Pharmacol. Rev.* **2000**, *52*, 673-751.
2. Murakami, A.; Ohigashi, H. *Oxidative Stress and Disease. Phytochemicals in Health and Disease.* **2004**, *12*, 187-211.
3. Lopez-Lazaro, M. *Curr. Med. Chem. – Anti-Cancer Agents* **2002**, *2*, 691-714.
4. Whitman, S.C.; Kurowska, E.M.; Manthey, J.A.; Daugherty, A. *Atherosclerosis* **2005**, *178*, 25.
5. Manthey, J. A.; Manthey, C. L.; Montanari, US Patent 6,184,246, 2001.
6. Lin, N.; Sato, T.; Takayama, Y.; Mimaki, Y.; Sashida, Y.; Yano, M.; Ito, A. *Biochem. Pharmacol.* **2003**, *65*, 2065-2071.
7. Yoshimizu, N.; Otani, Y.; Saikawa, Y.; Kubota, T.; Yoshida, M.; Furukawa, T.; Kumwi, K.; Kameyama, K.; Fujii, M.; Yano, M.; Sato, T.; Ito, A.; Kitajima, M. *Aliment Pharmacol. Therapy* **2004**, *20* (Suppl. 1), 95 - 101.
8. Murakami, A.; Nakamura, Y.; Torikai, K.; Tanaka, T.; Koshiba, T.; Koshimizu, K.; Kuwahara, S.; Takahashi, Y.; Ogawa, K.; Yano, M.; Tokuda, H.; Nishino, H.; Mimaki, Y.; Sashida, Y.; Kitanaka, S.; Ohigashi, H. *Cancer Res.* **2000**, *60*, 5059-5066.
9. Kandaswami, C.; Perkins, E.; Soloniuk, D.S.; Drzewiecki, G.; Middleton, E. Jr. *Cancer Lett.* **1991**, *56*, 147-152.

10. Ghai, G; Rosen, R.; Ho, C-T.; Chen, K.Y.; Telang, N.; Lipkin, M.; Huang, M.T.; Boyd, C.; Csiszar, K. WO Patent 01/21137, 2001.
11. Manach, C.; Scalbert, A.; Morand, C.; Remesy, C.; Jimenez, L. *Am. J. Clin. Nutr.* **2004**, *79*, 727-747.
12. Horowitz, R.M.; Gentilli, B. In *Citrus Science and Technology;* Nagy S, Shaw P. E., Veldhuis M. K., Eds.; Avi Publishing Company Inc.: Westport, CT, **1977**; pp 397.
13. Kohno, H.; Yoshitani, S.; Tsukio, Y.; Murakami, A.; Koshimizu, K.; Yano, M.; Tokuda, H.; Nihshino, H.; Ohigashi, H.; Tanaka, T. *Life Sci.* **2001**, *69*, 901-913.
14. Jayaprakasha, G. K.; Negi, P.S.; Sikder, S.; Rao, L.J.; Sakariah, K.K. *Z. Naturforsch [C]* **2000**, *55*, 1030-1034.
15. Chen, J.; Montanari, A.M. *J. Agric. Food Chem.* **1998**, *46*, 1235-1238.
16. Chen, J.; Montanari, A.M.; Widmer, W.W. *J. Agric. Food Chem.* **1997**, *45*, 364-368.
17. Gaydou, E. M.; Bianchini, J.; Randriamiharisoa, R.P. *J. Agric. Food Chem.* **1987**, *35*, 525-529.
18. Kawaii, S.; Tomono, Y.; Katase, E.; Ogawa, K.; Nonomura-Nakano, M.; Nesumi, H.; Yoshida, T., Sugiura, M. and Yano, M. *J. Agric. Food Chem.* **2001**, *49*, 3982-3986.
19. Manthey, J.A.; Grohmann, K. *J. Agric. Food Chem.* **2001**, *49*, 3268-3273.
20. Manthey, J.A.; Grohmann, K. *J. Agric Food Chem.* **1996**, *44*, 811-814.
21. Coussens, L.M.; Werb, Z. *Nature* **2002**, *420*, 860-867.
22. Surh, Y.J.; Chun, K.S.; Cha, H.H.; Han, S.S.; Keum, Y.S.; Park, K.K.; Lee, S.S. *Mutat. Res.* **2001**, *480-481*, 243-268.
23. Bremner, P.; Heinrich, M. *J. Pharm. Pharmacol.* **2002**, *54*, 453-472.
24. Adcock, I.M. *Monaldi Arch. Chest. Dis.* **1997**, *52*, 178-186.
25. Adderley, S.R.; Fitzgerald, D.J. *J. Biol. Chem.* **1999**, *274*, 5038-5046.
26. Evans, D.A; Hirsch, J.B; Dushenkov, S. J. Sci. Food Agric. **2006**, *86*, 2503-2509.
27. Chainani-Wu, N. *J. Altern. Complement Med.* **2003**, *9*, 161-168.
28. Chiang, L.C.; Ng, L.T.; Chiang, W.; Chang, M.Y.; Lin. C.C. *Planta. Med.* **2003**, *69*, 600-604.
29. Xagorari, A.; Papapetropoulos, A.; Mauromatis, A.; Economou, M.; Fotsis, T.; Roussos, C. *J. Pharmacol. Exp. Ther.* **2001**, *296*, 181-187.
30. Deby, C.; Goutier, R. *Biochem. Pharmacol.* **1990**, *39*, 399-405.
31. Aust, S.D.; White, B.C. *Advan. Free Rad. Biol. Med.* **1985**, *1*, 1-17.
32. Babbs, C.F. *Ann. Emergency Med.* **1985**, *14*, 777-783.
33. Fazilatun, N.; Nornisah, M.; Zhari, I. *Pharm. Biol.* **2005**, *43*, 15-20.
34. Lomnitski, L.; Bergman, M.; Nyska, A.; Ben-Shaul, V.; Grossman, S. *Nutr. Cancer* **2003**, *46*, 222-231.
35. Ishiwa, J.; Sato, T.; Mimaki, Y.; Sashida, Y.; Yano, M.; Ito, A. *J. Rheumatol.* **2000**, *27*, 20-25.

36. Murakami, A.; Shigemori, T.; Ohigashi, H. *J. Nutr.* **2005**, *135*, 2987S-2992S.
37. Ohshima, H.; Tatemichi, M.; Sawa, T. *Arch. Biochem. Biophys.* **2003**, *417*, 3-11.
38. Balkwill, F.; Mantovani, A. *Lancet* **2001**, *357*, 539-545.
39. Manthey, J.A.; Grohmann, K.; Montanari, A.; Ash, K.; Manthey, C. L. *J. Nat. Prod.* **1999**, *62*, 441-444.
40. Kawaii, S.; Tomono, Y.; Katase, E.; Ogawa, K.; Yano, M. *J. Agric. Food Chem.* **1999**, *47*, 128-135.
41. Miyazawa, M.; Okuno, Y.; Fukuyama, M.; Nakamura, S.; Kosaka, H. *J. Agric. Food Chem.* **1999**, *47*, 5239-5244.
42. Breinholt, V.; Lauridsen, S.T.; Dragsted, L.O. *Xenobiotica.* **1999**, *29*, 1227-1240.
43. Obermeier, M.T.; White, R.E.; Yang, C.S. Xenobiotica **1995**, *25*, 575-584.
44. Siess, M.H.; Leclerc, J.; Canivenc-Lavier, M.C.; Rat, P.; Suschetet, M. *Toxicol. Appl. Pharmacol*. **1995**, *130*, 73-78
45. Malterud, K.E.; Rydland, K.M. *J. Agric. Food Chem.* **2000**, *48*, 5576-5580.
46. Chaumontet, C.; Bex, V.; Gaillard-Sanchez, I.; Seillan-Heberden, C.; Suschetet, M.; Martel, P. *Carcinogenesis* **1994**, *15*, 2325-2330.
47. Pan, M.H.; Chen, W.J.; Lin-Shiau, S.Y.; Ho, C.-T.; Lin, J.K. *Carcinogenesis* **2002**, *23*, 1677-1684
48. Murakami, A.; Koshimizu, K.; Ohigashi, H.; Kuwahara, S.; Kuki, W.; Takahashi, Y.; Hosotani, K.; Kawahara, S.; Matsuoka, Y. *Biofactors* **2002**, *16*, 73-82.
49. Kawabata, K.; Murakami, A.; Ohigashi, H. Biosci. Biotechnol. Biochem. **2005**, *69*, 307-314.
50. Slambrouck,S.V.; Parmar, V.S.; Sharma,S.K.; Bondt, B.D.; Fore, F.; Coopman, P.; Vanhoecke, B.W.; Boterberg,T.; Depypere,H.T.; Leclercq, G.; Bracke,M.E. *FEBS. Letters* **2005**, *579*, 1665-1669.
51. Kurowska, E.M.; Manthey, J.A. *J. Agric. Food Chem.* **2004**, *52*, 2879.
52. Wilcox, L.J.; Borradaile, N.M.; De Dreu, L.E.; Huff, M.W. *J. Lipid Res.* **2001**, *42*, 725.
53. Guthrie, N.; Kurowska, E.M.; Manthey, J.A.; Horowitz, R.M. US Patent, 6,987,125, 2006.
54. Li, R.W.; Theriault, A.G.; Au, K.; Douglas, T.D.; Casaschi, A.; Kurowska, E.M.; Mukherjee, R. *Life Sci.* **2006**, *79*, 365-373.
55. Eguchi, A.; Murakami, A.; Ohigashi, H. *FEBS Lett.* **2006**, *580*, 3321-3328.
56. Ross, R. *The New England Journal of Medicine* **1999**, *340*, 115-126.
57. Yu, H. Ph.D. thesis, Rutgers University, New Brunswick, NJ, 2004.
58. Li, S. Ph.D. thesis, Rutgers University, New Brunswick, NJ, 2005.
59. Schneider, G.; Unkrich, G.; Pfaender, P.; *Archiv der Pharmazie und Berichte der Deutschen Pharmazeutischen Gesellschaft,* **1968**, *301*, 785-792.

60. Swift, L. J. *J. Agric. Food Chem.* **1967**, *15*, 99-101.
61. Tatum, J.H.; Berry, R.E. *Phytochemistry*, **1978**, *17*, 447-449.
62. Boehm, H.; Voelcker, P.E. *Arch. Pharm.* **1959**, *292*, 529-536.
63. Mizuno, M.; Linuma, M.; Ohara, M.; Tanaka, T.; Iwamasa, M. *Chem. Pharm. Pharm. Bull.* **1991**, *39*, 945-949.
64. Sugiyama, S.; Umehara, K.; Kuroyanagi, M.; Ueno, A.; Taki, T. *Chem. Pharm. Pharm. Bull.* **1993**, *41*, 714-719.
65. Prendergast, P.T. Patent, WO Patent 2,001,003,681, 2001.

Chapter 14

Isolation and Purification of Polymethoxyflavones as Substrates for Efficacy Studies

Shiming Li[1], Chih-Yu Lo[1], Slavik Dushenkov[2], and Chi-Tang Ho[1]

[1]Department of Food Science, Rutgers University, 65 Dudley Road, New Brunswick, NJ 08901
[2]WellGen Inc., 63 Dudley Road, New Brunswick, NJ 08901

Polymethoxyflavones (PMFs) isolated from members of the citrus genus are of particular interest in that they exhibit a wide spectrum of biological activities. Due to the current interest in the exploration of the beneficial health properties accociated with citrus-derived PMFs, there has been an increased interest in clinical trials amed at determining efficacy parameters.. Unfortunately, the supply of pure PMFs available for these *in vivo* studies is a limiting factor because of the difficulties related to large scale isolation of the required PMFs. As a result of this, an efficient and large scale separation process for the PMFs was needed. In this paper, we discuss the newly developed preparatve methods for efficient and large scale isolation of PMFs from sweet orange (*Citrus sinensis*) peel. These procedures employ chiral liquid chromatography and supercritical chromatography (SFC) technology.

There are numerous research reports concerning polymethoxyflavone's (PMFs) biological activities, including anti-inflammatory, anti-carcinogenic, and anti-atherogenic properties (*1-6*). The majority of the bioactivity studies of PMFs were performed *in vitro*, and these procedures require only small amounts (milligram scale) of materials as investigational product. Animal and Clinical trials such as pharmacokinetics, safety, metabolism, and efficacy studies utilizing specific PMFs have rarely been done, primarily because of the difficulties associated with isolation of suitable quantities of these unique PMF materials.

Although some PMFs are commercially available, the cost is often too high to permit these *in vivo* studies to be performed.. For example, 3,5,6,7,8,3′,4′-heptamethoxyflavone has been reported to exhibit *in vitro* potent anti-tumor activity and also to be a chemopreventive agent against nitric oxide carcinogenesis (*7,8*). However, more in-depth *in vitro* investigations and efficacy studies of 3,5,6,7,8,3′,4′-heptamethoxyflavone have not been initiated, because of its limited availability and high cost ($300/mg). An efficacy study in animals of lower species (mice, rats *etc.*) could easily consume grams of 3,5,6,7,8,3′,4′-heptamethoxyflavone at a price of $300,000/g. It could theoretically cost billions of dollars for clinical trial studies using this compound, a cost that puts this material beyond the pratical range for commercial applications. To make these biologically interesting PMFs available for the pharamaceutical and nutrition industries, a lower cost procedure was required, which will be described in the subsequent text

Analysis and Identification Method

Earlier studies related to the properties of PMFs have focused on the analysis and identification of PMFs in a variety of plant families, especially those from the citrus genus and a Variety of analytical methods have been applied to characterize different PMFs in citrus plants.

For example, using Gas Chromatography (GC) with a coated capillary column, Gaydou, et al. separated and identified 27 PMFs from three industrially-derived types of orange peel oils (*9*). In this study, they also found that sinesetin, nobiletin, and 3,5,6,7,8,3′,4′-heptamethoxyflavone are characteristic of orange peel oils, and are among the most abundant PMFs. The other three PMFs identified, tangeretin, tetra-*O*-methylscutellarein, and 3,5,6,7,3′,4′-hexamethoxyflavone, were found to be less abundant in the three samples they tested. Gas Chromatography/Mass Spectrometry (GC-MS) is another technique that was first used to charcaterize the PMFs extant in sweet orange, tangeretin and grapefruit (*10*). There were six predominant PMFs in cold pressed orange peel oils separated during this study and hydroxylated PMFs were identified in tangeretin oil. Monohydroxylated pentamethoxyflavone (M^+ = 388 amu), previously identified in tangeretin was also present in both orange and grapefruit oils, but only at lower concentrations. Due to its ease of use, robustness, and ability to process both volatile and non-volatile compounds, high performance liquid chromatography (HPLC) has become popular for citrus-derived PMF analysis and identification. The high resolution and accurate quantification of citrus PMFs and other related natural products by HPLC has been reviewed (*11*). HPLC has also been used for simultaneous separations of PMFs and flavone glycosides (*12*) as well as the simultaneous determination of four PMFs (nobiletin, tangeretin, 3,5,6,7,8,3′,4′-heptamethoxyflavone, and 5-demethyl-nobiletin) plus four other compounds (glycyrrhizin, hesperidin, honokiol and magnolol) from a multi-component traditional Chinese medicine (*13*). The detection of flavonoid compositions in 42 species and cultivars of the citrus genus by HPLC has also been described (*14*). It is reported that reverse phase HPLC coupled with mass spectrometry (MS) yielded enhanced identifications of PMFs (tangeretin and nobiletin) and other flavonoids (naringin, isonaringin, hesperidin, neohesperidin, hesperitin and naringenin) (*15*) and also enabled the qualitative and quantitative determination of the flavonoid content of extracts of Greek navel sweet orange peel (*16*). Additionally, the application of HPLC-MS/NMR (Nuclear Magnetic Resonance) was used as a complementary analytical tools for the reliable identification of PMFs in residues from molecular distillation of cold preesed peel oils of *Citrus sinensis* (*17*).

Supercritical Fluid Chromatography (SFC) was also applied in the analysis of PMFs. It has been reported that the SFC method is both rapid and quantitative (*18,19*). Milligram quantities of six PMFs were isolated from sweet orange peel using a SFC procedure (*19*). However, both HPLC and SFC are purely analytical methods used solely for the purpose of individual PMF identification.

Isolation and Preparation Procedures for PMFs

Purification and isolation procedures for PMFs were of little interest until recently when a separation method of PMFs based on high-speed counter-current chromatography was reported (*20*). Although this method was able to isolate some PMFs in multi-milligram quantities, it was laborious and time-consuming, which in turn limited its scalability and application for larger separations. As a result, the PMF supply for potential in vivo and clinical trial study remained very limited.

The first reported large scale separation of PMFs was a nobiletin isolation (*21*) procedure based on preparative, chiral HPLC. The cold pressed orange peel is first passed through a silica-gel column to remove orange peel oils and to divide the PMFs into six groups. Group V, mainly containing nobiletin and 5,6,7,4′-tetramethoxyflavone, is then loaded onto a Welk-O chiral column (Regis Welk-O 1 R,R 450 gram column). Using a mixture ethanol and hexanes as eluting solvents, more than 2 grams of nobiletin can be obtained from one preparation within 45 minutes. The application of this chiral preparative HPLC method not only opened a new era of nobiletin research for *in vitro* and *in vivo* studies because of the availability of PMFs in large amount. Secondarily, it demonstrated the general application of the utility of chiral preparative HPLC columns for PMF isolation. This is the first successful application of chiral chromatography in separating PMFs, and it demonstrated the technique's ablity to resolve these closely related PMFs which exhibit a high degree of chemical structural and physical property similarity

More general separation methods, useful for isolation of large quantities of PMFs, were recently developed using supercritical fluid chromatography (SFC) (*22*). Following a number of attempts to purify PMFs from sweet orange (*citrus sinenesis*) we developed an efficient and scalable SFC method for the large scale separation of four common PMFs. This process has the potential application to become the predominant technology for large scale PMF isolation. This SFC technology has a number of advantages over the other separation techniques in cost effectiveness, time efficiency, ease of automation. This is the first reported SFC application for the preparative separation of PMFs. It is also of significance because it has not only provided an efficient and large scale preparation of PMFs, but has also explored a new application of the SFC technology in the field of PMF research. The four PMFs isolated in this study were nobiletin, tangeretin, 3,5,6,7,8,3′,4′-heptamethoxyflavone, and 5,6,7,4′-tetramethoxyflavone.

Summary

Analytical techniques for PMF separation and identification have been successfully developed and employed by various research groups interested in PMFs. These techniques were not able to provide quantities suitable to provide sufficient investigational product for preclinical and clinical studies related to efficacy and bioavailabilty of PMFs. However, to meet the demand of gram or kilogram required, efficient and scalable separation methods of PMFs have been developed employing chiral preparative HPLC chromatography and preparative SFC separation techniques.

References

1. Manthey, J.A.; Grohmann, K.; Guthrie, N. *Curr. Medi. Chem.* **2001**, *81*, 35.

2. Middleton, E.; Kandaswami, C.; Theoharides, T.C. *Pharmacol. Rev.* **2000,** *52*, 673.
3. Lopez-Lazaro, M. *Curr. Med. Chem. – Anti-Cancer Agents*. **2002,** *2*, 691.
4. Whitman, S.C.; Kurowska, E.M.; Manthey, J.A.; Daugherty, A. *Atherosclerosis* **2005**, *178*, 25.
5. Yoshimizu, N.; Otani, Y.; Saikawa, Y.; Kubota, T.; Yoshida, M.; Furukawa, T.; Kumwi, K.; Kameyama, K.; Fujii, M.; Yano, M.; Sato, T.; Ito, A.; Kitajima, M. *Aliment Pharmacology Therapy* **2004,** *20 (Suppl. 1)*, 95.
6. Lin, N.; Sato, T.; Takayama, Y.; Mimaki, Y.; Sashida, Y.; Yano, M.; Ito, A. *Biochem. Pharmacol.* **2003,** *65*, 2065.
7. Iwase, Y.; Takemura, Y.; Ju-ichi, M.; Yano, M.; Ito, C.; Furukawa, H.; Mukainaka, T.; Kuchide, M.; Tokuda, H.; Nishino, H. *Cancer Lett.* **2001,** *163*, 7.
8. Iwase, Y.; Takemura, Y.; Ju-ichi, M.; Ito, C.; Furukawa, H.; Kawaii, S.; Yano, M.; Mou, X.Y.; Takayasu, J.; Tokuda, H.; Nishino, H. *Cancer Lett.* **2000,** *154*, 101.
9. Gaydou, E.M.; Berahia, T.; Wallet, J-C.; Bianchini, J-P. *J. Chromatogr.* **1991,** *549*, 40.
10. Stremple, P. *J. High Resol. Chromatogr.* **1998,** *21*, 587.
11. Heimhuber, B.; Galensa, R.; Herrmann, K. *J. Chromatogr.* **1998**, *439*, 481.
12. Mouly, P.; Gaydou, E.M.; Auffray, A. *J. Chromatogr. A* **1998,** *800*, 171.
13. Lay, H-L.; Chen, C-C. *J. Liq. Chromatogr. & Rel. Technol.* **2000,** *23*, 1439.
14. Nogata, Y.; Sakamoto, K.; Shiratsuchi, H.; Ishii, T.; Yano, M.; Ohta, H. *Biosci. Biotechnol. Biochem.* **2006,** *70*, 178.
15. He, X.; Lian, L.; Lin, L.; Bernart, M.W. *J. Chromatogr. A* **1997,** *791*, 127.
16. Anagnostopoulou, M.A.; Kefalas, P.; Kokkalou, E.; Assimopoulou, A.N.; Papageorgiou, V.P. *Biomed. Chromatogr.* **2005,** *19*, 138.
17. Weber, B.; Hartmann, B.; Stockigt, D.; Schreiber, K.; Roloff, M.; Bertram, H-J.; Schmidt, C.O. *J. Agric. Food Chem.* **2006,** *54*, 274.
18. Morin, P.; Gallois, A.; Richard, H.; Gaydou, E. *J. Chromatogr.* **1991,** *586*, 171.
19. Dugo, P.; Mondello, L.; Dugo, G.; Heaton, D.M.; Bartle, K.D.; Clifford, A.A.; Myers, P. *J. Agric. Food Chem.* **1996,** *44*, 3900.
20. Wang, X.; Li, F.; Zhang, H.; Geng, Y.; Yuan, J.; Jiang, T. *J. Chromatogr. A* **2005,** *1090*, 188.
21. Li, S.; Yu, H.; Ho, C.-T. *Biomed. Chromatogr.* **2006,** *20*, 133.
22. Li, S.; Lambros, T.; Wang, Z.; Goodnow, R.; Ho, C.-T. *Journal Chromatogr. B,* **2006,** doi:10.1016/j/jchromb.2006.09.010.

Chapter 15

Polymethoxyflavones: Metabolite Identification and Pathway

Shiming Li[1], Di Tan[1], Slavik Dushenkov[2], and Chi-Tang Ho[1]

[1]Department of Food Science, Rutgers University, 65 Dudley Road, New Brunswick, NJ 08901
[2]WellGen Inc., 63 Dudley Road, New Brunswick, NJ 08901

Metabolic studies of polymethoxyflavones (PMFs) have become increasingly interesting to the scientific community. This is especially so in the areas of identification of *in vivo* biotransformation products because the elucidation and biological activity studies of PMFs metabolites are leading to further exploration of PMF's wide-ranging bioactivity. The metabolic pathway of tangeretin has been given a detailed examination, whereas the metabolic fate of other PMFs needs to more fully explored. Our *in vivo* transformation study of nobiletin in mice resulted in the finding of demethylated nobiletin metabolites isolated from urine and plasma. By comparing the Liquid Chromatography/Mass Spectrometry (LC/MS) results and Supercritical Fluid Chromatography (SFC) profiles of the synthesized hydroxylated PMFs with that of the metabolite mixture, three demethylated metabolites were identified in mouse plasma and urine as 3′-demethylnobiletin, 4′-demethylnobiletin and 3′,4′-dihydroxy-5,6,7,8-tetra-methoxyflavone.

It is well known that exogenous substances can be metabolized by xenobiotic enzymes, which are compromised of phase I and phase II enzymes (*1*). Phase I enzymes, such as cytochrome P450s, attach hydrophilic functional groups to exogenous substances. The phase II enzymes, including glutathione transferase, provides a glutathione to compounds that are metabolized by phase I enzymes.

Substances such as most of the polycyclic aromatic hydrocarbons, aniline derivatives, and other pro-carcinogens are biologically inactive in their original form, but are readily activated by phase I enzymes and generate toxic carcinogens which can bind to cellular DNA and cause damage to DNA, RNA, lipids and proteins (*2*). However, the activated carcinogens can be inactivated by phase II enzymes and excreted from the body. One of the plasma membrane proteins, P-Glycoprotein, can confer multi-drug resistance on cells by virtue of its ability to exclude cytotoxic drugs in an ATP-dependent manner (*1*).

Nutrients and drugs must be in solution to be absorbed from the gastrointestinal tract. During the course of absorption, the exogenous substances are metabolized in the intestine, and later in the liver. Metabolism is the natural response from the host body to defend itself from foreign molecules. Generally, it is a detoxification process performed by the phase I and II enzyme systems in which the xenobiotic substances are transformed into more hydrophilic molecules which are easily excreted from the body. Phase I enzymes are involved in oxidation and reduction reactions as well as ester, amide and ether linkage hydrolysis , and as a result, convert these moieties to more hydrophilic groups such as -OH, $-NH_2$, -SH, and $-CO_2H$ *etc.* Phase II enzymes, also called transferases or conjugation enzymes, conjugate the hydrophilic groups (-OH, $-NH_2$, -SH, and $-CO_2H$ *etc.*) to even more hydrophilic groups such as glucuronic acid or sulphates. Consequently, the polarity and hydrophilicity of the bio-transformed molecules are significantly increased and the tendency for these molecules to be easily absorbed and excreted is very greatly increased (*3,4*). Figure 1 illustrates the metabolic pathway of lipophilic compounds in the human body. Following ingestion, lipophilic substances bind to blood cells, albumen, lipoproteins, lymph cells, and lymph proteins, *etc.* Then they undergo the phase I and phase II transformations, generating water soluble hydrophilic metabolites, which now circulate freely in the blood system. Some fractions of these metabolites are eventually excreted either in urine via the kidney or by the bile system through the gut to become fecealy incorporated. Smaller molecules tend to be excreted with the urine, whereas larger, conjugated metabolites are eliminated in from bile and the intestine by feces. Other metabolic fractions circulates in the blood and are distributed to body tissue and organs. These are the active sites for these modified biological agents. Therefore, an understanding of the blood and plasma metabolic transformations is of key importance in developing the relationship between the metabolite profile and the bioactivity of these transformed nutraceutical and pharmaceuticals species It is noted that

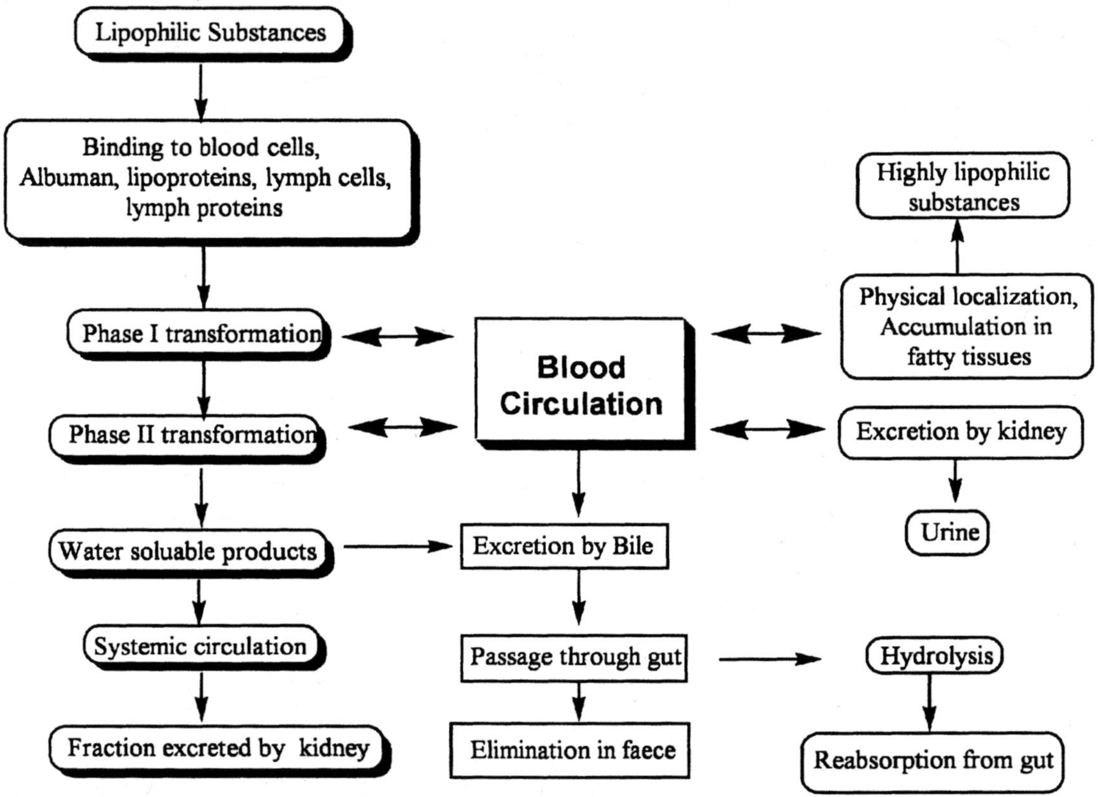

Figure 1. Metabolic pathway of lipophilic substances.

cytochrome P450 family of enzymes play a crucial role in activating and detoxifying xenobiotics in the process of phase I and phase II transformations. Various *in vitro* models consisting of different P450 enzymes have been developed to facilitate the understanding of *in vivo* metabolic mechanisms (*3*).

Metabolism Study of Flavonoids

Flavonoids are a naturally occuring class of polyphenolic compounds. The metabolism studies of polyphenolic compounds, such as the *in vitro* biotransformation of EGCG, curcumin, and chlorogenic acid, have been reviewed previously (*5*).

The antioxidant and biological activity of flavonoids are closely associated with the number and position of hydroxyl groups on the 3-ring skeleton structure of flavonoids. Biological activities of nutrients and pharmaceuticals generally depend on their bioavailability and sometimes more so on the bioactivity of the metabolites. For example, some diets may be rich in bioactive flavonoids but these flavonoids do not necessarily demonstrate the desired biological activity due to insufficient absorption of the flavonoids from the small intestine, extensive metabolization before reaching the site of action, or rapid elimination by kidney or bile. The biological activities and health properties of flavonoids like quercetin, apigenin, narigenin, and genistein have been extensively studied and there are numerous reports concerning their metabolism and pharmacokinetics (*6,7*). Metabolite identification, bioactivity, and metabolic studys of a bioactive molecule elucidate these metabolic pathways and also can identify which molecules are responsible for the biologically activity. It is essential to thoroughly understand the metabolism of a nutrient or a drug so that maximum health benefits can be developed.

A recent investigation of the pharmacokinetics and metabolism of apigenin (3′-hydroxy-5,7-dihydroxyflavone), a weak estrogenic flavonoid phytochemical present in aromatic plants (camomile, rosemary and parsley), celery, apple, honey, fennel and wheat germ *etc.*, identified the major metabolites in rat urine as glucuronated and sulfated derivatives of apigenin. This study also suggested slow metabolism of apigenin, related to its slow absorption and clearance rates. The pharmacokinetic data showed that apigenin excretion occurred mainly via the urinary pathway (*6*). These findings provide evidence for the hypothesis that there is an accumulation of apigenin in the body.

Metabolic Studies of PMFs

Hydroxylation or methoxylation of the benzo-pyrone skeleton imparts very different chemical and physical properties to the flavone structure. As a result, it is evident that the properties of PMFs are separate and distinct from those of common flavonoids with respective absorption and metabolism *in vitro* (*8*). There are many research reports regarding the important biological activities of PMFs, especially their *in vitro* bioactivity, which has been reviewed in the previous chapter, but there has been little investigation of their metabolic fate. More research in the field of PMF metabolism must be conducted if the knowledge of the PMF metabolic pathway is to be extended. It has been found that PMFs undergo *in vivo* biotransformation and produce metabolites with different bioactivities and pharmacological properties (*9*). As a result, the metabolite identification and metabolic fate of PMFs is of special interest.

As early as 1980, it was observed that the addition of tangeretin or nobiletin to human liver microsomes activated both the hydroxylation of the mutagens benzo[α]pyrene and the metabolism of aflatoxin B1 (*10*). Later, it was demonstrated that a nobiletin and tangeretin mixture induced liver mixed function oxidase (MFO) systems in rainbow trout in a dose-dependent manner (*11*). Yang *et al.* first reported the effects of tangeretin on cytochrome P450 activity in 1995, and they found that 7-ethoxyresorufin-*O*-deethylase (classified as CYP 1A) and nifedipine oxidase (CYP 3A4) in human liver microsomes were inhibited by tangeretin in a noncompetitive manner (*12*). However, a contradictory result from the immunoblot assay of rat liver samples showed that the flavone-like tangeretin increased the activity of the cytochrome P450 enzymes, ethoxyresorufin-*O*-deethylase, methoxyresorufin-*O*-demethylase, and pentoxy-resorufin-*O*-dealkylase. (*13*). Acetaminophen oxidation, catalyzed by rat liver P450 3A4, was stimulated by tangeretin, nobiletin, and other PMFs, but inhibited by 40-60% by myricetin and quercetin (*14*). These observations suggest an interaction between PMFs and liver cytochrome P-450 isozymes.

Cytochrome P450 (CYP) is the key enzyme system involved in the biotransformation of flavonoids through the catalysis of hydroxylation and demethylation reactions. This metabolic pathway of flavonoids is considered to be identical across species. The 3′ and 4′ positions on the B-ring of the flavonoid structure are the primary sites for biotransformation. The number and position of the hydroxyl and methoxy groups on the B-ring greatly influences the course of this metabolism (*15,16*). It has been observed that CYP 1A2 plays the predominant role in the hepatic metabolism of genistein and tangeretin among the different forms of CYP P450 enzymes, suggesting that inter-individual differences in the metabolism of PMFs may result from the differential activities of this sub-family of enzymes (*15*). Studies have shown that tangeretin was demethylated by interacting with human and rat liver microsomes of P450 system (*13,17*). It was also reported that CYP 1A2 is the major enzyme involved in the metabolism of tangeretin. Other isoforms such as CYP 3A4, CYP 2D6, and CYP2C9 may also play a minor role (*15*).

Metabolic Studies of Tangeretin

Nielson and coworkers carried out both *in vitro* and *in vivo* experiments to study the metabolic fate of tangeretin by analyzing various samples from corresponding experiments. In the *in vitro* experiment, tangeretin was incubated with Aroclor induced rat liver microsomes and three major metabolites of tangeretin were identified as 4'-demethyltangeretin, 3′,4′-dihydroxy-5,6,7,8-tetramethoxyflavone, and 5,4′-dihydroxy-6,7,8-trimethoxy-flavone (*16*). In their *in vivo* biotransformation study of tangeretin, the rats were repeatedly gavage fed. The samples of rat urine and feces were then collected and analyzed by HPLC-MS and proton NMR techniques (*18*). Both the major and minor metabolites were isolated and characterized. Ten compounds were characterized as the metabolites of tangeretin. The dominant metabolite in rat urine was 4'-demethyltangeretin and the other major metabolite 3′,4′-dihydroxyflavone. Other metabolites with intact 4′-methoxy groups and demethylated at various positions on A ring were also identified (*15,18*). As a result, the conclusion can be drawn that the 4′-methoxy group of tangeretin is the primary site for demethylation, while the 3′-position is the most vulnerable site for hydroxylation or oxidation by phase I enzymes. Additionally, urine analysis determined that 38% of the tangeretin metabolites were excreted as conjugates of glucuronates and sulfates (*15,18*).

Metabolic Studies of Nobiletin

Interest in nobiletin metabolism studies started several years ago. Ohigashi and his coworkers examined the *in vitro* biotransformation of nobiletin by treating nobiletin with rat liver S-9 mixture for 24 h. From HPLC analysis and a proton NMR – NOE study, they identified the major metabolite as 3′-demethylnobiletin (Figure 2). They also found that the demethylation rate of nobiletin was slow, with a half life greater than 24 h, much lower than that of ethoxycoumarin ($t_{1/2}$ = 2 h; reference *19*), suggesting that the half life of nobiletin in an *in vivo* system might be considerably longer. In contrast to the demethylation of PMFs, the 3′ hydroxy group on the B-ring of polyhydroxylated flavonoids (PHFs), such as quercetin, was found to be methylated in rat plasma, indicating that the active site of methylation and demethylation is at the 3'-position of certain PHFs and PMFs, respectively, and is one of the key steps in the metabolic transformations of PMFs and PHFs bearing methoxy or hydroxy groups on the B-ring (*8*). Ohigashi and his associates continued their experiments with male SD rats and identified the dominant metabolite as 3′-demethylnobiletin, plus two other mono-demethylated nobiletin, and two di-demethylnobiletin products (*20*). These results were consistent with their previous *in vitro* experimental results (Figure 2). Further, in their *in vivo* experiment, they detected the only metabolite of nobiletin, 3′-demethylnobiletin, in serum (*20*).

In contrast, a recent study of nobiletin biotransformation within the same species - the male SD rat fed nobiletin orally, Ohsawa's research group isolated and identified one major and two minor metabolites by three-dimensional HPLC equipped with a photodiode array detector from the rat urine. The structure of the major metabolite was characterized as 4′-demethylnobiletin by spectroscopic methods (*21*), not 3'-demethylnobiletin (Figure 2). One of the minor metabolites was also mono-demethylnobiletin, but not 5-demethylnobiletin, as determined by proton NMR. The other minor metabolite was di-demethylnobiletin, characterized by ESI-MS. This result is in conflict with the conclusion obtained from Ohigashi's group who identified the major metabolite of nobiletin as 3'-demethylnobiletin from the same genus of animal.

The biotransformation of nobiletin by the fungi *Aspergillus niger* and the antimutagenic activity of a metabolite were investigated (*22*). Following a 3-day incubation of nobiletin in an *A. niger* culture medium, only one metabolite was isolated and this was identified as 4′-demethylated nobiletin. Okino and Miyazawa (*22*) also investigated the antimutagenic activity of the metabolite, 4'-demethylnobiletin and found that it had suppressive effects on the gene expression for the SOS response to DNA damage in Salmonella typhimurium TA1535/pSK1002 induced by the chemical mutagens furylfuramide, MeIQ, and Trp-P-1. This was the first report of a study of the biological and pharmacological activity of 4'-demethylnobiletin. Earlier, Miyazawa *et al.* also reported that nobiletin exhibits inhibition activity against the SOS response and antimutagenic activity against chemical mutagens (*23*). In their comparison study, nobiletin and 4′-demethylnobiletin showed similar suppressive effects against furylfuramide, UV, Trp-P-1 and MeIQ, but 4′-demethylnobiletin exhibited greater suppressive effects than nobiletin against activated Trp-P-1 and MeIQ (*22*).

Recent Metabolic Study of Nobiletin

During the course of our metabolism studies on nobiletin, we conducted *in vivo* experiments with nobiletin on CD-1 mice (*24,25*). Since the quantity of metabolite samples available from mouse urine and plasma was limited, four potential metabolites of nobiletin were synthesized as standard compounds and used to compare metabolite mixtures during chromatography analysis. Various techniques were employed to separate and characterize the metabolites: Supercritical fluid chromatography (SFC), equipped with a chiral column, gave

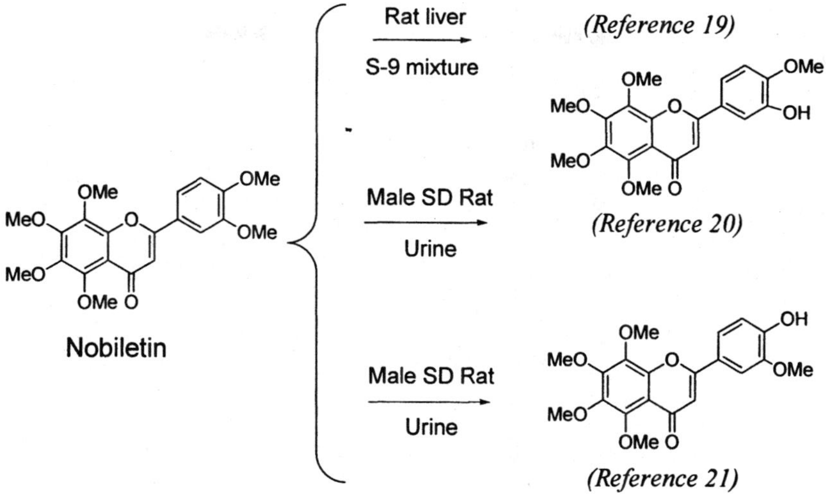

Figure 2. Major metabolites of nobiletin.

the best results in separating the nobiletin and its metabolites from mouse urine and plasma. The major metabolites from both mouse urine and mouse plasma were identified by comparing the SFC profiles of the metabolite mixture from mouse urine and plasma with those of standard compounds synthesized for this study (*24,25*).

Syntheses of Potential Nobiletin Metabolites

Based on the previous literature reports from Nielsen, Ohigashi, and Ohsawa, four potential metabolites were synthesized: 3′-demethylnobiletin, 4′-demethylnobiletin, 3′,4′-dihydroxy-5,6,7,8-tetramethoxyflavone, and 5-demethylnobiletin. The detailed synthetic procedures for 3′-demethylnobiletin and 4′-demethylnobiletin followed literature procedures with some modifications and they were published recently (*25*). The preparation of 3′,4′-dihydroxy-5,6,7,8-tetramethoxyflavone followed the synthetic route of 3′-demethyl-nobiletin, with the exception of the minor change of starting material to 3,4-dibenzyloxybenzaldehyde using 3,4-dibenzoylbenzaldehyde and 2′-hydroxy-3′,4′,5′,6′-tetramethoxyacetophenone. 5-Demethylnobiletin was obtained by a one-step selective demethylation reaction of nobiletin with a Lewis acid like BF_3. During the course of continued PMF isolation and characterization, 5-demethylnobiletin was also obtained from orange peel extracts (*24,25*).

Physical Properties of Synthesized Compounds

During analytical method development for the *(analytical or preparative?)* separation of potential metabolites of nobiletin, both normal phase silica gel HPLC and C18 reverse phase HPLC were employed. Distinct separation techniques were required for the homologues nobiletin, 3′,4′-dihydroxy-5,6,7,8-tetramethoxyflavone, 5-demethylnobiletin. 3′-demethylnobiletin or 4′-demethylnobiletin. The analytical HPLC profiles of nobiletin, 3′- and 4′-demethylnobiletin are illustrated in Figure 3. Under these analytical conditions, 3′- and 4′-demethylnobiletin were unseparable because they had nearly the same retention times on LC (1.25 min and 1.26 min). LC/MS/MS analysis of 3'- and 4'-demethylnobiletin showed molecular weights of 388 amu for both compounds and their fragmentation pattern from MS/MS was also identical (Figure 4). Therefore, separation of 3′-demethylnobiletin and 4′-demethylnobiletin could not be resolved using these chromatographic techniques.

Separation by Supercritical Fluid Chromatography (SFC)

The technique of SFC was employed to further attempt an analytical separation of 3′-demethylnobiletin and 4′-demethylnobiletin. The result obtained by the SFC experiment showed that there was a significant difference between the SFC chromatograms of 3′-demethylnobiletin and 4′-demethylnobiletin. During a 20 min elution (Figure 5), the separation of these two compounds was greater than 10 min, indicating that their behavior on SFC was significantly different from that demonstrated by C_{18} reverse phase analytical HPLC. The SFC retention time of 4′-demethylnobiletin (peak 1 in Figure 3) is 7.1 min whereas that of 3′-demethylnobiletin (peak 2 in Figure 1 is 17.9 min, under the same experimental conditions. Nobiletin and other potential metabolites were also screened on the same SFC conditions and resulted in good separations. The SFC experiment demonstrated that it is superior to other methods, and therefore it is feasible to use SFC technique in the identification of nobiletin metabolites and other difficult to separate compounds in the reverse phase HPLC.

Identification of Nobiletin Metabolites in Mouse Urine

The SFC profile of nobiletin metabolites in mouse urine was obtained. The major absorption peak in the mouse urine metabolites was found at 7.1 min (Figure 8), corresponding to 4′-demethylnobiletin (by comparison with of the SFC chromatogram of the urine sample to that of the 4′-demethylnobiletin standard (Figure 7). A minor absorption peak was eluted at 17.9 min (Figure 8), corresponding to 3′-demethylnobiletin (Figure 7). Thus, our SFC study of mouse urine sample concluded that the major metabolite is 4′-demethylnobiletin, and that 3′-demethylnobiletin was found to be only as a minor metabolite. Furthermore, efforts to quantify the urine levels of metabolites have been made and the concentration of the major metabolite, 4′-demethylnobiletin, in mouse urine was found to be 28.9 μg/mL under the conditions of this study.

Identification of Nobiletin Metabolites in Mouse Plasma

The SFC profile of nobiletin metabolites in mouse plasma was also obtained and is illustrated in Figure 7. By comparing the SFC profile of mouse plasma to that of nobiletin and previously synthesized standard compounds , two of the major metabolites of nobiletin were identified as 3′,4′-dihydroxy-5,6,7,8-tetramethoxyflavone and 4′-demethylnobiletin. 3′-Demethylnobiletin was seen

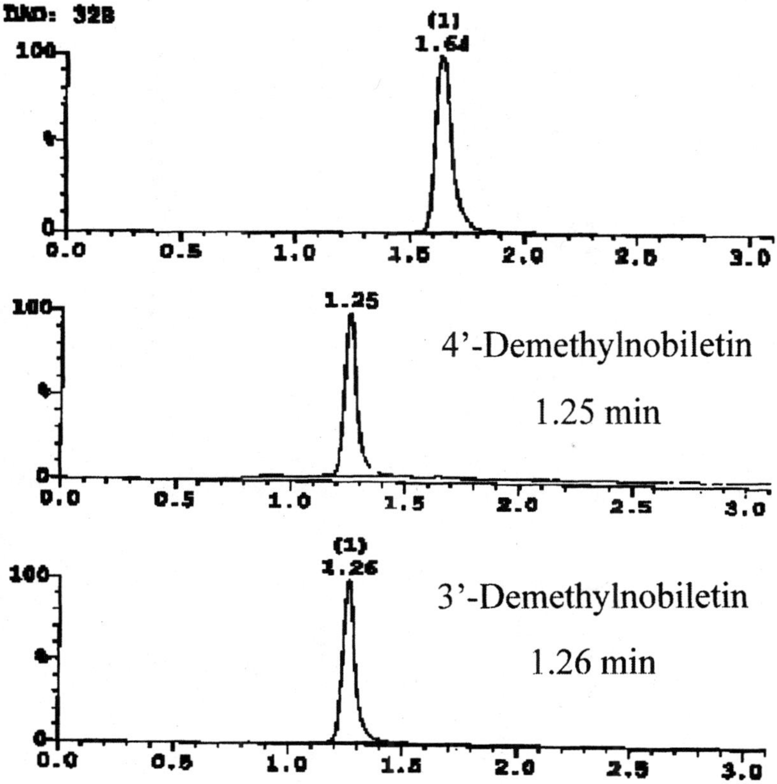

Figure 3. HPLC profile of nobiletin, 4′- and 3′-demethylnobiletin

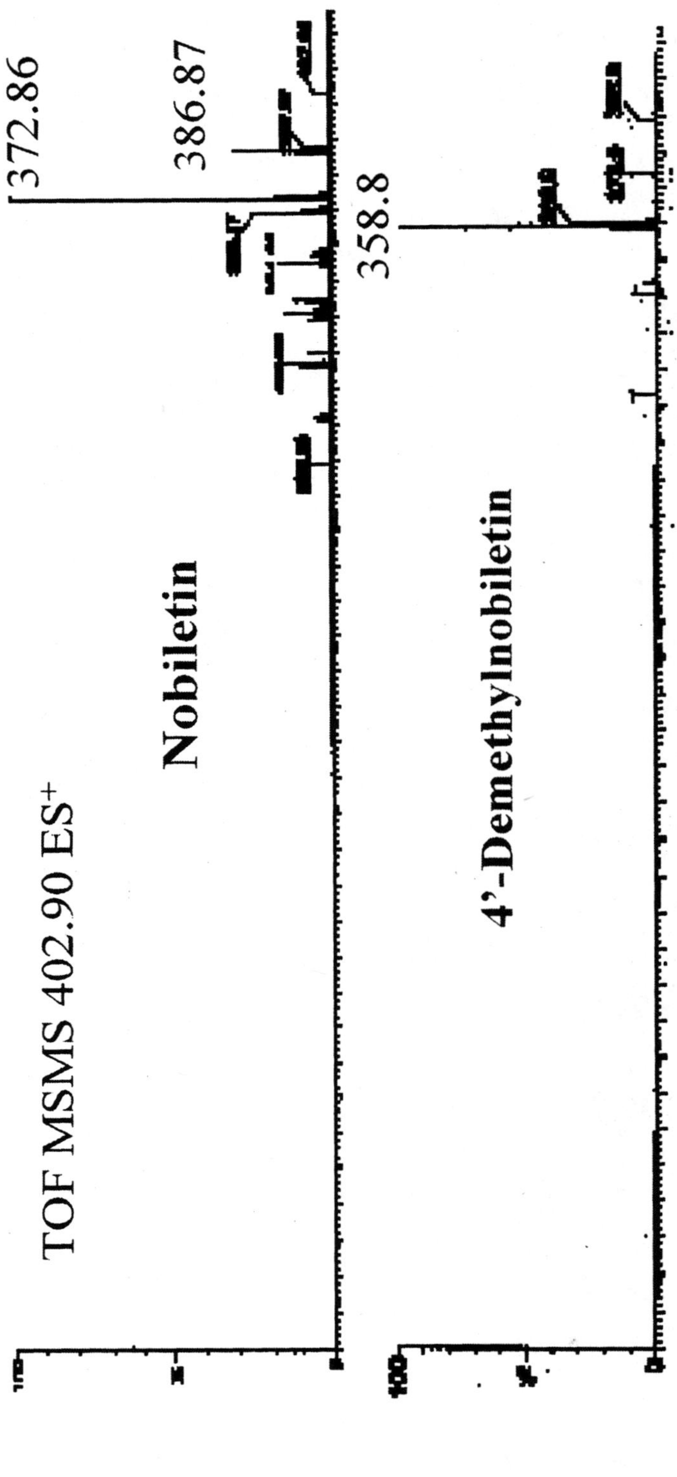
TOF MSMS 402.90 ES+
Nobiletin
372.86
386.87
4'-Demethylnobiletin
358.8

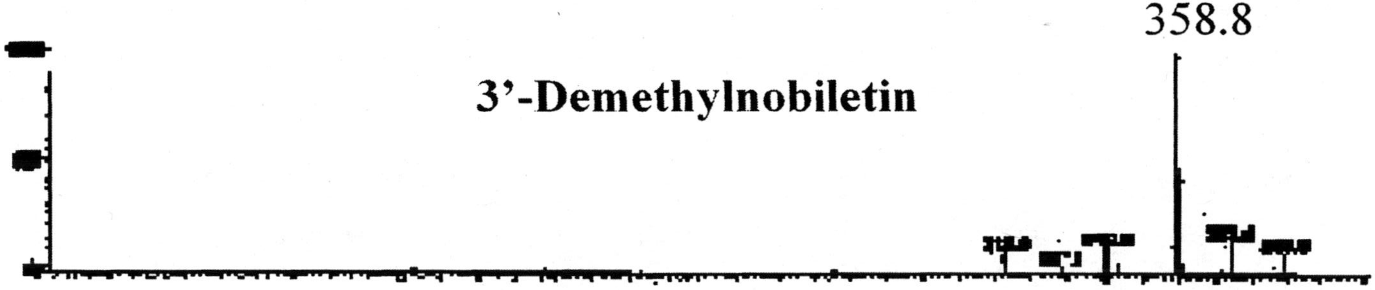

Figure 4. LC/MS/MS of nobiletin, 3'- and 4'-demethylnobiletin

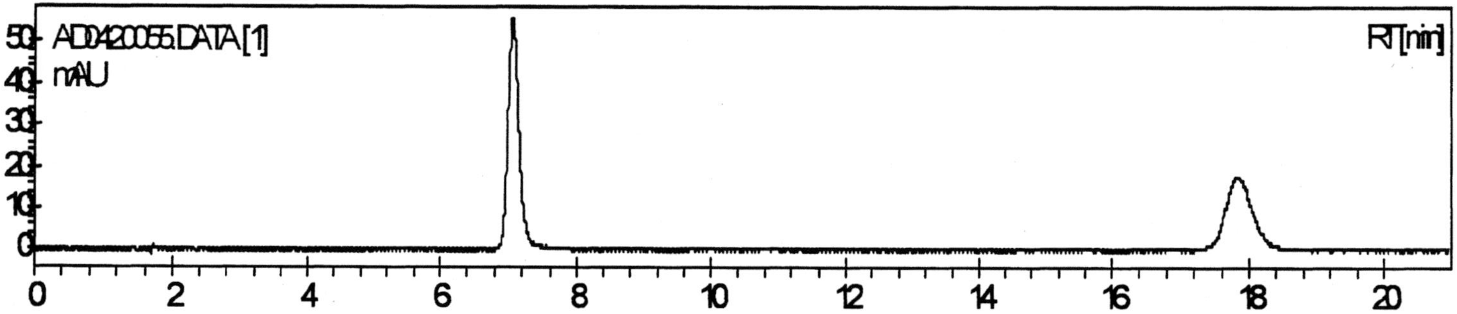

Figure 5. SFC separation profile of 3'- and 4'-demethylnobiletin

(Early eluting peak at 7.1 min is 4'-demethylnobiletin and late eluting peak at 17.9 min is 3'-demethylnobiletin; Chromatographic conditions: Chiralpak AD column, CO_2 pressure 100 bar, 20% methanol as modifier, temperature of 30 °C, flow rate 2.0 mL/min, UV wavelength 220 nm)

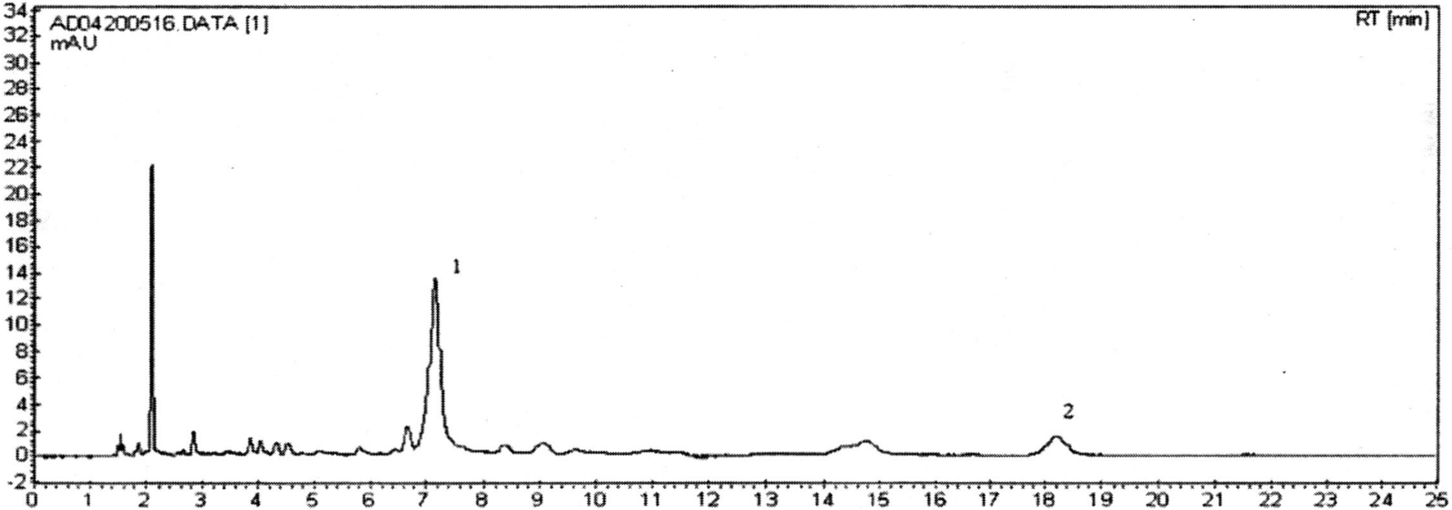

Figure 6. SFC profile of nobiletin metabolites in mouse urine (Peak 1 is 4'-demethylnobiletin and Peak 2 is 3'-demethylnobiletin)

as a minor metabolite. 5-Demethylnobiletin was not found in nobiletin plasma metabolites. Also, there was a recognizable amount of free nobiletin existing in mouse plasma. The identification of other metabolites in mouse plasma and the concentration measurements of the metabolites are in progress.

Conclusion

The metabolic study of PMFs indicates that the major metabolites are demethylated PMFs like mono- and di-demethylation products, and hydroxylation (oxidation) products detected in the tangeretin biotransformation pathway. The demethylation occurs mainly at the B-ring of C6-C3-C6 flavone skeleton with the formation of 4′-, 3′- and 3′,4′-demethylated PMFs. Therefore, a conclusion can be drawn from these preliminary results that the metabolic pathway of PMFs is mainly one of B-ring demethylation. Further mechanistic investigations into PMF biotransformations, especially in a variety of animal species and organs are required to have a more complete understanding of the whole PMF metabolic pathway. The biological activity of the identified metabolites is an exciting area which is largely uninvestigated.

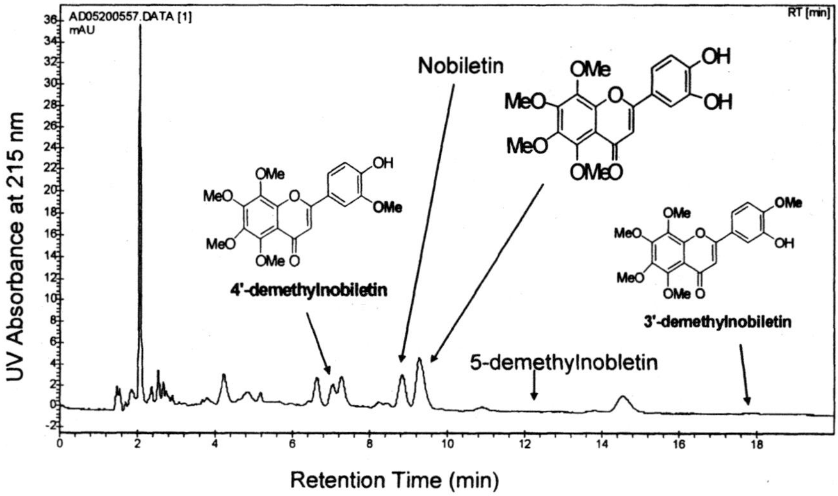

Figure 7. SFC profile of mouse plasma metabolites of nobiletin

References

1. Murakami, A.; Ohigashi, H. *Oxidative Stress and Disease. Phytochemicals in Health and Disease.* **2004**, *12*, 187-211.

2. Ohshima, H.; Tatemichi, M.; Sawa, T. *Arch. Biochem. Biophys.* **2003**, *417*, 3-11.
3. Yan, Z.; Caldwell, G.W. *Curr. Top Med. Chem.* **2001**, *1*, 403-425.
4. Iersel, M.L.P.S.; Verhagen, H.; Bladeren, P. J. *Mutation Res.* **1999**, *443*, 259-270.
5. Li, S.; Ho, C-T. In *Herbs: Challenges in Chemistry and Biology*; Wang, M.; Sang, S.; Hwang, L. S.; Ho, C-T., Ed.; ACS Symposium Series 925; American Chemical Society: Washington, DC, 2006; pp240-253.
6. Gradolatto, A.; Basly, J.P.; Berges, R.; Teyssier, C.; Chagnon, M.C.; Siess, M.H.; Canivenc-Lavier, M.C. *Drug Metabolism and Disposition.* **2005**, *33*, 49-54.
7. Middleton, E.; Kandaswami, C.; Theoharides, T. C. *Pharmacol. Rev.* **2000**, *52*, 673-751.
8. Murakami, A.; Kuwahara, S.; Takahashi, Y.; Ito, C.; Furukawa, H.; Ju-ichi, M.; Koshimizu, K.; Ohigashi, H. *Biosci. Biotechnol. Biochem.* **2001**, *65* (1), 194-197.
9. Sergeev, I. N.; Li, S.; Ho, C-T.; Dushenkov, S. *Life Sci.* **2006**, *80*, 245-253.
10. Conney, A.H.; Buening, M.K.; Pantuck, E.J.; Pantuck, C.B.; Fortner, J.G.; Anderson, K.E.; Kappas, A. *Ciba Found Symp.* **1980**, pp. 147-167.
11. Nixon, J. E.; Hendricks, J.D.; Pawlowski, N.E.; Pereira, C.B.; Sinnhuber, R.O.; Bailey, G.S. *Carcinogenesis* **1984**, *5*, 615-619.
12. Obermeier, M.T.; White, R.E.; Yang, C.S. *Xenobiotica* **1995**, *25*, 575-584.
13. Canivenc-Lavier, M.C.; Bentejac, M.; Miller, M.L.; Leclerc, J.; Siess, M. H.; Latruffe, N.; Suschetet. M. *Toxicol. Appl. Pharmacol.* **1996**, *136*, 348-353.
14. Li, Y.; Wang, E.; Patten, C.; Chen, L.; Yang, C.S. *Drug Metabolism and Disposition* **1994**, *22*, 566-571.
15. Breinholt, V.M.; Rasmussen, S.E.; Brosen, K.; Friedberg, T.H. *Pharmacol. Toxicol.* **2003**, *93*, 14-22.
16. Nielsen, S.E.; Breinholt, V.; Justesen,U.; Cornett, C.; Dragsted, L.O. *Xenobiotica* **1998**, *28*, 389-401.
17. Vyas, K.P.; Shibata, R.J.; Highet, H.J.; Yeh, P.E; Thomas, D.E.; Ryan, W.L.; Jerina, D.M. *J. Biol. Chem.* **1983**, *258*, 5649-5659.
18. Nielsen, S.E.; Breinholt, V.; Cornett, C.; Dragsted, L.O. *Food Chem. Toxicol.* **2000**, *38*, 739-746.
19. Murakami, A.; Kuwahara, S.; Takahashi, Y.; Ito, C.; Furukawa, H.; Ju-ichi, M.; Koshimizu, K.; Ohigashi, H. *Biosci. Biotechnol. Biochem.* **2001**, *65*, 194-197.
20. Murakami, A.; Koshimizu, K.; Ohigashi, H.; Kuwahara, S.; Kuki, W.; Takahashi, Y.; Hosotani, K.; Kawahara, S.; Matsuoka, Y. *Biofactors* **2002**, *16*, 73-82.
21. Yasuda, T.; Yoshimura, Y.; Yabuki, H.; Nakazawa, T.; Ohsawa, K.; Mimaki, Y.; Sashida, Y. *Chem. Pharm. Bull.* **2003**, *51*, 1426-1428.

22. Okuno, Y.; Miyazawa, M. *J. Nat. Prod.* **2004**, *67*, 1876-1878.
23. Miyazawa, M.; Okuno, Y.; Fukuyama, M.; Nakamura, S.; Kosaka, H. *J. Agric. Food Chem.* **1999**, *47*, 5239-5244.
24. Li, S. Ph.D. thesis, Rutgers University, New Brunswick, NJ, 2005.
25. Li, S.; Wang, Z.; Sang, S.; Huang, M-T.; Ho, C-T. *Mol. Nutr. Food Res.* **2006**, *50*, 291-299.

Chapter 16

Bioavailability of Polymethoxyflavones

Shiming Li[1], Yu Wang[1], Slavik Dushenkov[2], and Chi-Tang Ho[1]

[1]Department of Food Science, Rutgers, The State University of New Jersey, 65 Dudley Road, New Brunswick, NJ 08901–8520
[2]WellGen Inc., 63 Dudley Road, New Brunswick, NJ 08901

One of the most influential factors of bioavailability of a nutrient or a drug is absorption. Both solubility and permeability affect absorption. Sufficient absorption of a drug or nutrient is necessary to elicit a therapeutic effect or for a successful nutritional usage. There are a variety of procedures to measure solubility and permeability. Prior to this report, bioavailability studies have been performed with flavonoids of all subclasses except the flavones and it was found that isoflavones had the best bioavailability relative to other flavonols studied. In the course of this work, we performed *in vitro* bioavailability assays of several polymethoxyflavones (PMFs) by measuring their solubility and permeability and concluded that PMF bioavailability is generally good owing to their lipophilic nature of the multiple methoxy groups on the PMF structure. This report represents the first systematic *in vitro* study of PMF bioavailability.

Background of Bioavailability

Bioavailability of a drug or a nutrient is that amount of the drug or the nutrient which reaches the blood circulatory system and target tissues (*1*). The bioavailability is an overall effect of absorption, distribution, metabolism and excretion (ADME). Hence, its determination is the combined rate of these factors. Absorption describes a drug or a nutrient's ability to pass into the systemic circulation following oral administration. Metabolism is the rate that a drug or a nutrient is eliminated from the systemic circulation, following its initial absorption. Distribution describes how well a drug or a nutrient reaches the target tissues. Excretion is the rate at which a drug or a nutrient is excreted from the systemic circulation (*2*). Figure 1 is a graphic description of the ADME process. Only the "unbound fraction" of a drug or a nutrient remains to reach the tissue and is able to interact with the molecular target (*1,3*).

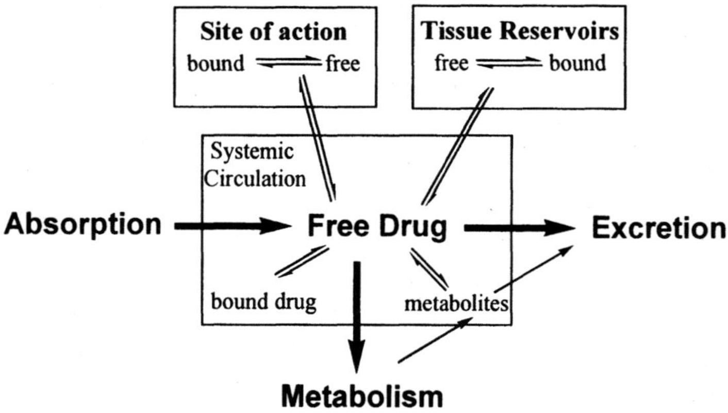

Figure 1. Bioavailability and ADME

One of the most influential factors of bioavailability of a nutrient or a drug is absorption, which can be defined as the product of solubility and permeability. Absorption is usually the first pharmacokinetic component studied (*3*). Solubility refers to aqueous solubility and is must include a specific pH range. A nutrient or a drug substance can only be absorbed from the gastrointestinal tract when it is in solution (*2*). Therefore, the fraction of drug absorbed into the portal vein is a function of the aqueous solubility of the substance. Permeability is a compound's ability to across through intestinal membranes into the blood circulation system (*4*). Both solubility and permeability affect absorption and they must be considered together, and not as independent parameters.

Compounds with high solubility and high permeability are usually well absorbed, whereas compounds with low solubility and low permeability are poorly absorbed. Compounds with mixed properties must be carefully characterized to ensure that they show sufficient absorption/permeability to elicit a therapeutic or nutritional use (*1*).

It is reasonably straightforward to measure solubility but more difficult to measure permeability because permeability is less defined than solubility (*4*). The US Food and Drug Administration (FDA) accepts three absolute permeability criteria: (i) mass balance, absolute bioavailability, or intestinal perfusion studies in human subjects - costly methods usually reserved for compounds that are well along in the approval process; (ii) *in vivo* intestinal perfusion studies in animals; and (iii) *in vitro* permeation experiments across a monolayer of cultured human intestinal cells (Caco-2 cells) (*2*).

Currently, the most productive methods for solubility measurements is LYSA (lyophilization solubility assay, reference *1*). LYSA is a high throughput solubility measurement included in the Multi-dimensional Drug Optimization (MDO) system. The LYSA method has a relatively high throughput (hundreds of compounds per day) in comparison to low throughput solubility assays like THESA (thermodynamic solubility assay).

Permeability represents the overall effects of influx and efflux in the body (Figure 2). Influx refers to a drug or nutrient being transported from the small intestine to the blood system by absorptive transporters such as multidrug resistence proteins (MRP1 and MRP3), ,whereas efflux is defined as a process whereby the drug or nutrient is pumped out from blood system to the intestine by a secretory transporter like breast cancer resistance protein (BCRP), P-glycoprotein and MRP2 (*5*). Cultured "intestine-like" cells such as monolayer cultured cells (Caco-2) have been used for many years to test drug and nutrient permeability. The Caco-2 cell monolayers are human carcinoma cell lines that have many enterocyte-like properties. They are used by the pharmaceutical industry to evaluate the oral absorption potential of drugs and/or to study their absorption mechanisms (*1*). Transport studies through Caco-2 cells provide information about: (i) intestinal permeability, (ii) transport mechanisms (paracellular or transcellular or active carrier), (iii) the role of intestinal metabolism, and (iv) the influence of P-glycoprotein efflux system (*6*). Caco-2 permeability studies can deliver valuable information in the early lead discovery of biologically active compounds and in the development phase for intestinal permeability and absorption prediction through membranes. Hence, Solubility and permeability are essential parameters in the prediction of absorption affecting the bioavailability of ingested substances (*1,6*).

Another type of permeability test known as PAMPA (Parallel Artificial Membrane Permeation Assay) was developed by Roche scientists in 1998 (*4,7*). PAMPA, performed in microplates, measures permeation of a compound through a phospholipid-coated filter medium that mimics intestinal cell

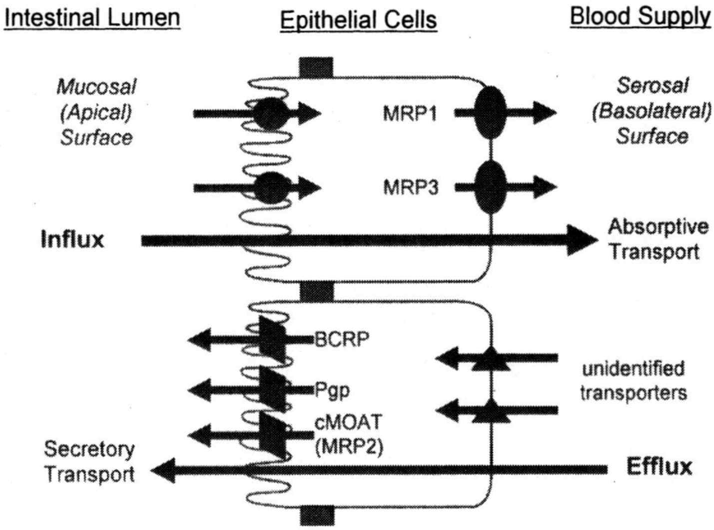

Figure 2. Influx and efflux pathways

structures. The idea behind the PAMPA assay is to predict human intestinal permeability. Among the three possible pathways through a membrane (paracellular, transcellular and active transport), 80%-90% of all small molecular drugs permeability is transcellular. Transcellular permeation is based on passive diffusion, driven by a concentration gradient between donor and acceptor. The small intestine (duodenum, jejunum and ileum) is the major absorption site with the largest absorption area. The PAMPA assay mimics these absorptive conditions using an artificial phospholipidmembrane (*4,7*). Many samples per day can be analyzed by the PAMPA system

Absorption and Bioavailability of Flavonoids

To elucidate the role of dietary flavonoids in human health, it is essential to know the concentrations and forms present in the plasma and tissues after ingestion of these flavonoids (*6*). Absorption from the small intestine is more efficient than that of the colon. Good absorption from small intestines leads to higher plasma values. Upon absorption from the small intestine, flavonoids are then conjugated with glucuronic acid or sulfate, or even *O*-methylated (*6*). Most flavonoids, except catechins, are usually present in the diet as β-glycosides. Glycosides were thought to be too hydrophilic to be well absorbed by passive diffusion in the small intestine (*6*). Research results have shown that the sugar moiety of quercetin plays an important role in increasing the bioavailability of quercetin (*8*). The plasma concentrations and kinetics after single dose administration of some flavonoids in humans are summarized in tabular form by Hollman (*9*). The table showed that catechins and EGCG are quickly absorbed. The bioavailability of catechins was similar. Overall, isoflavones showed the highest bioavailability among the studied flavonols. Although bioavailability studies have been performed with flavonoids of all subclasses, with the exception of the flavones (*9*), there seems to be a lack of understanding or consideration of the importance of bioavailability in the *in vitro* studies, which are subsequently used for the design of *in vivo* experiments. Bioavailability should be carefully considered in the design and interpretation of *in vitro* and *in vivo* experiments (*10,11*). It is important to understand the correlation between *in vitro* and *in vivo* bioavailability, ensuring that the latter can be predicted, based on the data from *in vitro* experiments.

Absorption and Bioavailability of PMFs

With the increased interests in exploring the health benefits of PMFs, the bioavailability study of PMFs has lagged behind. Ohigashi and his colleagues investigated the *in vitro* absorption of nobiletin and luteolin and found that nobiletin preferably accumulated in a differentiated Caco-2 cell monolayer while luteolin did not (*12*). The conclusion can be drawn from this *in vitro* experiment that nobiletin has a much higher permeability and a tendency to accumulate in the intracellular compartment relative to luteolin. The same research group conducted an *in vivo* study of nobiletin in SD male rats and concluded that nobiletin has the distinct property to accumulate in a wide range of organs including the stomach, small and large intestines, liver, and kidney during the 1 to 4 hour periods following single dose administration. In contrast, luteolin and its conjugates were predominantly detected in the gastrointestinal tract (*13*).

Recent *in vitro* Bioavailability Studies of PMFs

We performed a bioavailability study of some PMFs by measuring their solubility and permeability with the goal of further understanding the cause of their absorption and bioavailability properties. The high throughput screening procedure, LYSA, was used to measure the solubility of PMFs. Some permeability data were also obtained from both PAMPA and Caco-2 experiments. From the data it was found that the solubility of PMFs is fairly low, but their permeability is quite high (Table I). As a net result, the absorption profile of PMFs is high, possibly contributing to the high relative bioavailability of PMFs. Although multiple methoxy groups are hydrophobic, contributing to the poor solubility of PMFs the lipophilic nature of methoxy groups are closely related to the good permeability of PMFs (*14*).

Experimental Conditions and Reagents

Preparation of PMF samples: 5,6,7,3′,4′-pentamethoxyflavone, tangeretin, nobiletin, 3,5,6,7,8,3′,4′-heptamethoxyflavone, 5-hydroxy-6,7,8,3′,4′-pentamethoxyflavone, and 3-hydroxynobiletin were isolated from sweet orange peel. 3′-Hydroxy-5,6,7,4′-tetramethoxyflavone and Gardenin A were purchased from Indofine, Inc. (Hillsborough, NJ). 3,5,6,7,8,3′,4′,5'-octamethoxyflavone was synthesized from Gardenin A according to a published procedure (*14*). 3′-hydroxy-5,6,7,8,4′-pentamethoxyflavone, 4′-hydroxy-5,6,7,8,3′-pentamethoxyflavone, and 3′,4′-hydroxy-5,6,7,8-tetramethoxyflavone were synthesized according to a reported procedure (*14*). A total of 12 PMFs were weighed and dissolved in 0.1 mL of DMSO, which is the stock solution of PMFs (10 mM).

Reagents and solvents: Phosphate buffer (pH 6.5), EDTA solution, sterile PBS solution, cell culture media, trypsin, and non-essential amino acids were purchased from Fisher Scientific (Fairlawn, NJ).

LYSA Procedure

DMSO stock solutions of PMFs were diluted to 0.05, 0.1, 0.2 and 0.5 mM for preparation of a four point calibration curve. The OD (Optical Density) maxima (λ_{max}) was determined by direct UV measurement for each PMF. In a similar manner, sample solutions of PMFs were also prepared in duplicate from the DMSO stock solutions (10 mM). The DMSO was then removed with a centrifuge vacuum evaporator. After evaporation of DMSO, the solid PMFs were dissolved in a 0.05 M phosphate buffer (pH 6.5), stirred for one hour and then shaken for two hours. After standing overnight,, the solutions were filtered using a microtiter filter plate. The spectra of the filtrate and its 1/10 dilution were measured with the UV microplate reader and solubility was calculated for each PMF studied.

PAMPA Procedure

PAMPA is an automated assay which is based on a 96-well microplate. The permeation of drugs is measured using a "sandwich" construction (this needs to be fixed?) (*4,7*). A filterplate is coated with a phospholipid membrane and placed into a donor plate containing a drug/buffer solution. The filterplate is next filled with buffer solution (acceptor). The donor concentration is measured at the start of the experiment (reference, t_{start}) and compared with the donor and acceptor concentrations after a certain time t_{end}. The drug concentration analysis is based on UV spectroscopy. UV absorption of the samples, compared between the initial reading of the assay (t_{start}: reference) and at the end (t_{end}: donor, acceptor), allow the determination of a sample distribution between donor, membrane and acceptor. Once the permeation time (t_{end} - t_{start}) is known, a permeation rate constant can be calculated. The unit of this constant is 10^{-6} cm/s, is indicative of a kinetic value. The detailed experimental procedure of PAMPA measurement is illustrated in references *4* and *7*.

Caco-2 Experimental Procedure

Cell culture: Caco-2 cells were obtained from American Type Culture Collection (Rockville, MD). Caco-2 cells at passage number 28 to 40 were used in these studies. All cells were grown in Dulbecco's modified Eagle's medium containing 90% Dulbecco's modified Eagle's medium, 10% fetal bovine serum, 1% nonessential amino acids, 100 units/ml penicillin, and 100 μg/ml streptomycin. All cells were grown at 37 °C in a humidified atmosphere containing 5% CO_2. Culture medium was changed every other day, and cells were passed every 3 to 5 days by trypsinization with 0.05% trypin and 0.53 mM EDTA at 37 °C for 10 min (*15*).

Directional transport assays. Directional transport assays were performed essentially as described by Irvine, *et al.* (*16*). Briefly, Caco-2 cells were seeded in Transwell plates at a density of 6.25 X 10^4 cells/cm^2. The cells were then fed with fresh medium every other day and cultured for 21 to 25 days. Prior to the assay, cells were rinsed with transport medium (Hank's balanced solution, pH 7.4, containing 10 mM Hepes). Cells were equilibrated in the transport medium at 37 °C for 30 min. The transepithelial electrical resistance (TEER) of cell monolayers was measured using "chopstick" electrodes (World Precision Instrument, Sarasonta, FL). The TEER values were determined and corrected by subtracting the resistance of blank filters.

Result and Discussion

In this study we demonstrated the overall bioavailability of PMFs are good. The PMFs have excellent permeability but poor aqueous solubility.

In the LYSA assay, any number that is greater than 200 μg/mL is considered soluble. The number between 100 μg/mL and 200 μg/mL is considered medium

Table I. Solubility and Permeability Data of PMFs

Name	*Structure*	*LASA (µg/ml)*	*PAMPA*		*Caco-2 (P_{app})*		
			(cm/s x10^{-6})	Permea-bility	A to B (cm/s x10^{-6})	B to A (cm/s x10^{-6})	Ratio (A to B)/ (B to A)
Sinensetin	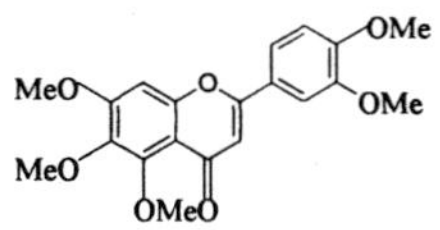	nd	1.02	H	nd	nd	nd
Tangeretin	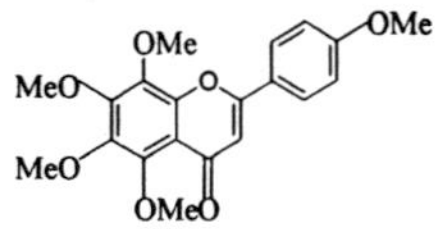	19	1.62	H	nd	nd	nd
Nobiletin	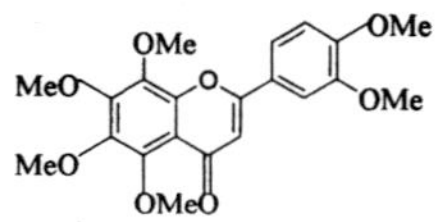	12*	1.38	H	nd	nd	nd
5,6,7,8,3',4',5'-Hepta-methoxy-flavone	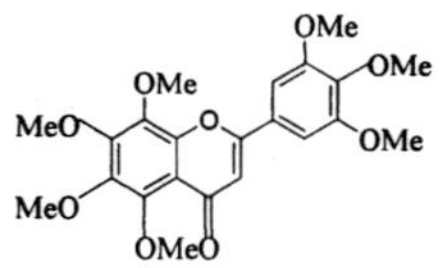	14	0.75	M	nd	nd	nd
3,5,6,7,8,3',4'-Hepta-methoxy-flavone	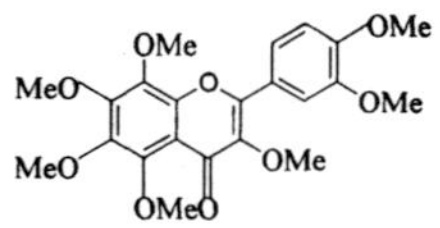	8	0.93	M	802	535	0.7
Gardenin A	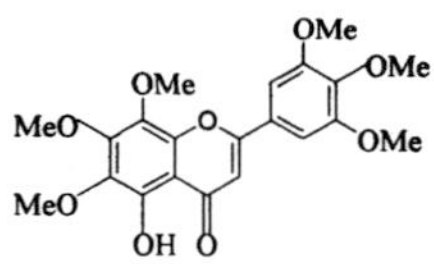	nd	0.9	M	nd	nd	nd
3'-Hydroxy-5,6,7,4'-tetra-methoxy-flavone	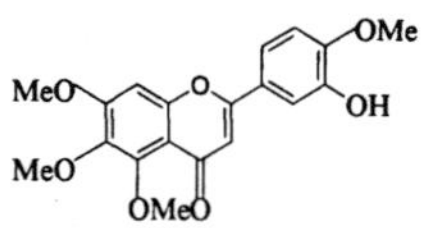	6	0.98	M	558	507	0.9

Table I. Solubility and Permeability Data of PMFs (*Continued*)

Name	*Structure*	*LASA* *(μg/ml)*	*PAMPA*		*Caco-2 (P_{app})*		
			(cm/s x10^{-6})	Permea-bility	A to B (cm/s x10^{-7})	B to A (cm/s x10^{-7})	Ratio (A to B)/ (B to A)
3'-Hydroxy-5,6,7,8,4'-penta-methoxy-flavone	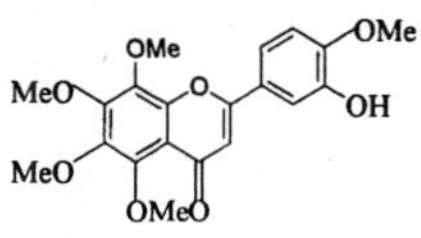	29	1.05	H	nd	nd	nd
4'-Hydroxy-5,6,7,8,3'-penta-methoxy-flavone	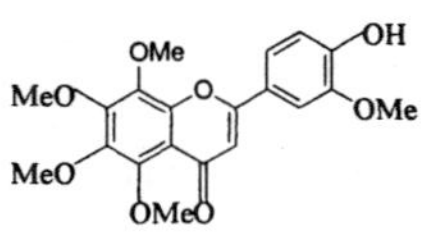	22	1.14	H	nd	nd	nd
3',4'-Dihydroxy-5,6,7,8-penta-methoxy-flavone	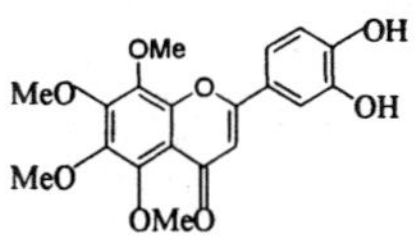	53	0.98	H	nd	nd	nd
3-Hydroxy-5,6,7,8,3',4'-hexa-methoxy-flavone	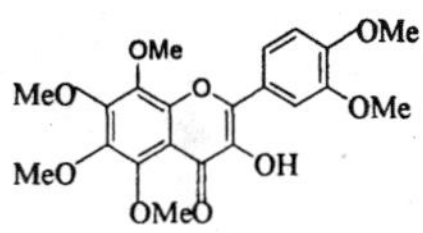	37	0.55	M	1700	1000	0.6
5-Hydroxy-6,7,8,3',4'-penta-methoxy-flavone	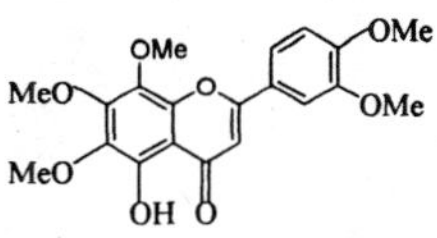	32	nd-	nd	nd	nd	nd

**Data measured from HPLC method; nd, determination in progress; H: high; M: medium.*

solubility. However, there is a very close relationship between solubility, permeability and potency. Solubility alone is insufficient to describe the bioavailability of an active compound in oral administration. Lipinski depicted the minimum acceptable solubility that is required for an orally active drug. He grouped the compounds in sets of three and showed that, at pH 6.5 or 7.0, the minimum thermodynamic aqueous solubility required for low, medium and high permeability values for a particular clinical dose. For example, to achieve oral absorption, a compound with medium intestinal permeability and a projected human dose of 1 mg/kg requires a minimum aqueous solubility of 52 µg/mL. For a highly permeable compound with a 0.1 mg/kg projected human dose, the required minimum solubility is only 1 µg/mL (*17*).

Solubility of PMFs

The LYSA data is shown in Table I and identifies that the overall solubility of PMFs is low. The solubility of hydroxylated PMFs is higher than their fully methoxylated counterparts. For example, the solubility of 5-demethylnobiletin (32 µg/mL) is better than that of nobiletin (12 µg/mL). In another example, the solubility of 3-hydroxynobiletin (37 µg/mL) is significantly higher than that of 3-methoxynobiletin (8 µg/mL). The more hydroxyl groups that the PMFs have, the better is their solubility . Thus, 3',4'-dihydroxy-5,6,7,8-tetramethylflavone has the highest solubility (53 µg/mL) among the 10 PMFs screened by LYSA. The sequence of solubility ranges from high to low: 3',4'-dihydroxy-5,6,7,8-tetramethylflavone > 5-demethylnobiletin > 3'-demethylnobiletin > 4'-demethylnobiletin > nobiletin. Usually, higher solubility values are obtained from LYSA readings. Another observation was that a hydroxyl group at the 3-position of the PMF C-ring tends to increase the PMF solubility dramatically: the solubility of 3-hydroxynobiletin (37 µg/mL) is much greater than 3'-methoxynobiletin (8 µg/mL). Interestingly, 3'-hydroxy-5,6,7,4'-tetramethoxy-flavone, with a LYSA solubility of 6 µg/mL, is nearly insoluble. The explanation of this interesting result may be the lack of an 8-methoxy group on the A-ring, in comparison with 3'-demethylnobiletin. The solubility of 5,6,7,3',4'-pentamethoxyflavone was not obtained primarily because its solubility is too low to detect in the LYSA assay.

PAMPA Data of PMFs

Table I also shows the passive permeability data obtained from the PAMPA assay. Overall, the permeability of PMFs is medium to high, meaning that PMFs can easily cross the phospholipid membrane. This may be due to the hydrophobicity imparted by their multiple methoxy groups. In contrast to the general flavonoids, which possess multiple hydrophilic hydroxyl groups, the PMFs have strong tendency toward lipophilicity because of the hydrophobic feature of methoxy groups.

Tangeretin has the highest permeability (1.62 x 10^{-6} cm/s), followed by nobiletin (1.38 x 10^{-6} cm/s), 4'-demethylnobiletin (1.14 x 10^{-6} cm/s), and 3'-demethylnobiletin (1.05 x 10^{-6} cm/s). All PMFs tested have or have close to high permeability with the exception of 3-hydroxynobiletin, whose permeability is in the medium range (0.55 x 10^{-6} cm/s). Surprisingly, 3',4'-dihydroxy-5,6,7,8-tetramethoxy-flavone has relatively high permeability (0.98 x 10^{-6} cm/s).

Usually, compounds that are highly soluble exhibit poor permeability and lipophilic compounds have high permeability but poor solubility. From Table I, we can see that PMFs have poor solubility in general. However, PMFs having relatively better solubility are also highly permeable to the membrane, which in turn contributes to their high absorption and bioavailability. These examples are tangeretin, 3',4'-dihydroxy-5,6,7,8-tetramethoxyflavone, 4'-demethylnobiletin, 3'-demethylnobiletin, and nobiletin.

Caco-2 Results

In analysis of Caco-2 data, the apparent permeability (P_{app}) from A (apical side) to B (basolateral side) represents the overall effects of transportation carried out by both absorptive transporters and secretory transporters. A compound exhibits good permeability when $P_{app} > 100 \times 10^{-7}$ cm/s. The apparent permeability (P_{app}) from B (basolateral side) to A (apical side) only measures the effects of secretory transport. If the ratio of P_{app} from B to A over A to B is greater than 3, there will be efflux liability, which means that the compound is pumped out too fast from blood circulation system, which will affect the concentration of the substance in the systemic circulation, and consequently affect the absorption and bioavailability of the compound.

There are only three PMFs whose Caco-2 data (Table I) are currently available. From the available Caco-2 data of PMFs assayed, we can see that the P_{app} for all PMFs from A to B are 508 x 10^{-7} cm/s for 3'-demethylsinenstin, 802 x 10^{-7} cm/s for 3-methoxynobiletin, and 1700 x 10^{-7} cm/s for 3-hydroxynobiletin. These values are significantly greater than 100 x 10^{-7} cm/s, indicating that the PMFs have superb permeability. For all compounds tested, the P_{app} ratios of B to A over of A to B were less than 3. They are 0.9 for 3'-demethylsinenstin, 0.7 for 3-methoxynobiletin, and 0.6 for 3-hydroxynobiletin. Therefore, there is no existence of an efflux-liability problem.

In consideration of absorption and bioavailability, the data set (LYSA, PAMPA and Caco-2) for each compound should be considered as a whole. To this extent, a conclusion can be drawn that PMFs are marginally soluble, with 3',4'-dihydroxy-5,6,7,8-tetramethoxyflavone, 4'-demethylnobiletin and 3'-demethylnobiletin having better solubility among the PMFs assayed. Tangeretin and nobiletin are highly permeable but have poor solubility. However, all of the PMFs assayed are highly permeable across the small intestine wall. Considering the solubility and permeability together, an overall good absorption of PMFs can be predicted. Further *in vivo* bioavailability explorations of PMFs are required to complete the understanding of PMF ADME characteristics.

References

1. Van de Waterbeemd, H.; Lennernas, H.; Artursson, P. *Methods Princ. Med. Chem.* 18. **2003**, Wiley-VCH Veriag GmbH & Co. KGaA, Weinheim, Germany,.
2. http://www.dddmag.com/feats/0208cem23.asp
3. Kararli, T.T. *Crit. Rev. Ther. Drug Carrier Syst.* **1989**, *6*, 39-86.
4. Kansy, M.; Senner, F.; Gubernator, K. *J. Med. Chem.* **1998**, *41*, 1007-1010.
5. Ritschel, W. A. *S.T.Pharma* **1987**, *3(2)*, 125-141.
6. Walle, T.; Walgren, R.; Walle, U.K.; Galijatovic, A.; Vaidyanathan, J. B. In: Rice-Evens, C.A. and Packer, L., editors. *Flavonoids in Health and Disease*. Second Edition. Marcel Dekker, Inc., London, **2003**, 349-361.
7. Kansy, M.; Fischer, H.; Kratzat, K.; Senner, F.; Wager, B.; Parrilla, I. *Helvetica Chimica Acta*, **2000**, 447-464.
8. Hollman, P.C.H.; van Trijp, J.M.P.; Buysman, M.N.C.P.; van der Gaag,M. S.; Mengelers, M.J.B.; de Vries, J.H.M.; Katan, M.B. *FEBS Lett.* **1997**, *418*, 152-156.
9. Hollman, P.C.H. *Pharmaceutical Biology* **2004**, *42*(supplement), 74-83.
10. Williamson, G.; Manach., C. *Am. J. Clin. Nutr.* **2005**, *81* (suppl), 243S-255S.
11. Manach, C.; Scalbert, A.; Morand, C.; Remesy, C.; Jimenez, L. *Am. J. Clin. Nutr.* **2004**, *79*, 727-747.
12. Murakami, A.; Kuwahara, S.; Takahashi, Y.; Ito, C.; Furukawa, H.; Ju-ichi, M.; Koshimizu, K.; Ohigashi, H. *Biosci. Biotechnol. Biochem.* **2001**, *65 (1)*, 194-197.
13. Murakami, A.; Koshimizu, K.; Ohigashi, H.; Kuwahara, S.; Kukí, W.; Takahashi, Y.; Hosotani, K.; Kawahara, S.; Matsuoka, Y. *Biofactors* **2002**, *16*, 73-82.
14. Li, S. *Dissertation*, Rutgers University, New Brunswick, New Jersey, USA, **2005**.

15. Guo, A.; Marinaro, W.; Hu, P.; Sinko, P. *Drug Metabolism and Disposition.* **2002**, *30*, 457-463.
16. Irvine, J. D.; Takahashi, L.; Lockhart, K.; Cheong, J.; Tolan, J. W.; Selick, H. E.; Grove, J. R. *J. Pharm. Sci.* **1999**, *88*, 28-33.
17. Lipinski., C. A. *J. Pharmacological and Toxicological Methods.* **2000**, *44*, 235-249.

Safety and Bioactivity

Chapter 17

Kava *(Piper methysticum)* Safety Concerns and Studies on Pipermethystine, an Alkaloid in Kava

Jin-Woo Jhoo[1,2], Catharina Y. W. Ang[1], Nan Mei[1], Tao Chen[1], Klaus Dragull[3], and Chung-Shih Tang[3]

[1]National Center for Toxicological Research, U.S. Food and Drug Administration, 3900 NCTR Road, Jefferson, AR 72079
[2]Department of Food Science and Technology in Animal Resources, Kangwon National University, 192–1 Hyoja–2, Chunchon, Kangwon 200–701, South Korea
[3]Department of Molecular Biosciences and Bioengineering, University of Hawaii, Honolulu, HI 96822

Kava (Piper methysticum) extract products have been associated with a number of severe hepatotoxicity cases. Factors influencing the reported kava-linked liver injuries could include the methods of kava preparation, genetic differences between populations, herbal-drug interactions, enzyme inhibition and minor toxic kava constituents in the kava extracts. This article provides highlights of recent findings on kava safety issues, and also presents results of our current studies on pipermethystine, a kava alkaloid, in commercial products and its potential mutagenicity. The level of pipermethystine was found to be negligible in most of the kava dietary supplement products analyzed; however, one product sold as a mixture of kava roots and leaves did contain detectable amounts of pipermethystine. *In vitro* mutagenicity tests indicated that pipermethystine was negative in Salmonella umu assay at a dose range up to 1000 μM and in the mouse lymphoma assay at concentrations lower than 2.5 μg/mL. However, it showed cytotoxicity in mouse lymphoma cells in a dose-response manner.

Kava, also has been referred to as kava kava, is *Piper methysticum* Forst. f. in the family Piperaceae. It is a perennial shrub grown in Pacific islands including Fiji, Tonga, Vanuatu, Samoa, Futuna, New Caledonia, and the Hawaiian islands (*1*). The rootstocks are commonly used in the preparation of traditional beverages with calm and relaxation effects for social and ceremonial occasions for centuries. Some of the events where kava is commonly used are weddings, funerals, religious activities, welcoming of visitors, and exchanging of gifts. Various kava preparations have also been used as folk medicines for treatments of a range of symptoms, including chills, headaches, gastrointestinal upset and skin disease. Lebot *et al.* (*1, 2*) described many different aspects about kava as used in the South Pacific region.

During the last 20 years, some Western manufacturers have become interested in making kava extracts as an alternative medicine for the treatment of mental disorders, nervous anxiety, tension, and restlessness (*3*). Upon the passage of the Dietary Supplement Health and Education Act in 1994, more herbal dietary products with perceived beneficial effects became readily available in the U. S. A. The sales of kava dietary supplements or related products were booming worldwide in the late 1990s and early in the year 2000. Consumers were led to believe that kava products were effective, non-addictive, and safe. Kava root extract capsules, kava root powder capsules and kava tea blends became one of the top-selling herbal dietary supplements for a number of years before 2002.

Within a few years, however, severe cases of hepatic toxicity, possibly associated with the consumption of herbal products containing kava were reported in several countries. Germany was the first country to ban kava products in 2002, and England, France, and Australia also removed kava products from the market in the following years. The liver injuries included hepatitis, cirrhosis, and liver failure. The U.S. FDA (*4*) issued an advisory to health care professionals and consumers regarding the potential risks of severe liver injury with use of kava supplements. Up to the present time, there have been 78 cases of hepatotoxicity presumed to be linked to kava ingestion (*5*), but it is still under debate as to whether the ban of kava products was justified or was an overreaction (*6*).

Many of the liver failure cases were associated with products containing acetone or alcohol extracts of kava (*7*). Proposed causes of liver injuries associated with kava include the commercial preparation methods (organic solvent extraction), genetic difference in consumers, inhibition of cytochrome P450 (CYP) enzymes by kavalactones, herbal-drug interactions, and the presence of toxic alkaloid, such as pipermethystine, in aerial parts of kava plant. Detailed discussions on the kava controversy and proposed causes of kava hepatotoxicity appeared recently (*5, 8-11*). The present article provides highlights of some key findings related to the chemical and safety aspects of kava, and also presents results of our current studies on the determination of pipermethystine, a kava alkaloid, in dietary supplement products and testing for cytotoxicity and mutagenicity of pipermethystine.

Factors Related to Kava Safety Issues

Composition and Active Constituents

Fresh kava rootstock contains about 80 % water. On a dry weight basis, the composition is approximately 12% water, 43% starch, 10% fiber, 3.2% sugars, 3.6% proteins, 3.2% minerals, and 15% kavalactones which can vary from 3 to 20%. Many factors influence the composition, including the age of plant, cultivars, location and growing conditions (*1, 12*). Kavalactones, also have been referred to as kavapyrones, are regarded as pharmacologically active constituents responsible for the anti-anxiety and mild sedative action. They are present as a mixture of more than 18 different forms that appear to have synergistic effects. Six kavalactones (Figure 1) are present in highest concentrations and they account for 96% of the lipid extract. These major kavalactones are (+)-methysticin, (+)-dihydromethysticin, (+)-kavain, (+)-dihydrokavain, yangonin, and desmethoxyyangonin (*13*).

The total kavalactone content is generally highest in the roots, followed by stumps and basal stems (*12*), and the lowest content is in the stems and leaves (*1*). The distribution of various kavalactones is different among different cultivars and kava plant parts. Dihydrokavain and dihydromethysticin are the major constituents in leaves and stems. Kavain and methysticin are the major components of the roots and rhizome (*14, 15*). The ratios of kavalactones may be important to biological effects.

Other constituents from kava rootstock include trace pyrrolidine alkaloids, three chalcones (flavokavains, A, B, C) (*13, 14*), and a different alkaloid, pipermethystine, in kava leaves, which is specific to *P. methysticum* (*16*). There are also alcohols (*e.g.* dihydrokavain-5-ol), ketones (cinnamalaketone), essential oils, and methylene (*e.g.* dioxy-3, 4-cinnamalaketone) (*13, 14*).

Methods of Kava Preparation for Consumption

Traditional Methods

In the South Pacific islands, rootstocks are used to prepare kava beverages. The traditional methods involve chewing, masticating, grating, grinding or

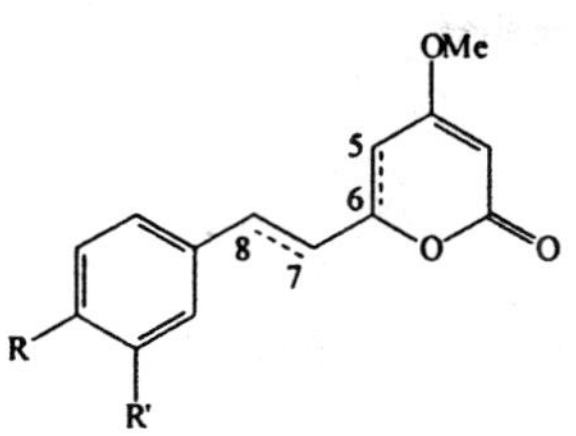

	R	R'	C_5-C_6	C_7-C_8
Kavain	H	H	—	=
Dihydrokavain	H	H	—	—
Methysticin	—O—CH_2—O—		—	=
Dihydromethysticin	—O—CH_2—O—		—	—
Yangonin	—OMe	H	=	=
Desmethoxyyangonin	H	H	=	=

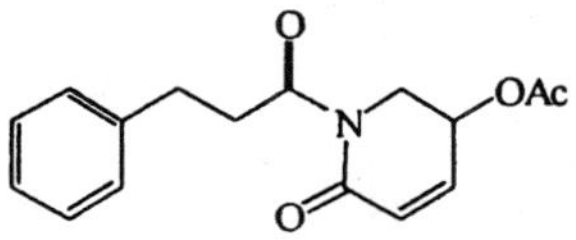

Pipermethystine

Figure 1. Structures of six major kavalactones and pipermethystine.

pounding fresh or dried kava rootstocks and stumps, followed by immersing and mixing the processed mass in cold water (*1, 17*). The solids are strained through a strainer made of certain types of plant barks or leaves. The aqueous infusions of kava rootstock made by these methods are actually emulsions of suspended kava resin in water. Kavalactones are released into the water as a suspension rather than as a true solution (*1*). The kava beverages made by the traditional methods are generally referred to as aqueous infusions or water extracts.

Industrial Extraction Methods

Kava pharmaceuticals or dietary supplements are commonly made using organic solvent, such as acetone and ethanol in a mixture with water. An ethanol-water mixture can be used to extract kava and gives crude extracts containing about 30% kavalactones, while an acetone-water can be used to result in concentrated extracts containing 70% kavalactones (*18*). The solvents are then dried and the extracts may or may not be standardized to certain concentrations of total kavalactones in the final forms of capsules, tablets, or tinctures. The dietary products made from organic solvent extraction are referred to as herbal products or dietary supplements. However, some capsules may contain only the powdered kava roots, not the concentrated extracts.

It is well documented that the compositions of aqueous infusions and organic extracts are very different (*19, 20*), the former contains more polar components and the latter contains more non-polar constituents. Because the polarity of different kavalactones are different, the ratios of the individual kavalactones are dependent upon the preparation methods (*19, 20*), and the difference in kavalactone ratios may contribute to different biological effects in consumers.

Aqueous Infusion and Its Effects on Indigenous Populations

Some indigenous populations, who are frequent or excessive users of kava beverages (prepared by the traditional methods), have not developed liver problems (*21*). Approximately 10 % Caucasian populations have a genetic CYP 2D6 deficiency that could contribute to the hepatotoxicity of patients who ingest kava extract products (*8, 22*), while Polynesians do not have this deficiency (*8*). A number of studies, however, have also shown various other symptoms possibly caused by the heavy consumption of kava.

General Adverse Effects

Some adverse health effects were found in heavy users, such as those reported in Australian aborigines, Arnhem Land (*23, 24*), and Fiji (*25*). These adverse effects included general poor health, a "puff" face, scaly rash, headache, chest pain, indigestion, (*23-25*) mild and reversible gastrointestinal disturbance, and kava dermatopathy (scaly skin eruption) (*26*). A more recent survey by Tavana *et al.* (*17*) in Savaii, Samoa also reported that heavy drinkers might experience dizziness, dry pale skin, weight loss, and upset stomach. But the symptoms were reversible. Their data, collected by interviewing with traditional healers and biomedical practitioners, suggest that kava-related hepatotoxicity is lacking in native populations in Savaii, Samoa. No liver problems were reported for patients who had a history of drinking only kava without alcohol.

Effects on Liver Functions

Very heavy users of kava can consume 400 g/week of kava root (*24*) that may contain 40 g total kavalactones (based on 10% kavalactones in roots) and might show weight loss and abnormal liver functions. A survey by Russmann *et al.* (*26*) demonstrated that heavy kava drinkers in New Caledonia showed elevated γ-glutamyl transferase in 23/27 and minimally elevated transaminases in 8/27. These abnormal enzyme functions were not regarded as a sign of liver injury, but rather as an indication of CYP45O enzyme induction.

A study by Clough *et al.* (*27*) showed that the liver function changes in users of aqueous kava extracts appear to be reversible. In the case of herbal product hepatotoxicity, the aminotransferase and alkaline phosphatase levels were especially high and not reversible. More recent kava users developed higher levels of liver enzymes γ-glutamyl transferase and alkaline phosphatase, but not alanine aminotransferase or bilirubin. The abnormalities in liver function usually return to normal within 1-2 weeks. No evidence for irreversible liver damage was reported.

Correlations to Other Health Effects

On the positive side of health effects, populations with more kava consumption may have less cancer incidences. Steiner (*28*) published a statistical correlation and concluded that there was a close inverse relationship between kava consumption and age-standardized cancer incidence rates for all sites. Countries or cities compared were Vanuatu, Fiji, Western Samoa, Micronesia, New Caledonia, Hawaii/ Hawaiians, New Zealand/Maoris and Los Angeles. Another case-control study by Clough *et al.* (*29*) showed that there was no association between kava use and ischaemic heart disease in Aboriginal communities in eastern Arnhem Land Australia.

Organic Solvent Extracts and Kava Toxicity

Kava and other herbs including garlic, ginkgo, echinacea, ginseng and St. John's wort have the potential to modulate the activity of drug-metabolizing enzymes (notably cytochromes P450 isozymes) and/or drug transporter P-glycoprotein. All of these products participate in potential pharmacokinetic interactions with anticancer drugs (*30*).

Enzyme Inhibition and Herbal-drug Interactions

Several studies have indicated that kava extract and kavalactones inhibited human liver microsome enzymes. Mathews, *et al* (*31*) reported that whole kava extract caused significant inhibitory effects on the activities of human liver microsomes cytochromes P450 enzymes (CYP1A2, 2C9, 2C19, 2D6, 3A4 and 4A9/11). However, individual kavalactones had varied effects on different P450 enzymes. Some kavalactones also formed 455 nm metabolic intermediate complexes after incubation with human liver microsomes and NADPH, but kavain and desmethoxyyangonin did not. These data indicate that kava has a high potential for causing drug interactions through inhibition of P450 enzymes responsible for the majority of the metabolism of pharmaceutical agents. Unger *et al.* (*32*) found that kavaprones exhibited inhibitory effect on CYP3A4 and dihydromethysticin was the main inhibitory component of the ethyl acetate extract of kava.

Organic Solvent Extracts vs. Aqueous Infusions

Using 2 cell lines to study the effects of kava, Zou *et al.* (*33*) reported that the parent compounds of each of the four test samples (methysticin, yangonin, desmethoxyyangonin and ethanolic extract of root) were primarily responsible for the observed cell toxicity and that CYP 1A1, 2A6, 2E1, and 3A4 or epoxide hydroxylase did not appear to be involved. In *in vitro* studies, kava was not activated to toxic metabolites by enzymes.

Aqueous kava extracts had no effect on liver function tests in rats administered in daily dosages of 200 or 500 mg active kavalactones/kg body weight for 2 or 4 weeks (*19*). The data showed that alanine aminotransferase, aspartate aminotransferase, alkaline phosphatase and lactate dehydrogenase, and malondialdehyde were not elevated, and in some other cases, the enzymes were even significantly reduced, suggesting lack of toxic effect by aqueous kava extracts on the liver. The study also included the analysis of kavalactone content as extracted by different solvents. The ratios of different kavalactones were different for the different extraction methods; kavain and methysticin were predominant in water extracts while the proportions of these two compounds were less in ethanol extracts. Dihydrokavain and dihydromethysticin levels were higher in non-polar solvent extracts. Whether some or all of the kavalactones are hepatotoxic is still unclear; and whether large amounts of these kavalactones or other compounds extracted are responsible for the liver damage remains to be resolved.

Cote *et al.* (*20*) reported a significant difference in the ratio of the major kavalactones between commercial kava extracts and traditional kava beverage, and found a significant difference in their ability to inhibit of P450 enzymes (CYP3A4, 1A2, 2C9, 2C19). The inhibition was more pronounced for the commercial preparation. Although all of the extracts inhibited human CYP3A4, 1A2, 2C9 and 2C19 in a low micromolar range, the aqueous extract was the least potent for inhibiting all these P450s. However, it is not known if the inhibition of P450 is responsible for the hepatotoxicity. Nevertheless, they concluded that if the hepatotoxicity reported for the commercial caplet was the result from P450 inhibition, the traditional extract should also be hepatotoxic at high doses.

Toxicological Studies on Pipermethystine

The reported kava hepatotoxicity may result from minor toxic constituent(s) in kava extracts. Dragull *et al.* (*34*) isolated and identified three piperidine alkaloids. They also found that pipermethystine was concentrated in kava stem peelings and leaves; 3α, 4α-epoxy-5β-pipermethystine and awaine were new alkaloids; and the epoxide existed only in cultivar Isa among 11 cultivars examined. None of the compounds was detectable in commercial root powders from Fiji, Tonga or Hawaii. The pipermethystine content was 0.06 to 0.85 % in stem peelings and 0.32 to 2.43% (dry weight basis) in leaves. However, no commercial dietary supplement products were evaluated. Nerurkar *et al.* (*35*) reported that pipermethystine significantly decreased cellular ATP levels, mitochondrial membrane potential, and induced apoptosis as measured by the release of caspase-3 after 24 hr of treatment. Their data suggest that pipermethystine is capable of causing cell death, probably in part by disrupting mitochondrial function.

As a follow up of the above studies, we collected 12 commercial dietary supplement products from local markets or *via* internet in the USA and analyzed for their pipermethystine content. We also evaluated *in vitro* cytotoxicity and mutagenicity of pipermethystine that was purified at the University of Hawaii.

Experimental

Sample Preparation

Four products of capsule type, five of root powder type and two tincture products of kava-containing dietary supplements were collected in local markets or *via* Internet. For the preparation of solid samples, 1 g of finely ground sample powders (pre-grind in a coffee grinder as needed) was extracted with 40 mL of ethyl acetate with 30 min sonication. The extract was filtered, and the filtrate was evaporated *in vacuo.* The resulting residue was dissolved in 2 mL of ethyl acetate for GC-FID analysis. For liquid samples, 5 g of sample solution was dried under reduced pressure. The resulting residue was dissolved in 4 mL of ethyl acetate with sonication. The solution was centrifuged and the clear supernatant was used for GC-FID analysis.

GC-FID Analysis and GC-MS Confirmation

The method was modified from Dragull *et al.* (*34*). A Hewlett-Packard (HP) 5890 system equipped with FID detector and 6890 auto-injector was employed. Prepared sample solutions were separated on a DB-XLB capillary column (20 m × 0.18 mm, 0.18 μm, J&W Scientific, Folsom, CA). The column temperature initially was held at 150 °C for 1 min, raised to 300 °C at 5 °C/min and held 2 min. The total time was 33 min for each run. Injector and detector temperatures were maintained at 250 °C and 280 °C, respectively. The injection volume was 1 μL in the splitless mode. Quantification of pipermethystine in the sample was conducted by relating peak area to that of external standard curves that were constructed with a range from 10 μg/mL to 1000 μg/mL (10, 25, 50, 75, 100, 250, 500, 750 and 1000 μg/mL of pipermethystine). The calibration curve was linear over these concentrations, and the correlation coefficient was $r^2 > 0.995$. The identification of pipermethystine was performed by GC-FID analysis by comparing the retention time with authentic standard and confirmed by GC-MS analysis. In order to minimize the MS interference, pipermethystine in sample was first separated from major kavalactones by HPLC. The eluate containing the pipermethystine peak was collected and further purified with SPE (Sep-Pak cartridge, C-18) for the GC-MS analysis. The MS spectrum of sample was compared with that of the authentic compound.

Umu Test

This method was based on the report by Oda *et al.* (*36*). *S. typhimurium* TA 1535/pSK1002 was grown overnight in LB-broth supplemented with 50 μg of ampicillin/mL. The culture was then diluted 50-fold with fresh TGA medium (1% trypton, 0.5% NaCl, 0.2% glucose with ampicillin at 50 μg/mL). The culture media were further incubated until a bacterial optical density of 0.25 to 0.30 at 600 nm was reached. A 2.4 mL portion of bacterial culture was divided into test tubes, then 0.1 mL of test compound was added along with either 0.5 mL of 0.1M phosphate buffer (pH 7.4) or microsome mixture (0.25mM NADP, 5mM glucose-6-phosphate, 0.5 unit glucose-6-phosphatedehydrogenase, 3 mM $MgCl_2$, and microsomes in 1 mL) for metabolic activation. Rat liver microsome was prepared from Spargue-Dawley rats which were pre-treated with 3-methylcholanthrene (3-MC). After further incubation at 37 °C for 2 hrs, the β-galactosidase activity was measured with procedure reported by Miller (*37*). The absorbance at 420 nm and 550 nm was measured. 1-Nitropyrene (330 ng/mL) and 2-amino-3-methylimidazol [4,5-*f*] quinoline (IQ; 330 ng/ml) were used as positive controls in the systems without or with metabolic activation, respectively. DMSO was used as the negative control in both systems. The galactosidase activity was calculated using the following equation: units=$[[A_{420}-(1.75 \times A_{550})] / (t \times v \times A_{600})] \times 1000$, where, t is time of the reaction in minutes and v is volume (mL) of culture used in this assay.

Mouse Lymphoma Assay (MLA)

The $Tk^{+/-}$-3.7.2C heterozygote of the L5178Y mouse lymphoma cell line was utilized for this experiment. MLA experiment was performed according to the procedures of Chen and Moore (*38*). The test compound, pipermethystine, was dissolved in DMSO and added to the suspended cells at concentration range from 0.5 to 3 µg/mL. The cell cultures were incubated for 4 hr, washed twice with fresh medium and then re-suspended in fresh medium. The culture flasks were placed in a humidified incubator at 37 °C in the presence of 5% CO_2 and maintained in log phase growth for a 2-day expression period. The cells were counted and the densities were adjusted using fresh medium.

For the mutant selection, trifluorothymidine (TFT, 3 µg/mL) was added to the cell culture and cells were seeded into four 96-well flat-bottom microtiter plates using 200 µL per well and a density of 2000 cells/well. For the determination of plating efficiency, cultures containing 8 cells/mL were prepared with dilution, and 200 µL this culture was aliquoted per well into two 96-well flat-bottom microtiter plates. The colonies were counted after 11 days of incubation at 37 °C in a humidified incubator with 5% CO_2 in air. Mutation frequencies (MF) and relative total growth (RTG) were determined according to the method of Chen and Moore (*38*).

Results and Discussion

Pipermethystine Content in Dietary Supplement Products

The results of analysis of pipermethystine and a brief description of the products are shown in Table I. These products varied widely within and between product types as far as labeling for kavalactone content. Analytical results showed that pipermethystine was not detectable in 11 of 12 products tested (Table I). However, one product (No. 5, Table I) composed of kava root chips and leaves was found to contain pipermethystine at 126µg/g.

The finding that pipermethystine was present in samples containing kava leaves but not in other samples of root powders agreed with the report of Dragull *et al.* (*34*), showing the presence pipermethystine in kava leaves and stems but not roots. Since all kava parts might be used for commercial extracts (*34*) and human exposure to pipermethystine would then occur, we evaluated the potential cytotoxicity and mutagenicity of pipermethystine in this study.

Umu Test

This test system is based on the SOS response using strain *Salmonella typhimurium* TA1535/pSK 1002 that contains a *umu*C'-'*lac*Z gene. Mutagenic chemicals induce the SOS response that activates *umu* region of the plasmid, leading to production of *lac*Z gene product, β-galactosidase whose activity can be measured by colorimetry.

Table I. Pipermethystine Content of Dietary Supplement Products in U. S. Market

Product No.	*Type*	*Description/Comment*	*Pipermethystine µg/g*
1	Capsule	Extract of kava rhizome and root, 60 mg kavalactones per capsule	--[a]
2	Capsule	Kava, 30% kavalactones, 250 mg per capsule	--[a]
3	Capsule	450 mg per capsule, kava root	--[a]
4	Capsule	500 mg per capsule, kava root extract	--[a]
5	Powder and chips	Kava tea – root chips and leaves	126
6, 7, 8	Powder	Kava beverage powder, from three different origins, one supplier	--[a]
9, 10	Powder	Kava root powder of different cultivars, different suppliers	--[a]
11	Tincture	Rhizome and root, certified Organic grain alcohol (85-90%) and kava extractives	--[a]
12	Tincture	Organic kava root, grain alcohol (50-60%)	--[a]

[a] Not detectable; <90 µg/g.

Our data showed that in either system (with or without metabolic activation) the positive control induced a significant increase of β -galactosidase activity, while pipermethystine was negative for the induction of the SOS response at the treatment concentration range from 1 to 1000 µM (Figure 2). Based on the *umu* test, DNA damaging effects of pipermethystine may be negligible. It was interesting to note that pipermethystine treatment inhibited cell growth in a dose-dependent manner. Because the cell growth rate is one of the factors in calculating the *umu* units, the results showed lower *umu* units by treatments with high concentrations of pipermethystine as compared to the negative control.

Mouse lymphoma assay (MLA)

By this assay, cells deficient in *thymidine kinase* (*Tk*) due to a mutation in the gene that results in $Tk^{+/-}$ to $Tk^{-/-}$ are resistant to the cytotoxic effects of the pyrimidine analogue TFT. Thus, *Tk* proficient cells are sensitive to TFT, which

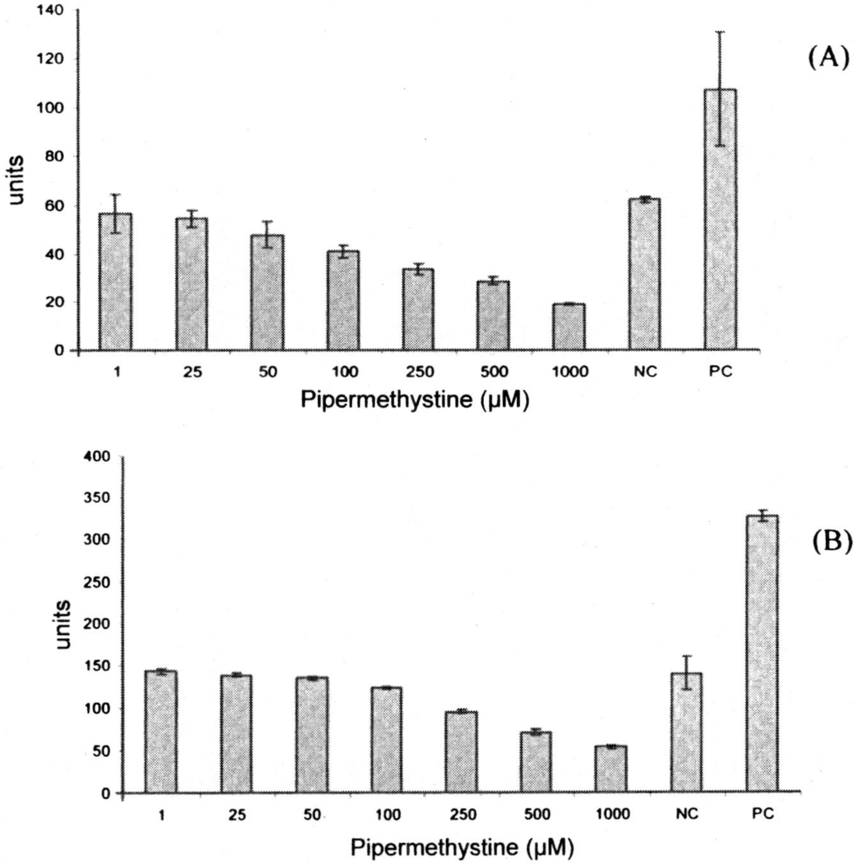

Figure 2. Mutagenicity study of pipermethystine by umu test, (A) in buffer solution without metabolic activation, PC: positive control (1-nitropyrene, 330 ng/mL), NC: negative control (DMSO); (B) with metabolic activation by microsomes, PC: 2-amino-3-methylimidazo[4,5-f] quinoline (IQ), 330 ng/mL, NC: DMSO, n=3.

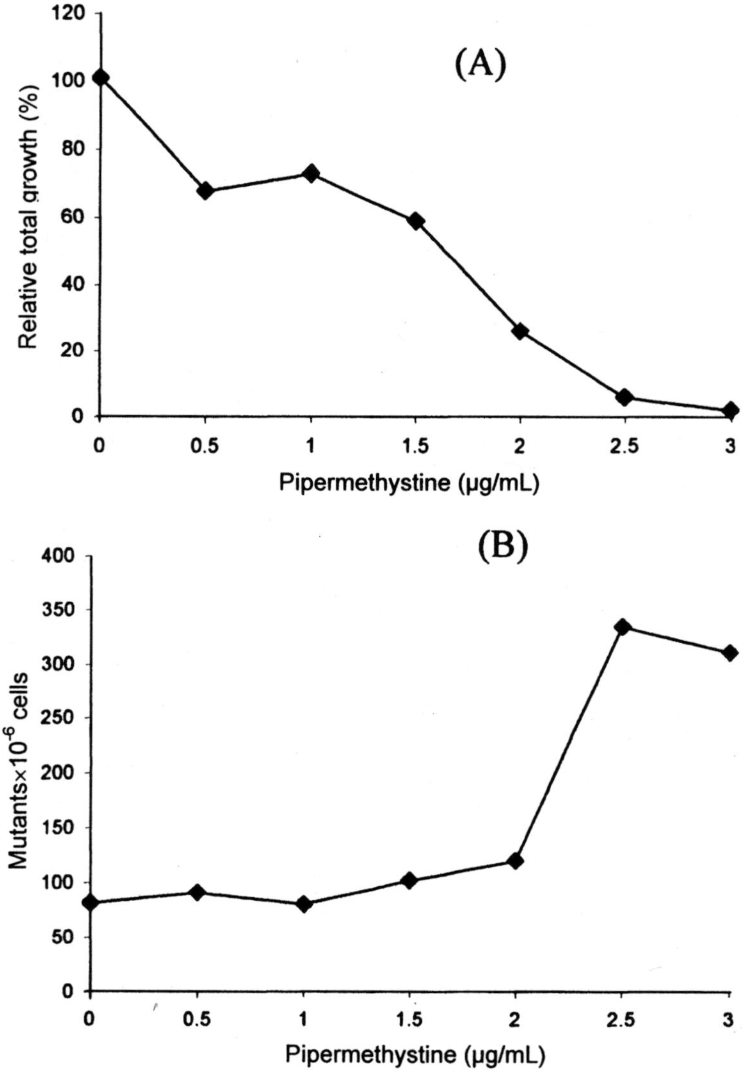

Figure 3. Mutagenicity study of pipermethystine by mouse lymphoma assay; (A) influence of pipermethystine concentration on relative total growth; (B) effect ofpipermethystine concentration on mutant cells

can cause the inhibition of cellular metabolism and prevent further cell division. Therefore, mutant cells are able to proliferate in the presence of TFT, whereas normal cells that have normal *Tk* activity are not. In our study, mutagenicity of pipermethystine was tested with a concentration range from 0.5 μg/mL to 3 μg/mL. The relative total growth (RTG) and the mutant frequencies (MF) with pipermethystine are shown in Figure 3.

Pipermethystine displayed strong cytotoxicity on mouse lymphoma cells. The RTGs for the pipermethystine treatments at 2 and 2.5 μg/mL were 26% and 6%, respectively (Figure 3a). Furthermore, the treatment at 3.0 μg/mL resulted in an RTG of 2% due to the high cytotoxicity and low plating efficiency.

The MFs did not increase over the control in the dose range from 0 to 2.0 μg/ml. Although the MFs were higher at the treatment concentrations of 2.5 and 3.0 μg/ml than the control, we cannot claim that pipermethystine is mutagenic in MLA because the RTGs were 6% and 2% at the two treatments, respectively. A high MF can be produced due to high cytotoxicity when RTG is smaller than 10% (*38*).

Conclusions

Many factors can contribute to kava biological functions and the reported hepatotoxicity. In addition to the genetic factors of individual consumers or populations, other factors, including the plant age, variety and parts, as well as the preparation methods, have a significant influence on the chemical composition of kava beverages or products. Consequently, the chemical composition plays a highly important role in determining the biological end-point. Kavalactones have been considered as the active constituents for the relaxation effect. The ratios of different kavalactones and other minor constituents may also contribute to the beneficial or adverse health effects. Pipermethystine was found in one commercial product tested in this study. It exhibited cytotoxicity, although it was not mutagenic in *in vitro* genotoxic tests. Accurate, sensitive and reliable analytical methods are indispensable for product quality control to ensure that potentially toxic components, such as pipermethystine, are below tolerance levels, and that the proportions of active constituents are appropriate in the final products.

References

1. Lebot, V.; Merlina, M.; Lindstrom, L. *Kava The Pacific Drug*; Yale University Press: New Haven, CT, 1992, pp. 255.
2. Lebot, V.; Merlin, M.; Lindstrom, L. *Kava-The Pacific Elixir: the definitive guide to its ethnobotany, history, and chemistry;* Healing Art Press: Rochester, VT, 1997.

3. Volz, H.P.; Kieser, M. *Pharmacopsych.* **1997**, *30*, 1-5.
4. U.S. Food and Drug Administration. Letter to health-care professionals. http://www.cfsan.fda.gov/~dms/supplement.html. March 25, 2002.
5. Clouatre, D. L. *Toxicol. Lett.* **2004**, *150*, 85-96.
6. Stickel, F.; Baumuller, H.-M.; Seitz, K.; Vasilakis, D.; Seitz, G.; Seitz, H. K.; Schuppan, D. *J. Hepatol.* **2003**, *39*, 62-67.
7. Curie, B. J.; Clough, A. R. *Med. J. Aust.* **2003**, *178*, 421-422.
8. Anke, J.; Ramzan, I. *Planta Med.* **2004**, *70*, 193-196.
9. Bauer, R.; Kopp, B.; Nahrstedt, A. *Planta Med.* **2003**, *69*, 979-972.
10. Whitton, P. A.; Lau, A.; Salisbury, A.; Whitehouse, J.; Evans, C. S. *Phytochemistry* **2003**, *64*, 673-679.
11. Schmidt, M. *J. Altern. Complement. Med.* **2003**, *9*, 183-188.
12. Siméoni, P.; Lebot, V. *Biochem. Syst. Ecol.* **2002**, *20*, 413-424.
13. He, X.; Lin, L.; Lian, L. *Planta Med.* **1997**, *63*, 70-74.
14. Bilia, A. R.; Scalise, L.; Bergonzi, M. C.; Vincieri, F. F. *J. Chromatogr. B* **2004**, *812*, 203-214.
15. Smith, R.M.; Thakrar, H.; Arowolo, A.; Shafi, A.A. *J. Chromatogr. A.* **1984**, *283*, 303-308.
16. Smith, R. M. *Tetrahedron* **1979**, *35*, 437-439.
17. Tavana, C.; Stewart, P.; Snyder, S.; Ragone, D.; Fredrickson, K.; Cox, P. A.; Borel, J. *HerbalGram* **2003**, *59*, 28-32.
18. Denham, A.; McIntyre, M.; Whitehouse, J. *J. Altern. Complement. Med.* **2002**, *8*, 237-263.
19. Singh, Y. N.; Devkota, A. K. *Planta Med.* **2003**, *69*, 496-499.
20. Cote, C. S.; Kor, C.; Cohen, J.; Auclair, K. *Biochem. Biophy. Res. Comm.* **2004**, *322*, 147-152.
21. Stevinson, C.; Huntley, A.; Ernst, E. *Drug Safety* **2002**, *25*, 251-261.
22. Russmann, S.; Lauterburg, B. H.; Helbling, A. *Ann. Inter. Med. Lett.* **2001**, *135*, 68-69.
23. Cawte, J. *Aust. N. Z. J. Psychiatry* **1986**, *20*, 70-76.
24. Mathews, J. D.; Riley, M. D.; Fejo, L.; Munoz, E.; Milns, N. R.; Gardner, I. D.; Powers, J. R.; Ganygulpa, E.; Gununuwawauy, B. J. *Med. J. Aust.* **1988**, *148*, 548-555.
25. Kava, R. *Pacific Health Dialog* **2001**, *8*, 115-118.
26. Russmann, S.; Barguil, Y.; Cabalion, P.; Kritsanida, M.; Duhet, D.; Lauterburg, B.H. *Eur. J. Gastroenterol. Hepatol.* **2003**, *15*, 1033-1036.
27. Clough, A. R.; Bailie, R. S.; Curie, B. J. *J. Toxicol. Clin. Toxicol.* **2003**, *41*, 821-829.
28. Steiner, G. G. *Hawaii Med. J.* **2000**, *59*, 420-422.
29. Clough, A. R.; Wang, Z.; Bailie, R. S.; Burns, C. B.; Curie, B. J. *J. Epidemiol. Comm. Health* **2004**, *58*, 140-141.
30. Sparreboom, A.; Cox, M. C.; Acharya, M. R.; Figg, W. D. *J. Clin Oncol.* **2004**, *22*, 2489-2503.

31. Mathews, J. M.; Etheridge, A. S.; Black, S. R. *Drug Metabol. Disp.* **2002**, *30*, 1153-1157.
32. Unger, M.; Holzgrabe, U.; Jacobsen, W.; Cummins, C.; Benet, L. Z. *Planta Med.* **2002**, *68*, 1055-1058.
33. Zou, L.; Harkey, M. R.; Henderson, G. L.; Dike, L. E. *Planta Med.* **2004**, *70*, 289-292.
34. Dragull, K.; Yoshida, W. Y.; Tang, C.S. *Phytochemistry* **2003**, *63*, 193-198.
35. Nerurkar, P. V; Dragull, K; Tang, C.-S. *Toxicol. Sci.* **2004**, *79*, 106-111.
36. Oda, Y.; Nakamuro, S.; Oki T.; Kato T.; Shinagawa H. *Mut. Res.* **1985**, *147*, 219-229.
37. Miller, J. H. *Experiments in Molecular Genetics*; **1972**. Cold Spring Harbor Laboratory: Cold Spring Harbor, NY, p 352-355.
38. Chen, T., Moore, M.M. *Optimization in Drug Discovery: In Vitro Methods*; **2004**. Humana Press: Totowa, NJ, p. 337-352

Disclaimer: The views presented in this article do not necessarily reflect those of the U.S. Food and Drug Administration, Kangwon National University or University of Hawaii.

Chapter 18

PurePowder®: A New Process for Sterilization and Disinfecting of Agricultural and Botanical Herbal Products

Kan He[1], Zhongguang Shao[1], Naisheng Bai[1], Tangsheng Peng[2], Marc Roller[3], and Qunyi Zheng[4]

[1]Department of Research and Development, Naturex, Inc., South Hackensack, NJ 07606
[2]SGS, 291 Fairfield Avenue, Fairfield, NJ 07004
[3]Naturex, Inc., 84911 Avignon Cedex 9, France
[4]Kent Financial Service, Inc., 211 Pennbrook Road, Far Hills, NJ 07931

Herbal powders comprise a large portion of the botanical products sold in the dietary supplement market. These powders are generally contaminated with microbial organisms found in botanical raw materials either harvested from farms or collected from the wild. Currently, ethylene oxide and gamma irradiation are the dominant modes used to sterilize botanical powders. However, these technologies have recently been subject to safety concerns due to their potentially harmful effects on human health. We report here a new process as an alternative to sterilizing and disinfecting botanical powders using hydrogen peroxide. The mechanism of the sterilization and disinfection is the use of hydrogen peroxide whereby contaminated products are brought into contact with nascent oxygen, resulting in the oxidization and destruction of the contaminating microorganisms.

Botanical products, including potentially contaminated herbs, received a substantial boost when the U.S. Congress passed the Dietary Supplement and Health Education Act in 1994 (*1-3*). Herbal products that are sold in the marketplace consist of either a powder form, derived directly from the ground plant, or an extract. There is less of a concern for microbial contamination on the extract because sterilization has taken place during extraction. Plant powders normally carry a great number of bacteria and molds that often originate in soil. Practices of harvesting, handling, and production may cause additional contamination and microbial growth. Government regulations require low microbial limits for final herbal products that are usually subjected to disinfectant treatments by means of fumigation with ethylene oxide, thermal treatment with steam, and irradiation with γ-rays or high-energy electrons.

Both gamma and electron-beam radiation processes are used commercially for sterilization (*4*). The process is generally considered safe when carried out under controlled conditions and in suitable facilities. However, today's consumers are becoming increasingly reluctant to accept a product that has been irradiated. For safety concerns, herbal supplements are not allowed to be irradiated in the European Union. However, certain constituents of herbal supplements may legally be irradiated provided that the final products indicate their presence on labeling. Ethylene oxide (ETO) is also used by the herbal industry as a sterilant because of its potency in destroying microorganism. Safety concerns are due to the flammable and explosive properties of ETO, as well as it's being a probable human carcinogen. Furthermore, toxic emission, toxicity of by-product ethylene chlorohydrin, and residues of ETO may present health hazards to people. The use of ETO has been forbidden within countries of the European Union and in Japan. Super-heated steam sterilization has the advantage of being an established, effective and well-understood technology. Steam is incompatible with thermo-labile components that comprise a large portion of the materials used in herbal products. To heat-resistant (thermo-phyllic) bacterial strains, steam is less effective.

Hydrogen peroxide (H_2O_2) has been used in many industries as a disinfectant. The hydrogen peroxide solution is a germicide that is active by virtue of the fact that it releases nascent oxygen. It is a very short acting element since this release occurs rapidly. Nascent oxygen is oxygen in atomic form characterized by excessive chemical activity. This form is known to be one of the most reactive chemical species, weaker than the element fluorine and hydroxyl radical, but stronger than ozone, perhydroxyl radical, permangnate, hypobromous acid, chlorine dioxide, chlorine, etc. Hydrogen peroxide is very unstable and is easily decomposed to form oxygen and water by increasing the temperature or by reducing agents. Thus, the use of hydrogen peroxide is predicted as safe, effective, economical and environmentally friendly. By using hydrogen peroxide as an oxidizer, we have developed a new technique, the PurePowder® process, which is based on an oxidation process to achieve complete elimination of microbial contamination (*5*). The process reported herein can be an alternative method to the abovementioned methods.

Materials and Procedures

Test for Residual Hydrogen Peroxide

Hydrogen peroxide, at a concentration of 35%, was purchased from Arkema, Inc., (Philadelphia, PA) and was used in the entire experiment. Residual hydrogen peroxide was tested using EM Quant Peroxide Test kits (EM Science, Gibbstown, NJ). 0.5-1 g of powder was added into 2-4 mL of distilled water (or ethanol or acetone, in the case of organic solvent). The solution was then mixed and sonicated for 15 minutes at room temperature. After sonication, the solution was filtered through a 0.45 μm filter for further use. For this test the strip was dipped into the solution for 1 second. The test strip was removed, the excess liquid was shaken off and the reaction zone was compared with the color scale after 15 seconds. In the case of organic solvents, the test strip was dipped into the solution for 1 second. The test strip is moved slightly to and fro for 3-30 seconds until the solvent has evaporated from the reaction zone. The rest of the procedure is the same as aqueous solutions.

Test for Active Components and Chemical Profiles

The active marker components in herbal products were tested using high performance liquid chromatography (HPLC) technique and run on a Hewlett Packard model 1100 equipped with an autosampler, UV/VIS detector, and Hewlett Packard ChemStation software. HPLC conditions included the use of a Phenomenex, Prodigy ODS (5 μm, 4 ID x 125 mm) column or equivalent C-18 column. Gel permeation chromatography (GPC) was performed on PL aquagel-OH 30, (8 μm, 7.8 ID x 30). The proton nuclear magnetic resonance (NMR) spectra were acquired on a Varian Unity Inova 400 system.

Test for Microorganisms

Total aerobic count (TAC) and yeast & mold (Y&M) count were tested by using the BioMérieux Bactometer method. One gram of sample was added to 99 mL of phosphate dilution buffer (pH 7.2 ± 0.2) (for 1000 cfu/g cutoff), or 10 gram of sample was added to 90 mL of phosphate dilution buffer (for 100 cfu/g cutoff). The mixture was shaken to mix well and a pipette was used to transfer 0.1 mL of this dilution into duplicated wells of a prepared General Purpose Media Bactometer module (for TAC) and a prepared Yeast and Mold Media Bactometer module (for Y&M). The TAC module was incubated in a 35 °C bactometer chamber for 24 hours and the Y&M module was incubated in a 25 °C bactometer chamber for 48 hours. The growth of the microorganism was monitored by the Bactometer Processing System. The presence or absence of *Salmonella* and *Escherichia coli* was tested following the procedures published in USP 24. The Enterobacerial Count was tested following the procedures published by Pharmacopeial Forum (Vol. 25 (2), page 7761).

General Procedure for PurePowder® Process

To the herbal crude powder, about 2-15% of the weight of herbal powder of hydrogen peroxide (H_2O_2, 35% concentration) was applied. After the dispersion of the H_2O_2 solution, the mixing powder was heated for about 10-30 minute at 60-90°C. Samples were taken for microbial and hydrogen peroxide residue testing.

Anti-DPPH Test

The solution of α,α-diphenyl-β-picrylhydrazyl (DPPH) was purchased from Sigma Chemical Co. (St. Louis, MO). A 12.2 mg of DPPH solution was mixed into 100 mL of methanol as stock solution and was stored in a refrigerator and kept from light before using. About 1-2 g of herbal powder was weighed and sonicated in 25 mL of methanol for 1 hour. DPPH stock solution was added to the methanol extract to make the final DPPH concentration of 0.5 mM. The mixture was shaken vigorously and left to stand in the dark at room temperature for 30 minutes. The absorbance of the resulting solution was measured using a spectrophotometer read at 517 nm (Cary 300 Bio UV-visible spectrophotometer). The percentage of anti-DPPH was calculated as the difference of the absorbances at 517 nm of DPPH and sample and divided by absorbance of DPPH.

Results and Discussion

This sterilization process has proven to be effective in the reduction of microbial contamination from a total aerobic count levels from as high as 900,000 to as low as less than 10 cfu/g. Similarly, the yeast and mold count is reduced from as high as 400,000 to less than 10 cfu/g. The process also completely eliminates *Escherichia coli* and *Salmonella* contamination (Table I). The hydrogen peroxide residue was below the testing limit, which was less than 0.5 ppm after the process was completed.

It is generally recognized that, since most of the microorganisms contaminating herbal powders reside on their surfaces, the inner parts do not have to be exposed to heat, gas or radiation for decontamination. In the current method, hydrogen peroxide is applied onto the surface of plant material and the reaction conditions are adjusted to decompose the hydrogen peroxide to release the nascent oxygen or hydroxyl radical, which chemically react with the surrounding molecules causing damage to biological cells, including those of the contaminating microorganisms or insects. Microorganisms are more sensitive to be attacked by hydrogen peroxide while the chemical components inside the plant cell are relatively stable from being oxidized. For example, *Eleutherococcus senticosus* was used as "adaptogen" in the US dietary supplemental market. Phenylpropane derivatives, eleutherosides B and E serve as markers in the standardized Eleuthero product. It was observed that the contents of eleutherosides B and E before and after sterilization with H_2O_2 in Eleuthero powder were similar and the HPLC profiles were identical. This indicates that the chemical compositions have not been significantly changed through this sterilization process (Figure 1).

The US native plant, *Echinacea purpurea*, gained worldwide popularity as an immune stimulant. Phenolic compounds, chicoric acid and caffeoyal tartaric acid are two major components used as markers in the standardized *Echinacea purpurea* product. Similar to Eleuthero powder, no significant changes were found in the contents of chicoric acid and caffeoyal tartaric acid in Echinacea powder before and after H_2O_2 processing (Figure 2).

Valeriana officinalis has a long history of being prescribed for sleep disorder. Sesquiterponoids, valerenic acids are the markers of quality control of Valerian products. No significant degradation was detected in the contents of valerenic acids through H_2O_2 treatment (Figure 3).

The above examples demonstrate that hydrogen peroxide attacks the microorganism on the surface of the plant more than the chemical components inside the plant. Similar examples containing varieties of natural products can also be found without significant degradation. These included triterpene glycosides in ginsneg, black cohosh, astragalus, tetraterpenoid β-carotene in carrot, sequiterpene parthenolide in feverfew, iridoids harpagoside in devil's claw, flavonoids in ginkgo biloba, milk thistle, passion flower, anthraquinones pseudohypericin, hypericin in St. John's wort, fatty acids in saw palmetto, maca, alkaloids in goldenseal and yohimbe, etc. In some cases there were some color changes, which were noticed in the current sterilization process if the condition was not well controlled, i.e. over using hydrogen peroxide and overheating. In the case of powders milled to a very fine mesh, some degradation of chemical components was observed with the probable explanation that due to a fine

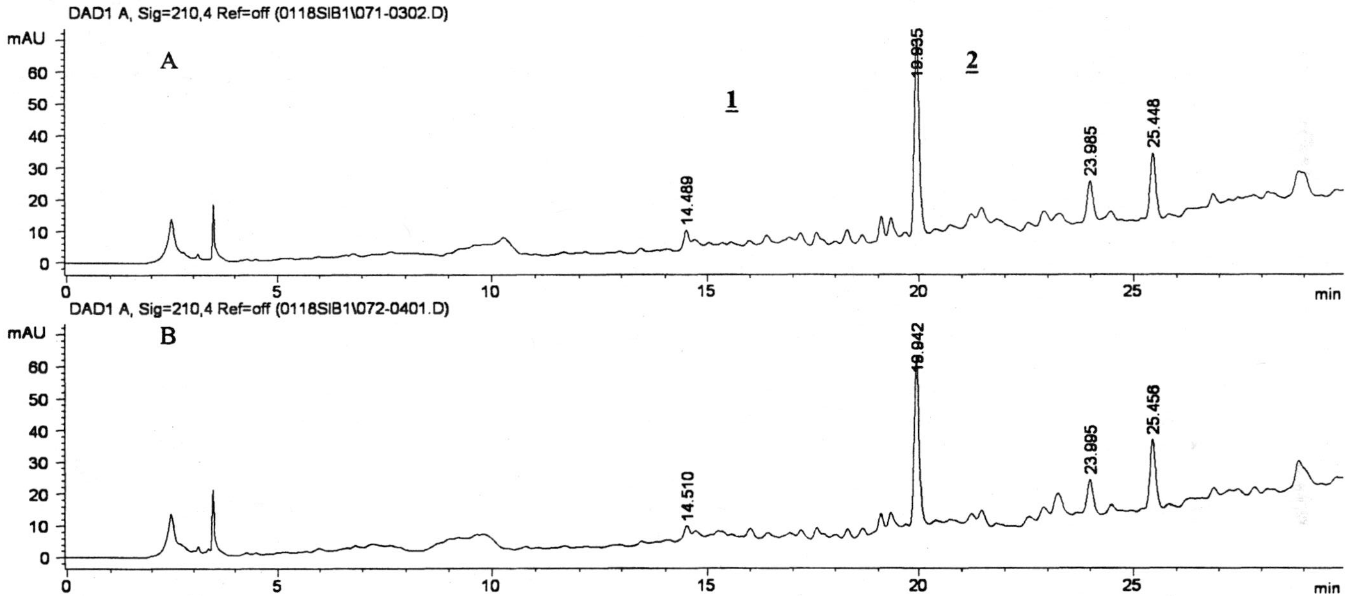

Figure 1. Comparison of HPLC chromatograms before (A) and after sterilization (B). 1 eleutheroside B, 2 eleutherosides E.

Table 1. Results of Microbial Analysis on Some Examples of Herbal Powder before and after Hydrogen Peroxide Processing

Herbal Name	Process	TAC[a]	E&S[b]	Y&M[c]
Black cohosh root & rhizome	Before	>10,000	P[d]	>4,000
powder (*Cimicifuga racemosa*)	After	<200	N[e]	<200
Echinacea leaf powder	Before	>100,000	P	>20,000
(*Echinacea purepurea*)	After	<1,000	N	<100
Eleuthero root powder	Before	>80,000	P	>20,000
(*Eleutherococcus senticosus*)	After	<1,000	N	<100
Ginger rhizome	Before	>480,000	P	>220,000
(*Zingiber officinalis*)	After	<200	N	<200
Green tea leave powder	Before	>100,000	P	>10,000
(*Camellia sinensis*)	After	<100	N	<20
Guarana seed powder	Before	>25,000	P	>25,000
(*Paullinia cupana*)	After	<1,000	N	<100
Milk thistle seeds powder	Before	>112,000	P	>116,000
(*Silybum marianum*)	After	<200	N	<200
Olive leave powder	Before	>5,000	P	>2,000
(*Olea europaea*)	After	<200	N	<200
Panax ginseng root powder	Before	>52,000	P	>32,000
(*Panax ginseng*)	After	<100	N	<100
Psyllium husk powder	Before	>900,000	P	>400,000
(*Plantago ovata*)	After	<100	N	<20
Saw palmetto berry powder	Before	>16,000	P	>4,000
(*Serenoa repens*)	After	<200	N	<200
St. John's wort flower powder	Before	>4,000	P	>4,000
(*Hypericum perforatum*)	After	<200	N	<200
Valerian root powder	Before	>140,000	P	>55,000
(*Valeriana officinalis*)	After	<100	N	<100

[a]Total Aerobic Count, [b]Yeast & Mold, [c]*Escherichia coli* and *Salmonella*, [d]Positive, [e]Negative.

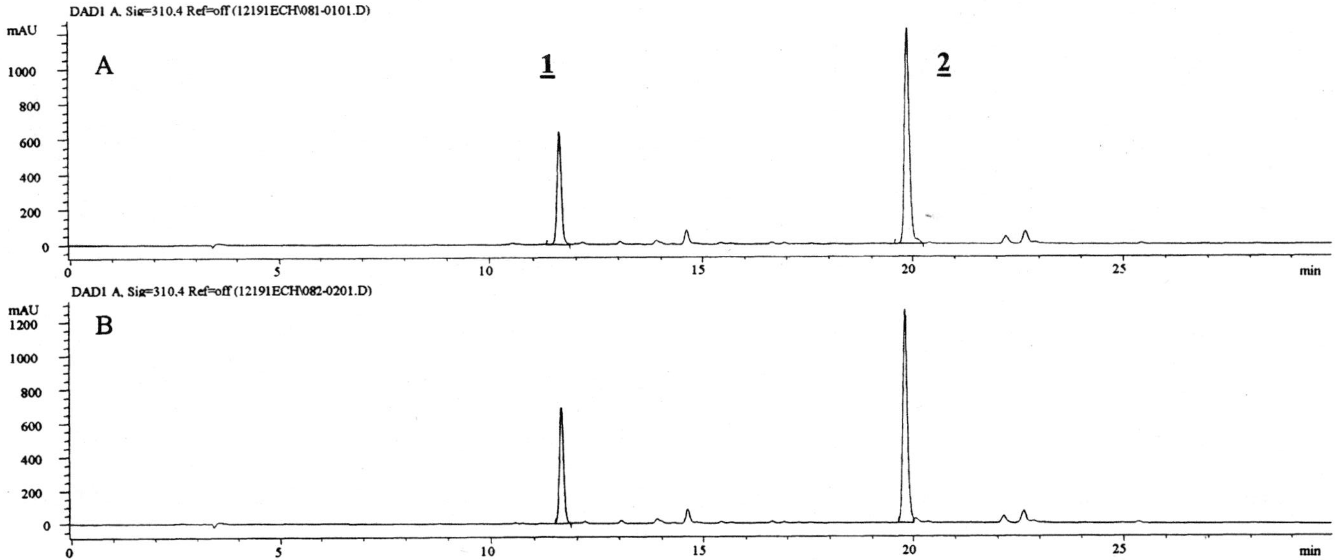

Figure 2. Comparison of HPLC chromatograms before (A) and after sterilization (B). 1 caffeoyal tartaric acid, 2 chicoric Acid.

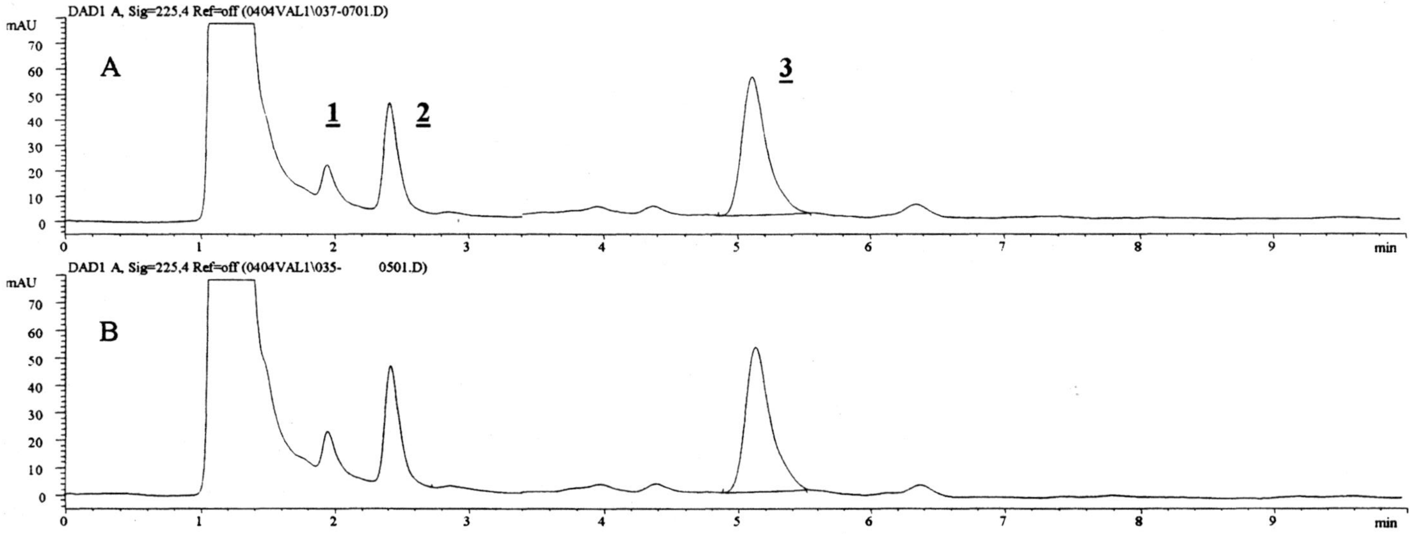

Figure 3. Comparison of HPLC chromatograms before (A) and after sterilization (B). 1 hydroxyvalerenic acid, 2 acetoxyvalerenic acid, 3 valerenic acid.

milling, more plant cell were broken leading to the leakage of chemical components.

The efficiency of sterilization relied on the condition of the decomposition of the hydrogen peroxide applied. For example, in the processing of Eleuthero powder, a different concentration of hydrogen peroxide, with and without heat, was compared. High temperature at 70-80°C can effectively remove microbial contamination. This is understandable since more hydrogen peroxide molecules are converted to release oxygen radicals (Table II).

Table II. Sterilization Results at Different Concentration of Hydrogen Peroxide and Temperature for Eleuthero Powder.

	Untreated	3%HP r.t.	3%HP heated	5%HP r.t.	5%HP heated	7%HP r.t.	7%HP heated
TAC	>1000	>1000	<1000	>1000	<1000	>1000	<1000

HP: hydrogen peroxide. r.t.: room temperature. heated: 70-80°C.

The USP product Psyllium (*Plantago ovata* or *P. psyllium*) is another good example which demonstrates the effectiveness of PurePowder® process. Psyllium contains mucilaginous polysaccharides that are active components that absorb excess water while stimulating normal bowel elimination. Recent interest in psyllium has arisen due to its high fiber content with cholesterol-lowing activity. Psyllium started with a very high load of microbial contamination. ETO is still the most widely used method of treating Psyllium. There is always the concern with the ETO residual levels. Some of the companies have developed cycles using gamma irradiation. It is a very narrow treatment window to get sufficient kill without damaging the Psyllium. Ozone has not effectively worked in the treatment of Psyllium because of issues with gas penetration in a static chamber. Steam sanitization of Psyllium works well to sanitize the husk but there is some evidence of degradation of the husk (*6-7*).

Treating Psyllium powder with different concentrations of hydrogen peroxide and different temperatures was tested. The result showed that both high concentrations of hydrogen peroxide at room temperature and low concentrations but with high temperature (70°C) can sufficiently kill microorganism (Table III).

In the comparison of polysaccharide property in Psyllium before and after the processing, it was found that the water absorption value, tested by swelling volume according to Unite States Pharmacopoeia procedure to evaluate the polysaccharide polymer, was the same. The signals from Psyllium polysaccharides showed similar spectra in proton nuclear magnetic resonance (NMR) (Figure 4). The retention time, chromatogram fingerprints, and the peak

Table III. Sterilization Results at Different Concentrations of Hydrogen Peroxide and Temperature for Psyllium Powder.

Psyllium	Untreated	5% HP at r.t. & heat	7% HP at r.t.	7% HP with heat	15% HP at r.t.
TAC	>1000	>1000	~1000	<1000	<1000

HP: hydrogen peroxide. r.t.: room temperature. heat: 70-80°C.

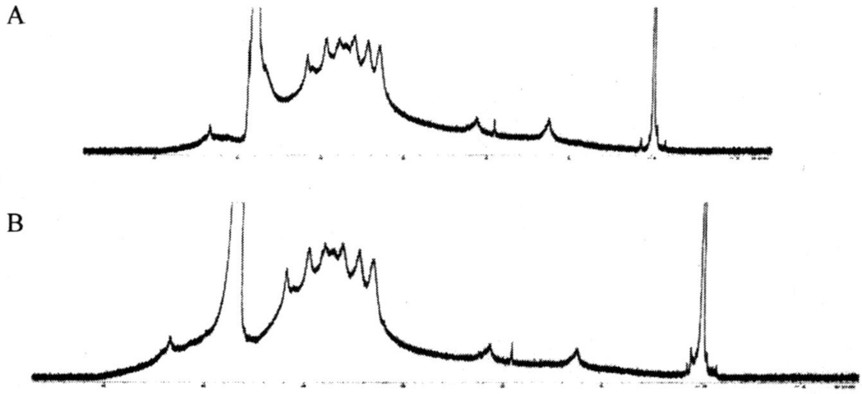

Figure 4. Polysaccharide signal profiles of ^{1}H-NMR spectra for psyllium. A: Untreated psyllium. B: Treated psyllium with H_2O_2 at 70-80°C.

areas of polysaccharides were also similar on the gel permeation chromatography (GPC), which was used to evaluate polysaccharide size and quantities (Figure 5). All these results indicate that the active component of polysaccharides in Psyllium has not changed after the treatment by the hydrogen peroxide solution.

Due to the highly oxidative activity of hydrogen peroxide, it is always a concern whether or not H_2O_2 will destroy the antioxidant capability of certain herbal products, such as green tea, ginger, and olive leaves. An anti-DPPH testing was used to evaluate the changes of their antioxidant activities (*8*). The anti-DPPH results showed 86.1% and 86.0% of inhibitory for green tea, 87.3% and 87.9% for ginger, and 90.3% and 90.0% for olive leaves, respectively. The unchanged anti-DPPH results of these three herbal products indicated that the hydrogen peroxide process did not cause the loss of antioxidant property. This was also confirmed by the comparison of HPLC fingerprints of the powder before and after processing (Figures. 6-8). Catechins including EGCG in green

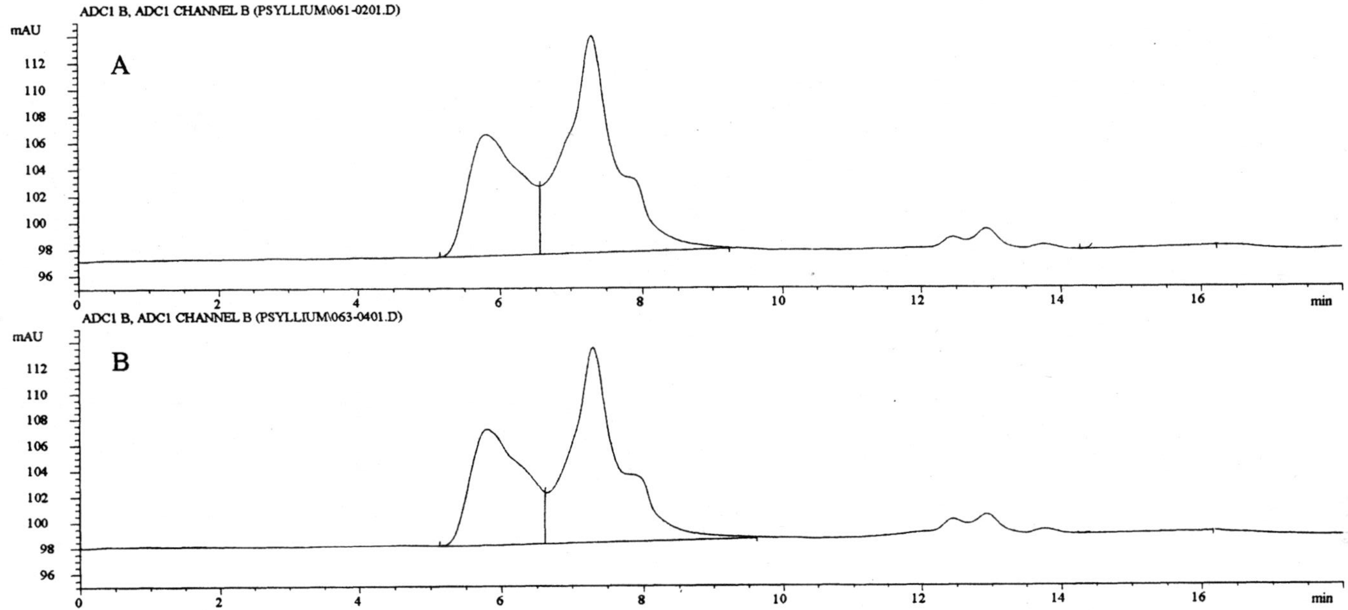

Figure 5. Polysaccharide profiles of GPC chromatograms for psyllium. A: Untreated psyllium. B: Treated psyllium with H_2O_2 at 70-80°C.

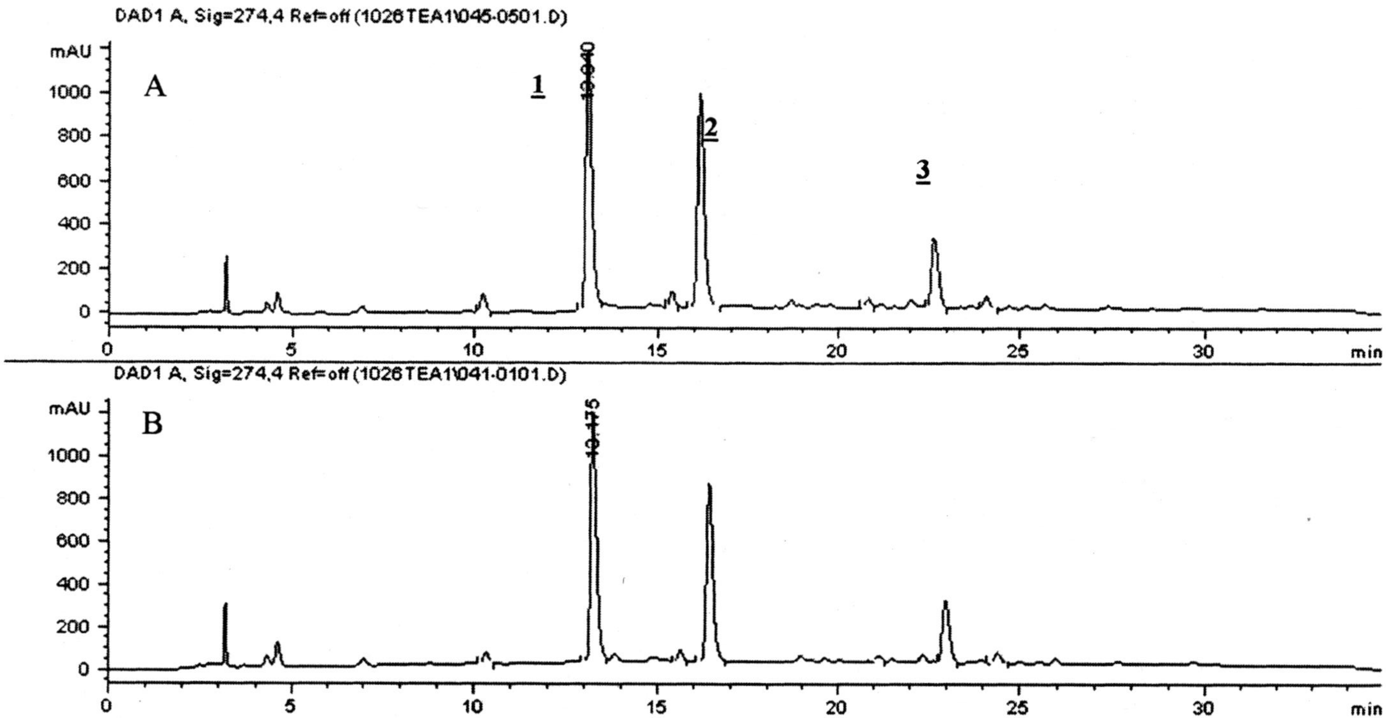

Figure 6. Comparison of HPLC chromatograms for green tea. A: Untreated green tea. B: Treated green tea with H_2O_2 at 70-80 °C. 1 caffeine, 2 (-) EGCG, 3 (-)-ECG.

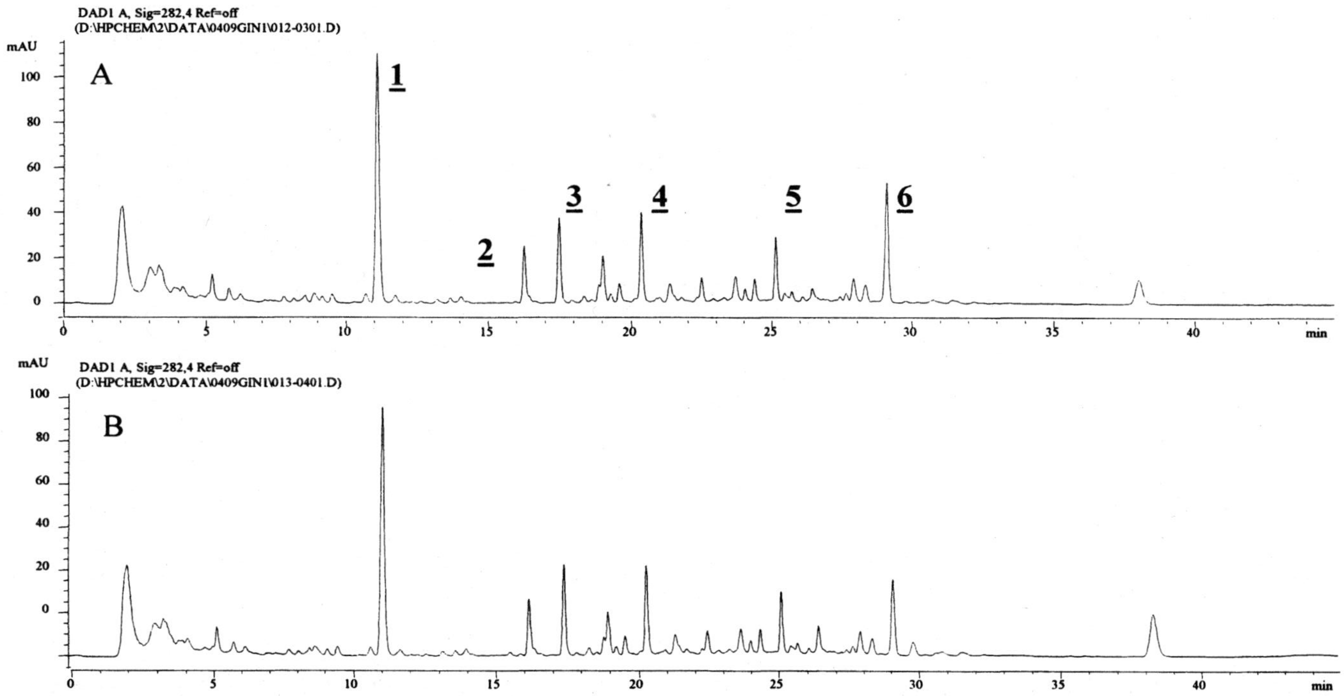

Figure 7. Comparison of HPLC chromatograms for ginger. A: Untreated ginger. B: Treated ginger with H_2O_2 at 70-80°C. 1 6-gingerol, 2 8-gingerol, 3 6-shogaol, 4 6-gingerdione, 5 8-gGingerdione, 6 10-gingerdiol.

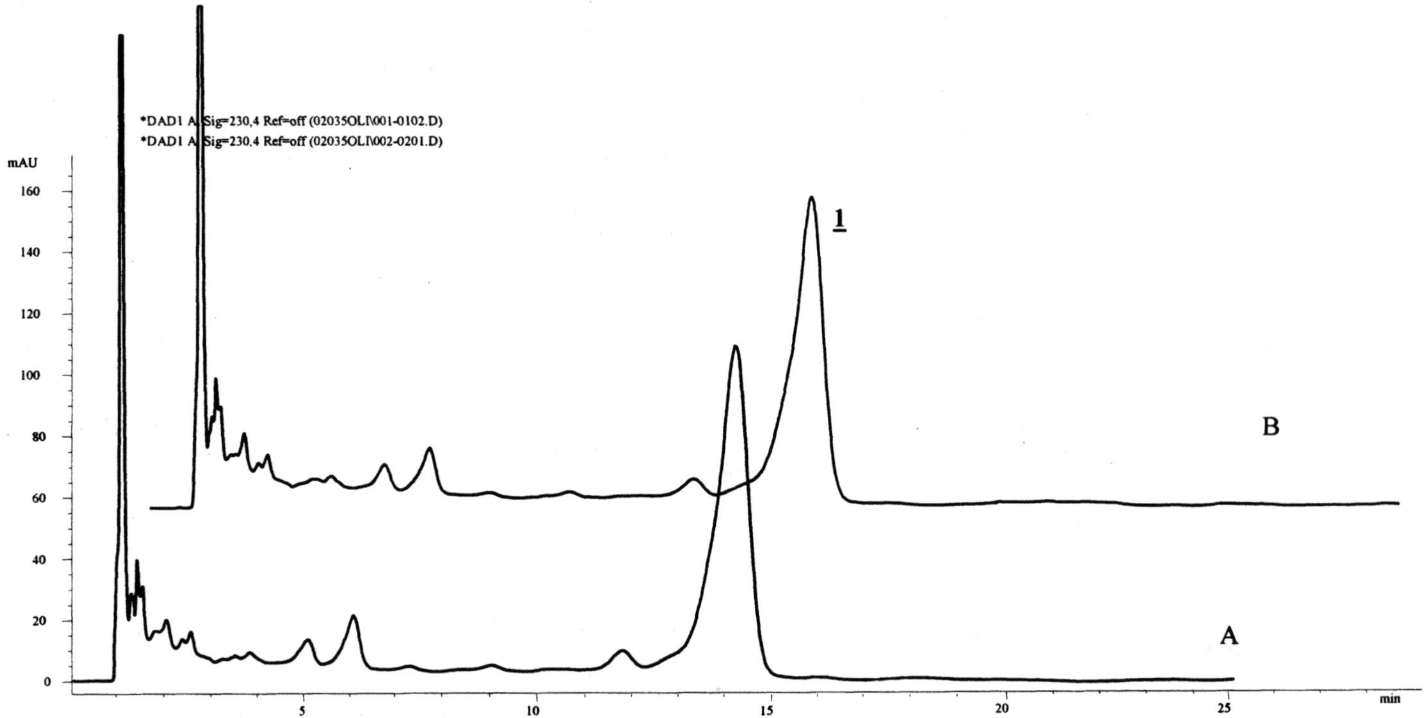

Figure 8. Comparison of HPLC chromatograms for olive leaf. A: Untreated olive leaf. B: Treated olive leaf with H_2O_2 at 70-80°C. 1 Oleuropein

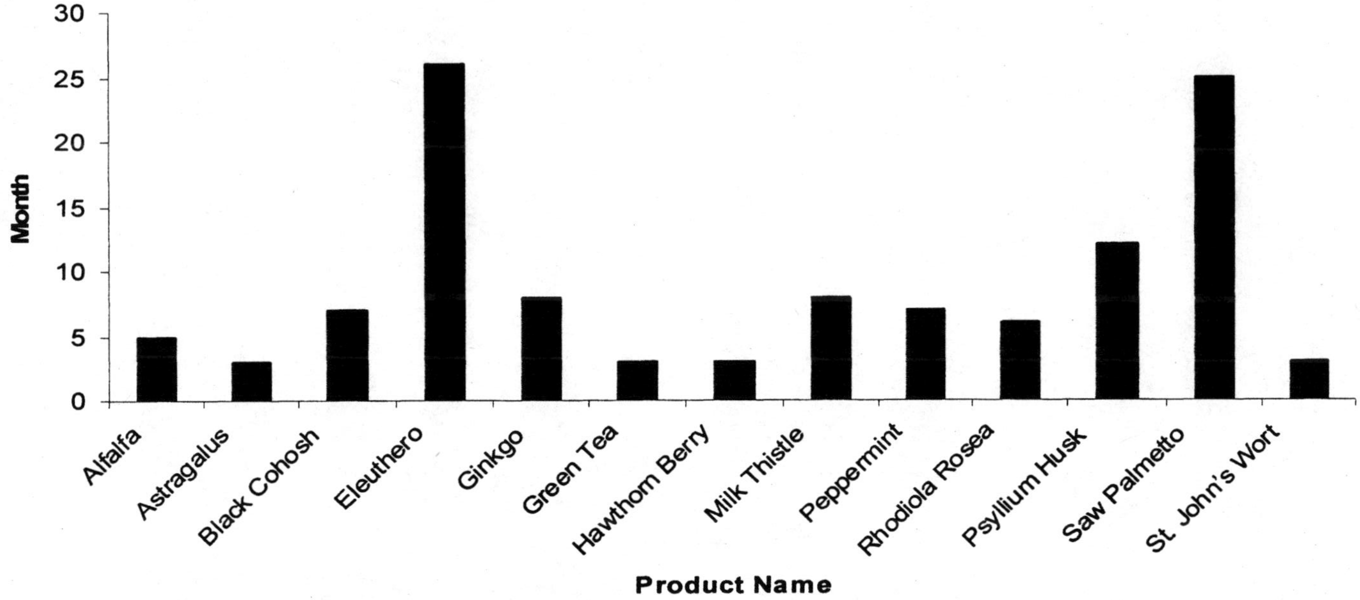

Figure 9. Result of microbial stability after hydrogen peroxide processing.

tea, gingerols in ginger, and oleuropein in olive leaf are well known for their antioxidant capability. The insignificant degradation of these components demonstrated that the current process maintained herbal chemical properties when it achieved successful sterilization of herbal product.

The current sterilization process eliminates the microorganism completely, which can be demonstrated by accessing the microbial stability analysis of the reoccurrence of microorganism growth. No microorganisms were detected for all the sterilized products. The longest period was 25 months (Figure 9).

Conclusion

PurePowder® has achieved successful sterilization on dozens of different botanical species without significantly damaging the chemical components inside the herbal powder. Stability testing shows an effective kill of microorganism that lasts at minimum 2 years under proper storage conditions. Food grade hydrogen peroxide is used in this process to generate free radicals. Following the decomposition pathway, hydrogen peroxide will break down into water and oxygen. Extensive validation testing demonstrates that no residues from this process are left in the product itself, nor in the surrounding environment.

References

1. Kelly, J.P.; Kaufman, D.W.; Kelley, K.; Rosenberg, L.; Anderson, T.E. Mitchell, A.A. *Arch Intern Med.* **2005**, *165*, 281-286.
2. Brevoort, P. *HerbalGram*. **1998**, *44*, 33-46.
3. Molyneaux, M. *HerbalGram*. **2006**, *70*, 68-69.
4. Morehouse, K.M.; Komolprasert, V. In *Irradiation of Food and Packaging Recent Development*, Komolprasert, V.; Morehouse, K.M. Eds. American Chemical Society, Washington, DC, 2004, pp. 1-11.
5. He, K.; Cui, B.L.; Shao, Z.G.; Koether, N.I.; Madis, V.; Zheng, Q.Y. US Patent No. US 6,682,697 B2, Jan 27, **2004**.
6. Cimiluca, P. Personal communication
7. Hukill, S. Personal communication.
8. Chen, C.W. and Ho, C.T. *J. Food Lipids.* **1995**, *2*, 35-46.

Chapter 19

Chemopreventive Effects of Dibenzoylmethane on Mammary Tumorigenesis

Chuan-Chuan Lin[1], Chi-Tang Ho[2], and Mou-Tuan Huang[3]

[1]Department of Food Science, China Institute of Technology, Taipei 115, Taiwan
[2]Department of Food Science, Rutgers, The State University of New Jersey, 65 Dudley Road, New Brunswick, NJ 08901–8520
[3]Laboratory for Cancer Research, School of Pharmacy, Rutgers, The State University of New Jersey, Piscataway, NJ 08854–8020

Dibenzoylmethane (DBM), a β-diketone structural analogue of curcumin, has been reported to exhibit chemopreventive activities in mammary tumorigenesis during the past few years. The underlying mechanisms might be complex and have not been well characterized, especially its function at the molecular level. In this report, we overview the recent mechanistic studies of DBM on the multiple stages of mouse mammary carcinogenesis from both chemical and biological aspects.

The Chemistry and Biology of Dibenzoylmethane

Dibenzoylmethane (DBM), a β-diketone structural analogue of curcumin, has been reported to exhibit anti-tumorigenic and chemopreventive activities during the past few years (*1-7*). Both DBM and its derivatives have been used as sun-screening agents (*8*). In biological aspects, DBM inhibits the mutagenicity and nucleic acid binding of chemical carcinogens *in vitro* (*9-11*). It modulates the Phase I/Phase II metabolic systems, induces apoptosis in various cancer cells, and, as a metal-chelator, exerts beneficial effects for the ischemic diseases (*1,3,5,12*). Talalay's group indicated that DBM and its structural derivatives are potent inducers of Phase 2 detoxification enzymes (*5*).

DBM shares a similar structural feature with another naturally-occurring dibenzoylmethane derivative, licodione, [1-(2,4-dihydroxyphenyl)-3-(4-hydroxyphenyl)-1,3-propanedione], a biosynthetic intermediate isolated from cultured *Glycyrrhiza echinata* L. cells (one licorice species) (*13*). Several isoprenoid-substituted dibenzoylmethanes were isolated from different licorice species (*14, 15*). Licorice, a sweet-tasting Glycyrrhiza (leguminosae) root, has long been used as a flavoring agent and recently as an anti-ulcer, anti-inflammatory agent in Eastern and Western countries (*16-18*). In addition to the main saponin triterpene constituent, glycyrrhizin, several bioactive flavonoid components, e.g. retrochalcones and dibenzoylmethane derivatives, have been isolated and tested for their biological activities (*16*). Among them, the licochalcone A from Xin-Jiang licorice showed potent inhibitory effect on DMBA/TPA-induced tumorigenesis (*17*). Due to the antitumor and anti-inflammatory effects of both curcumin and licochalcone, the structural analogues dibenzoylmethane and its derivatives have received more attention, especially its potential use as a chemopreventive agent (*6, 7*). The studies of the chemopreventive effects of DBM both *in vivo* and *in vitro* during the past few years are summarized in Table I and Table II.

The Multiple Stages of Mouse Mammary Carcinogenesis

In both rat and mouse models, DBM has been reported to have inhibitory effects toward carcinogen-induced mammary tumorigenesis (*2, 6, 7*). The typical animal model used for the induction of formation of mouse mammary tumors is shown in Figure 1. The mechanisms involved in this carcinogenic process are also proposed. 7,12-dimethylbenz[a]anthracene (DMBA) is treated orally for 5 weeks and the formation of mammary tumor would occur after the following few weeks. In the initiation stage, the DNA mutation is induced by the bioactivated metabolite of DMBA; in the promotion stage, estrogen is believed to act as an endogenous promoter in this case because estrogen can bind to the estrogen-responsible receptor for induction of cell proliferation. The chemopreventive agents would exert effects on the modulation of both DMBA and estrogen metabolisms or they might act as anti-estrogenic agents in this anti-carcinogenic process.

Inhibitory Effect of Dietary DBM on Mammary Tumorigenesis

Studies from Huang's laboratory indicated that while dietary curcumin had little or no effect on DMBA-induced breast tumorigenesis in mice, dietary DBM

Table I. Chemopreventive Effects of DBM *in vivo*

Biological activity	*Reference*
Inhibit DMBA-induced mouse and rat mammary tumorigenesis	*6, 7*
Inhibit lymphomas/leukemias in Sencar mice	*6*
Inhibit mammary DMBA-DNA adducts formation in mice and Rats	*2, 7*
Increase Phase II enzymatic activity in rat liver	*4*
Inhibit proliferation of mammary gland in mice	*2*
Decrease several biomarkers related to fat and lower serum estrogen in mice	*2*

Table II. Chemopreventive Effects of DBM *in vitro*

Biological activity	*Reference*
Induce Phase 2 detoxification enzymes in murine hepatoma cells	*5*
Inhibition of carcinogen-DNA adduct formation in MCF-10F	*4*
Modulate AhR function and expression of cytochromes P450 1A1, 1A2, and 1B1 in HepG2	*3*
Inhibit DMBA metabolism and the formation of DMBA-DNA adducts	*2*
Competitive binding to estrogen receptors with [^{3}H]-estradiol	*2*
Induce HIF-1 alpha and increases expression of VEGF	*12*
Induce cell cycle deregulation in various human cancer cells	*1*

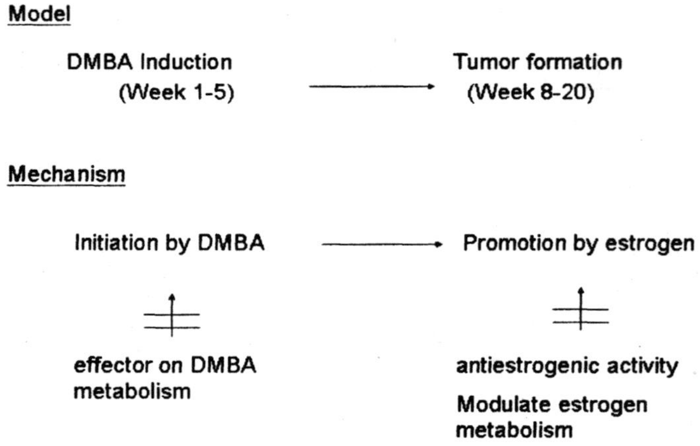

Figure 1. Typical mouse model for DMBA-induced mammary tumorigenesis and possible ways of inhibition

exhibited remarkable inhibitory effects on DMBA-induced mammary tumorigenesis in Sencar mice (*6*). Oral administration of 1 mg of DMBA to female Sencar mice (6 weeks old) once a week for 5 weeks, 68% of mice developed mammary tumors (average 1.08 tumors per mouse) at 20 weeks after the first dose of DMBA. Feeding 1% DBM in the diet at 2 weeks before DMBA treatment until the end of the experiment, inhibited both the multiplicity and incidence of mammary tumors by 97%. It was of interest to find that 2% curcumin diet had no effect on the DMBA-induced mammary tumor formation in mice. Similar results from another group also showed that DBM, but not curcumin, when added to diets fed to female rat, inhibited the formation of DMBA-DNA adducts *in vivo* and DMBA-induced rat mammary tumorigenesis (*7*).

In another animal study, we try to examine the inhibitory effect of dietary DBM on cell proliferation of mammary gland and uterus in Sencar mice. Immature female Sencar mice (26 days of age) were fed with 1% DBM in the diet until the first estrous phase of the estrous cycle. The average bromodeoxyuridine labeling index in the uterus (including epithelium and stroma) and mammary gland were decreased by 40% and 53%, respectively, compared to the corresponding control diet mice (*2*). This result showed a primary indication of possible antiestrogenic activity of DBM in Sencar mice.

This report summarizes recent mechanistic studies of DBM on the multiple stages of mouse mammary carcinogenesis from chemical and biological aspects (*30, 31*).

The Metabolic Fate of DBM and its Implication on DMBA-induced Mouse Tumorigenesis

DMBA, an effective carcinogenic initiator, is metabolically activated by cytochrome P450 oxidase to electrophilic diol-epoxide intermediates, which subsequently interact with DNA to form DMBA-DNA adducts (*19, 20*). In our previous *in vivo* and *in vitro* studies, DBM inhibited DMBA metabolism and formation of DMBA-DNA adducts in a dose-dependent manner (*2, 21*). Investigation of the underlying mechanism regarding the involvement of β-diketone moiety of DBM on mammary tumorigenesis is important. The β-diketone functionality in curcumin has been shown to exhibit antioxidative activity on tert-butylhydroperoxide-induced lipid peroxidation of erythrocyte membrane ghosts (*22*). Talalay also reported the potency of DBM as an inducer of Phase II detoxification enzymes, in part due to the β-diketone functionality (*5*). However, not any report published so far has concerned the influence of the β-diketone group on cytochrome P450 Phase I metabolizing enzymes, since DMBA needs to be oxidatively metabolized to bioactive carcinogen (*19*).

We examined the metabolic fate of DBM by NADPH-dependent cytochrome P450 enzymes in mouse liver microsomes. The identification of a major reductive DBM metabolite as well as several minor metabolites in β-diketone moiety from incubation with mouse liver microsomes *in vitro* is presented. Meanwhile, the possible metabolic pathway of DBM is proposed in Figure 2. These may provide partially the explanation from chemical aspects of the role of DBM as a modulator of the cytochrome P450 reductase that is required for the function of oxidase to metabolize DMBA. This might also result in the inhibition of DBM on DMBA-induced tumorigenesis.

The Inhibitory Effect of DBM on Estradiol-induced Mammary Proliferation

Overexpressions of oncogenes induced by estradiol (E_2) have been suggested to lead to the mammary tumorigenesis in animal and proliferation of human cancerous cells (*23-26*). Specifically, the synergistic effects of oncogenic expressions, e.g., *c-myc*, *ras*, *bcl-2*, and telomerase were observed in both transgenic mice and cultured cells, resulting in abrupt cellular proliferation and tumor formation (*27-29*). Consistently, estrogen response elements (EREs), which are required for gene expressions, have been identified in either promoter positions or coding sequences in several oncogenes (*23-26*).

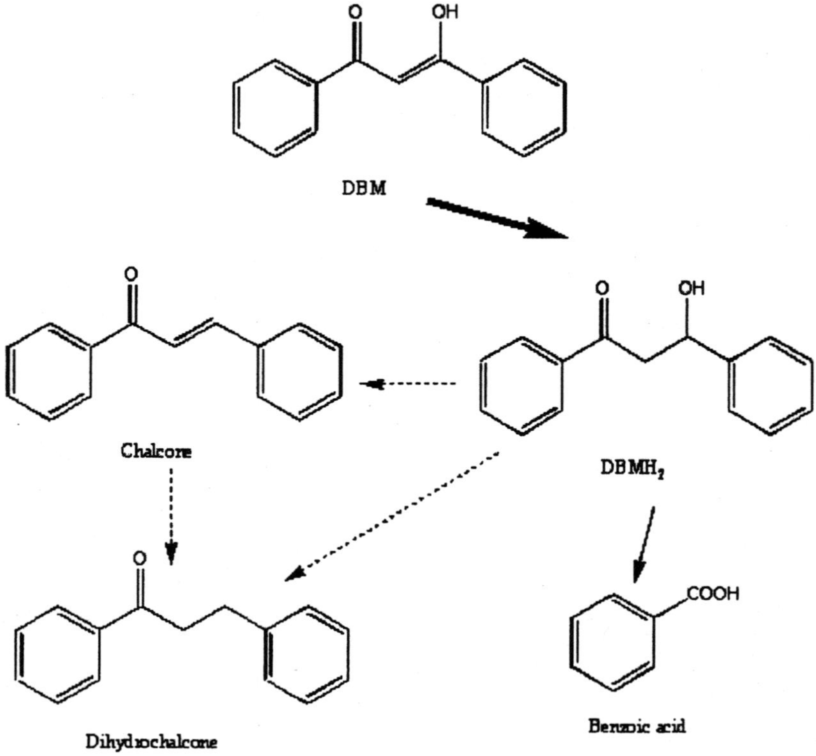

Figure 2. Proposed metabolic pathway of DBM by mouse liver microsomes

To examine how DBM affects E_2-dependent cell proliferation, the expressions of four oncogenes, *bcl-2*, *c-myc*, *Ha-ras* and *hTERT*, with their EREs having been identified in estrogen receptor-positive MCF-7 cells, were examined by quantitative RT-PCR, a technique for quantitative analysis of gene expression with high specificity. In a time-course study, MCF-7 cells were treated with 100 nM of E_2 for several time intervals and the results indicated that E_2-induced *c-myc*, *Ha-ras* and *bcl-2* reached their maximum expression levels after induction for 2 hrs, whereas *hTERT* required 8 hrs to aggrandize its climax (data not shown). As shown in Figure 3A, the expression levels of *hTERT*, *c-myc*, *Ha-ras* and *bcl-2* in E_2-treated cells were increased by 4.6, 4.1, 2.4, and 5.4 fold, respectively, compared to the E_2-untreated control. Treatment of 10 μM of DBM together with 100 nM E_2 reduced the expression of these four oncogenes to the basal levels, for *c-myc*, where a lesser extent of attenuation was observed. The inhibitory pattern of DBM was also compared to that of tamoxifen, a SERM that antagonizes the estrogenic action of E_2 in MCF-7 cells. Both DBM and tamoxifen exhibited similar patterns of inhibition as shown in Figure 3. Tamoxifen decreased the expressions of *hTERT*, *c-myc*, and *bcl-2*. However, tamoxifen did not show significant reduction of expression in *Ha-ras*. These results suggest that DBM inhibits the expression of E_2-regulated oncogenes, which might attribute to its inhibitory effect on E_2-stimulated cellular proliferation.

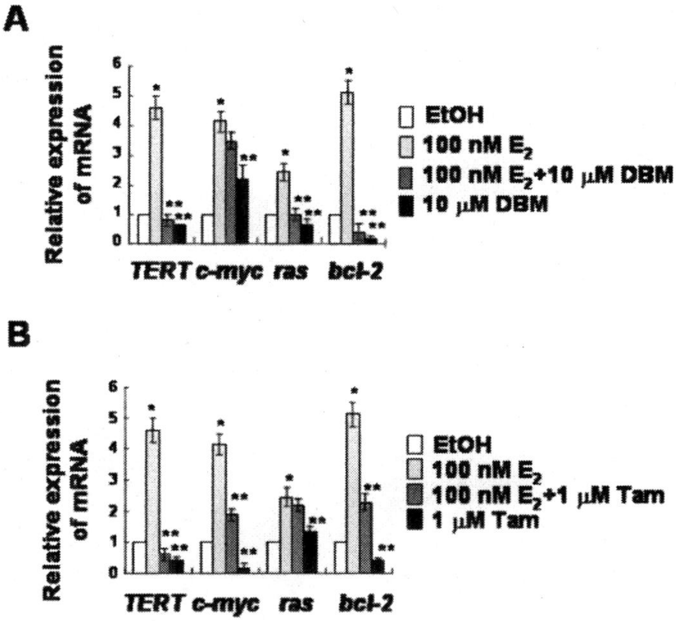

Figure 3. DBM and tamoxifen (Tam) affected the expressions of the E2-ER-ERE-dependent oncogenes in MCF-7 cells. MCF-7 cells were treated with EtOH, 100 nM E2, 100 nM E2+10 μM DBM, and 10 μM DBM (A) or EtOH, 100 nM E2, 100 nM E2+1 μM Tam, and 1 μM Tam (B) for 8 hrs to detect hTERT expression or 2 hrs to others. RNA was extracted and cDNA was then synthesized according to the standard protocol. The expression of hTERT, c-myc, Ha-ras and bcl-2 were measured by quantitative PCR analysis. Data are presented as the mean ± S.D. from triplicate determinations. Significant differences (P<0.05) are indicated by asterisks. (Adapted with permission from reference 31. Copyright 2006 Oxford University Press.)

To further investigate the underlying mechanism of DBM as an anti-estrogenic agent, chromatin immunoprecipitation (ChIP) was employed to examine the DBM on the ER binding to EREs within the regulatory regions of the target genes. *hTERT* and *bcl-2* promoters both harbor imperfect palindromic E_2-responsive elements (at -2,677 for *hTERT* and at +195 and +276 for *bcl-2*), which interact directly with ER. *c-myc* has been reported to contain a 116-bp GC-rich non-classical ERE (at +25 to +141). Our results showed that DBM alone or together with E_2 attenuated ERα binding to the responsive elements in these three genes in comparison with the E_2-treated control (Figure 4A). In real-time PCR analysis used to quantify the signals from ChIP, treatment of DBM exhibited at least 50% decrease in the binding efficiency of ER to the EREs (Figure 4B). Especially, more than 50% decrease in the binding efficiency was observed in the ERE of *c-myc*. Therefore, DBM significantly attenuates the binding of ER to the EREs.

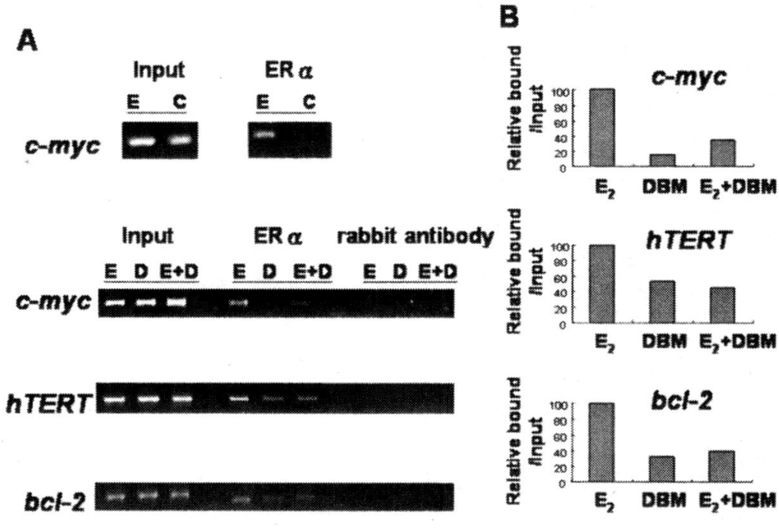

Figure 4. DBM attenuated the binding of ER to the EREs of c-Myc, hTERT, and bcl-2. (A) MCF-7 cells were treated with EtOH (C), 100 nM E2 (E), 10 μM DBM (D), or both (E+D) for 2 hrs to detect c-myc and bcl-2 or 8 hrs to detect hTERT. ChIP assays were performed using the antibody against ERα or an unrelated rabbit antibody. The final DNA extracts were amplified using pairs of primers that cover the regions of EREs as described in the text. The data were representatives of two or more experiments from independent immunoprecipitations. (B) Relative amounts of immunoprecipitated c-myc, hTERT, and bcl-2 ERE sequences compared to input chromatin were determined using quantitative PCR. (Adapted with permission from reference 31. Copyright 2006 Oxford University Press.)

Finally, we examined the anti-proliferative effect of DBM in the mammary glands of immature mice induced by E_2. First, female Sencar mice were given a single i.p. injection of varying concentration of DBM together with 0.3 µg of estradiol. While E_2 stimulated in a dose-dependent manner (Figure 5A), DBM inhibited E_2-induced incorporation of BrdU into mammary DNA in immature female Sencar mice in a dose-dependent manner (Figure 5B). In another set of mouse model, female Sencar mice were i.p. injected with 10 µmol of DBM, tamoxifen or vehicle control before injection of 0.3 µg of estradiol once a day for two days. BrdU (50 mg/kg body weight) was injected i.p. 2 hr after the second administration of estradiol and all mice were sacrificed one hour later. Figure 5C showed that DBM and tamoxifen reduced the BrdU incorporation in Sencar mice by 52% and 40%, respectively. These results demonstrate that DBM, like tamoxifen, inhibits the E_2-iduced mammary gland proliferation in a mouse model.

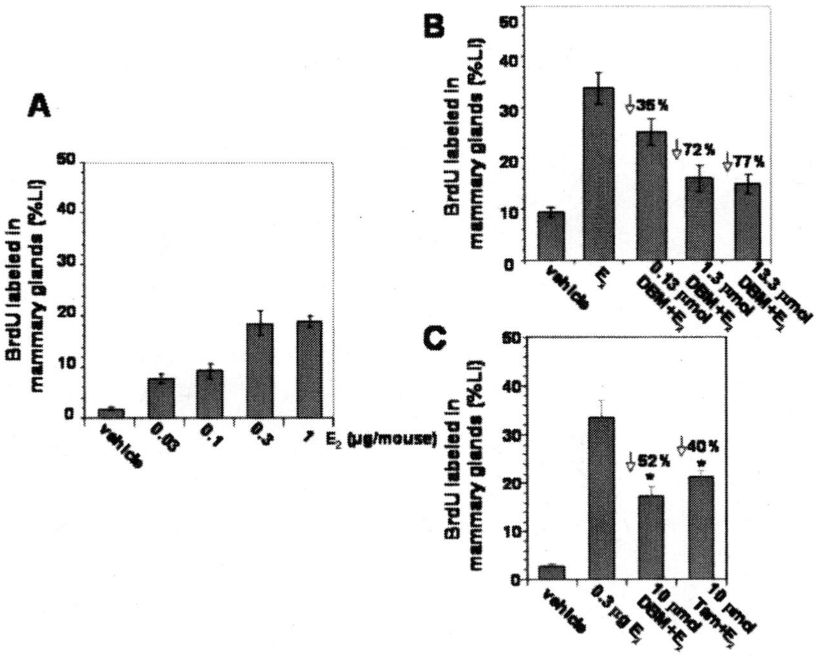

*Figure 5. DBM inhibited the E2-induced mammary gland proliferation in a mouse model. (A) Dose-response of E2 on BrdU incorporation. Female (B) Dose-response of DBM on E2-induced BrdU incorporation. (C) Inhibitory effects of DBM and tamoxifen on the incorporations of BrdU into mammary glands. Data are the mean ± S.E. from 5 mice. *, significantly different from the E2 injection alone. (Adapted with permission from reference 31. Copyright 2006 Oxford University Press.)*

Conclusion

Our previous study indicated that dietary DBM inhibits the incidence of 7,12-dimethylbenz[a]anthracene (DMBA)-induced mouse mammary tumorigenesis and decreases the number of breast tumors per mouse as well as the formation of DMBA-DNA adducts in mammary glands (*2*). *In vitro* competitive estrogen receptor binding studies demonstrated a direct binding between DBM and estrogen receptor *in vitro* (*21*). Further *in vivo* proliferation studies implicated the potential role of DBM as an anti-estrogenic agent (*2*). The underlying mechanisms might be complex and have not been well characterized, especially its function at the molecular level. In this report, we demonstrate that DBM strongly inhibits the E_2-induced proliferation in both the human breast cancer cell line MCF-7 as well as in a mouse model, reduces the expression of *bcl-2*, *c-myc*, *Ha-ras* and, *hTERT* and acts as an anti-estrogenic agent by attenuating the ER-ERE binding within the regulatory regions of these oncogenes. Taken together, our data suggest that DBM serves as a valuable model molecule for chemopreventive and anti-estrogenic drugs in tamoxifen-resistant cases.

References

1. Jackson, K.M.; DeLeon, M.; Verret, C.R.; Harris, W.B. *Cancer Lett.* **2002**, *178*, 161-165.
2. Lin, C.C.; Lu, Y.P.; Lou, Y.R.; Ho, C.-T.; Newmark, H.H.; MacDonald, C.; Singletary, K.W.; Huang, M.T. *Cancer Lett.* **2001**, *168*, 125-132.
3. MacDonald, C.J.; Ciolino, H.P.; Yeh, G.C. *Cancer Res.* **2001**, *61*, 3919-3924.
4. Singletary, K.; MacDonald, C. *Cancer Lett.* **2000**, *155*, 47-54.
5. Dinkova-Kostova, A.T.; Talalay, P. *Carcinogenesis* **1999**, *20*, 911-914.
6. Huang, M.T.; Lou, Y.R.; Xie, J.G.; Ma, W.; Lu, Y.P.; Yen, P.; Zhu, B.T.; Newmark, H.; Ho, C.-T. *Carcinogenesis* **1998**, *19*, 1697-1700.
7. Singletary, K.; MacDonald, C.; Iovinelli, M.; Fisher, C.; Wallig, M. *Carcinogenesis* **1998**, *19*, 1039-1043.
8. Nogueira, M.A.; Magalhaes, E.G.; Magalhaes, A.F.; Biloti, D.N.; Laverde, A.; Pessine, F.B.; Carvalho, J.E.; Kohn, L.K.; Antonio, M.A.; Marsaioli, A.J. *Farmaco.* **2003**, *58*, 1163-1169.
9. Wang, C.Y.; Lee, M.-S.; Nagase, H.; Zukowski, K. *J. Natl. Cancer Inst.* **1989**, *81*, 1743-1747.
10. Wang, C.Y.; Lee, M.-S.; Zukowski, K. *Mutat. Res.* **1991**, *262*, 189-193.

11. Choshi, T.; Horimoto, S.; Wang, C.Y.; Nagase, H.; Ichikawa, M.; Sugino, E.; Hibino, S. *Chem. Pharm. Bull.* **1992**, *40*, 1047-1049.
12. Mabjeesh, N.J.; Willard, M.T.; Harris, W.B.; Sun, H.Y.; Wang, R.; Zhong, H.; Umbreit, J.N.; Simons, J.W. *Biochem. Biophys. Res. Commun.* **2003**, *303*, 279-286.
13. Furuya, T.; Ayabe, S.-I.; Kobayashi, M. *Tetrahedron Lett.* **1976**, *29*, 2539-2540.
14. Demizu, S.; Kajiyama, K.; Hiraga, Y.; Kinoshita, K.; Koyama, K.; Takahashi, K.; Tamura, Y.; Okada, K.; Kinoshita, T. *Chem. Pharm. Bull.* **1992**, *40*, 392-395.
15. Fukai, T.; Nishizaira, J.; Nomura, T. *Phytochem.* **1994**, *35*, 515-519.
16. Shibata, S.; Saitoh, T. *J. Indian Chem. Soc.* **1978**, *55*, 1184-1191.
17. Shibata, S. In *Food Phytochemicals for Cancer Prevention II.* Ho, C.-T.; Osawa, T.; Huang, M.T.; Rosen, R.T., Eds.; American Chemical Society: Washington, DC, 1994; pp. 308-321.
18. Mizutani K. *Food Phytochemicals for Cancer Prevention II.* Ho, C.-T.; Osawa, T.; Huang, M.T.; Rosen, R.T., Eds.; American Chemical Society: Washington, DC, 1994; pp. 322-328.
19. Wood, A.W.; Levin, W.; Chang, R. L.; Yagi, H.; Thakker, D. R.; Lehr, R.E.; Jerina, D.M.; Conney, A.H. In *Polynuclear Aromatic Hydrocarbons.* Jones, P.W.; Leber, P., Eds.; Ann Arbor Science Publishers: Michigan, 1979; p 531.
20. Hall, M.; Grover, P.L. In *Handbook of Experimental Pharmacology*; Cooper, C.S.; Grover, P.L., Eds.; Springer-Verlag: New York, 1990; Vol 94/I, p 327.
21. Lin, C.C.; Ho, C.-T.; Huang, M.T. *Proc. Natl. Sci. Counc. Rep. China B.* **2001**, *25*, 158-165.
22. Sugiyama, Y.; Kawakishi, S.; Osawa, T. *Biochem. Pharmacol.* **1996**, *52*, 519-525.
23. Perillo, B.; Sasso, A.; Abbondanza, C.; Palumbo, G. *Mol. Cell Biol.* **2000**, *20*, 2890-2901.
24. Kyo, S.; Takakura, M.; Kanaya, T.; Zhuo, W.; Fujimoto, K.; Nishio, Y.; Orimo, A.; Inoue, M. *Cancer Res.* **1999**, *59*, 5917-5921.
25. Pethe, V.; Shekhar, P.V. *J. Biol. Chem.* **1999**, *274*, 30969-30978.
26. Dubik, D.; Shiu, R.P. Mechanism of estrogen activation of c-myc oncogene expression. *Oncogene* **1992**, *7*, 1587-1594.
27. Hahn, W.C.; Counter, C.M.; Lundberg, A.S.; Beijersbergen, R.L.; Brooks, M.W.; Weinberg, R.A. *Nature* **1999**, *400*, 464-468.
28. Sinn, E.; Muller, W.; Pattengale, P.; Tepler, I.; Wallace, R.; Leder, P. *Cell* **1987**, *49*, 465-475.

29. Lodish, H.; Berk, A.; Zipursky, L.; Matsudaira, P.; Baltimore, D.; Darnell, J. In *Molecular Cell Biology*; W. H. Freeman and Company: New York, 1999, p. 1055.
30. Lin, C.C.; Wei, G.J.; Huang, M.T.; Ho, C.-T. *J. Food Drug Anal.* **2005**, *13*, 284-288.
31. Lin, C.C.; Tsai, Y.L.; Huang, M.T.; Lu, Y.P.; Ho, C.-T.; Tseng, S.F.; Teng, S.C. *Carcinogenesis* **2006**, *27*, 131-136.

Chapter 20

Antiinflammatory and Anticancer Activities of Garcinol

Mou-Tuan Huang[1], Yue Liu[1], Vladimir Badmaev[2], and Chi-Tang Ho[3]

[1]Department of Chemical Biology, Laboratory for Cancer Research, School of Pharmacy, Rutgers, The State University of New Jersey, Piscataway, NJ 08854–8020
[2]Sabinsa Corporation, 70 Ethel Road West, Unit 6, Piscataway, NJ 08854
[3]Department of Food Science, Rutgers, The State University of New Jersey, 65 Dudley Road, New Brunswick, NJ 08901–8520

Garcinol was isolated from fruit rinds of *Garcinia indica*, and it contains one phenolic catechol group and 3 beta-diketone groups. Due to its novel chemical structure, it possesses strong antioxidant and anti-inflammatory activities. Topical application or oral administration of garcinol to CD-1 mice markedly inhibited TPA-induced ear inflammation in a dose-dependent manner. Topical application of garcinol to ears of CD-1 mice also markedly inhibited TPA-induced up-expression of pro-inflammatory cytokine IL-6 protein levels in ears in dose-dependent fashion. Oral administration of garcinol inhibited UVB-induced ear inflammation, and blocked up-expression of pro-inflammatory cytokine IL-1 beta and IL-6 protein levels in ears. Topical treatment of garcinol to backs of mice strongly inhibited TPA-induced skin tumor promotion in mice previously initiated with 7,12-dimethylbenz[a]anthracene (DMBA) both the numbers of skin tumors per mouse and percent of mice bearing with skin tumors.

Introduction

Garcinol that is a polyisoprenylated benzophenone derivative was isolated from the fruit rinds of *Garcinia indica* (*1-4*). Subsequently, in 1981 Krishnamurthy *et al* established the chemical structure of garcinol (*1-2*) and then revised by Bakana *et al* (*3*). The chemical structure of garcinol is shown in Figure 1. Garcinol contains one ortho-hydroxyl group (catechol group) and three beta-diketone groups. It has a conjugated double bond that exhibits strong antioxidant property (*1-6*). Yamaguchi *et al* demonstrated that garcinol had strong antioxidant property by using several *in vitro* assay systems (*6*). Electron spin resonance spectrometry (ESR) studies indicate garcinol had free radical scavenging activity. In the hypoxanthine oxidase assay system showed that garcinol was able to suppress superoxide anion as similar degree as alpha-tocopherol. Garcinol also was shown to suppress superoxide anion, hydroxyl radical and methyl radical in the H_2O_2/NaOH/DMSO system assay. In additional, Garcinol suppressed hydroxyl radical more strongly than DL-alpha-tocopherol in the Fenton reaction system. Taken together, garcinol is a potent free radical scavenger and is able to scavenge both hydrophilic and hydrophobic free radicals including reactive oxygen species. Garcinol competitively inhibited xanthine oxidase activity with an IC_{50} value of 52 μM (*7*). Liao *et al* also demonstrated that garcinol inhibited LPS-induced formation NO free radical in astrocytes (*7*). Garcinol inhibited formation NO free radical in the mouse RAW 264.7 macrophage cells in cultures (*7-9*). In this report, we demonstrated that garcinol had strong anti-inflammatory activity, and inhibited TPA- or UVB-induced ear inflammation and up-expression of pro-inflammatory cytokine IL-1 beta and IL-6 protein levels in ears of mice as well as was a potent inhibitor of TPA-induced skin tumor promoting activity in DMBA-initiated mice.

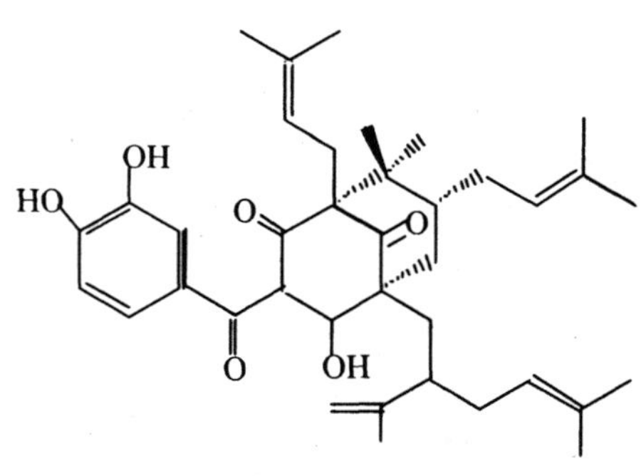

Garcinol

Figure 1. Chemical structure of garcinol.

Materials and Methods

Materials

7,12-Dimethylbenz[a]anthracene and 12-*O*-tetradecanoylphorbol-13-acetate were purchased from Sigma-Aldrich (St. Louis, MO). Acetone and 10% formalin-phosphate buffer were purchased from Fisher Scientific (Springfield, NJ).

Animals

Female CD-1 mice (3-4 weeks old for ear inflammation experiment, or 7-8 weeks old for skin tumor experiment were purchased from Charles River Laboratories (Kingston, NY). The mice were kept in our animal facility for at least one week before use. Mice were fed a Purina Laboratory Chow 5001 diet from Ralston-Purina Co. (St Louis, MO) and water *ad libitum* and kept on a 12 h light/12 h dark cycle. The dorsal area of each mouse was shaved with electric clippers at least 2 days before treatment with TPA or DMBA. Only mice that did not show signs of hair regrowth were used.

Measurement of Mouse ear Inflammation

Measurement of mouse ear inflammation was done according to the following procedure. Both ears of female CD-1 mice (3-4 weeks old; 5 - 6 mice per group) were treated topically with 20 μL acetone, 0.5 nmol TPA in acetone or test compound in acetone 10 min prior to TPA (1 nmol) treatment. Six hours later the mice were sacrificed by cervical dislocation, and 6-mm in diameter ear punches biopsies were taken and weighed. The increase in weight of ear punches was directly proportional to the degree of inflammation.

Tumor Studies on Mouse Skin

The dorsal area of female CD-1 mice (8 weeks old) was shaved with electric clippers. For studies on the inhibitory effect of topical application of garcinol on TPA-induced skin tumor promotion, the mice (30 mice per group) were treated topically with 200 nmol of 7,12-dimethylbenz[a]anthracene (DMBA) in 100 μL of acetone. A control group of mice received 200 μL of acetone alone. After 1 week the mice were treated topically with 200 μL of acetone or TPA (5 nmol) in acetone. Garcinol (0.2 or 10 μmol) in acetone were treated topically on dorsal area 10 min prior to TPA treatment twice weekly for 10 or 14 weeks. Skin tumors greater than 1 mm in diameter were counted and recorded every 2 weeks. All skin tumors were examined histopathologically.

Preparation of Garcinol.

Garcinol was isolated and purified from the dried fruit rinds of *Garcinia indica* (Kokum) according to the procedure as described by Yamaguchi *et al* (*5*). The dried fruit rinds of *G. indica* was extracted with ethanol. The extract was fractionated by preparative ODS (Octadecyl silica) column chromatography eluted stepwise with 70-80% (v/v) ethanol monitored at UV 254 nm, and the main fraction absorption at 254 nm eluted at 80% (v/v) ethanol were concentrated and dried by the rotary evaporator under 50 °C. The dried material was resolved in hexane and solution was under 5 °C for 2 days. Yellow amorphous powder was collected from the solution and washed with cold hexane on a glass filter. After drying in a vacuum desiccator, the amorphous was solubilized in hot acetonitrile and recrystallized at room temperature. Pale yellow crystal needles (garcinol) were obtained from solvent.

Results

Anti-inflammatory Activity

Both *in vitro* and *in vivo* tests have been used to evaluate the anti-inflammatory activities of garcinol (*7-9*).

The In vitro Cell Culture Anti-inflammatory Assay

Inhibition of lipopolysaccharide (LPS)-induced up-expression of iNOS and COX-2 in the mouse RAW264.7 macrophage cell culture has been widely used as the *in vitro* anti-inflammatory assays in cell culture system. Garcinol strongly inhibited LPS-induced up-expression of inducible nitric oxide synthase (iNOS) and cyclooxygenase (COX-2) in astrocytic and in the mouse RAW264.7 macrophage cell cultures (*7-9*). Since inflammation is associated with up-expression of iNOS and COX-2 mRNA and protein levels.

Inhibitory Effect of Topical Application of Garcinol and Curcumin on TPA-induced Edema of Mouse Ears

We have tested the anti-inflammatory activity of garcinol in a mouse ear inflammation model. Garcinol had a strong anti-inflammatory activity in TPA-induced edema of mouse ear test. The results are shown in (Table I). A single topical application of 1 nmol of 12-*O*-tetradecanoylphorbol-13-acetate (TPA) on ears of CD-1 mice resulted in increase in ear inflammation by increase an average weight of ear punches from 7.4 mg (acetone control group) to 15.1 mg (TPA-treated group) per punch. Topical application of 0.2 or 0.5 μmol of garcinol together with TPA (1.0 nmol) treatment resulted in suppressing TPA-induced ear inflammation by 39 or 78%, respectively. Topical application or 0.2 or 0.5 μmol of curcumin together with TPA (1.0 nmol) on ears of CD-1 mice inhibited TPA-induced ear inflammation by 58 or 88%, respectively (Table I). Our results suggested that both garcinol and curcumin had potent anti-inflammatory activity in the mouse ear inflammation model.

Table I. The Effects of Topical Application of Garcinol and Curcumin on TPA-induced Edema of Mouse Ears in CD-1 Mice

Treatment	*Average weight of ear punches (mg)*	*% Inhibition*
Acetone	7.44±0.36*	--
TPA	15.05±0.82	-
Garcinol (0.1 μmol) + TPA	12.14±0.81*	39%
Garcinol (0.5 μmol) + TPA	9.25±0.66*	78%
Curcumin (0.1 μmol) + TPA	10.20±0.80*	58%
Curcumin (0.5 μmol) + TPA	8.41±0.33*	88%

Both ears of mice (4 weeks old; 5 mice per group) were treated topically with acetone, TPA in acetone or TPA + test compound in acetone. The mice were sacrificed 6 h after TPA treatment. Ear punches (6-mm in diameter) were taken and weighed. Data are expressed as the mean±SE.

*Statistically different from the group 2 TPA alone ($P < 0.05$) as determined by the Student' t test.

Inhibitory Effect of Oral Administration of Garcinol on TPA-induced Edema of Mouse Ears

In addition, oral administration of garcinol to female CD-1 mice markedly inhibited TPA-induced ear inflammation in a dose-dependent fashion, whereas oral administration of curcumin had little or no effect on TPA-induced ear inflammation (data not shown). These observations suggested curcumin had poor bioavailability, whereas garcinol had good bioavailability by oral administration.

Inhibitory Effect of Topical Application of Garcinol on TPA-induced Persistent Inflammation and Up-expression of Pro-inflammatory Cytokine IL-1 beta and IL-6 Protein Levels in Mouse Ears

Topical application of TPA (0.8 nmol) once a day for 4 days induced persistent ear inflammation by increase an average of weight of ear punches from 7.4 mg (acetone-treated group) to 17.5 mg per punch (TPA-treated group) (Figure 2A), up-expression of cytokine IL-1 beta protein levels in ears from average of 0.54 pg/mg (acetone-treated group) to 64.7 pg/mg (TPA-treated group) wet tissue weight (Figure 2B), and up-expression of IL-6 cytokine protein levels in ears from 2.0 pg/mg to 8.0 pg/mg (Figure 2C), respectively.

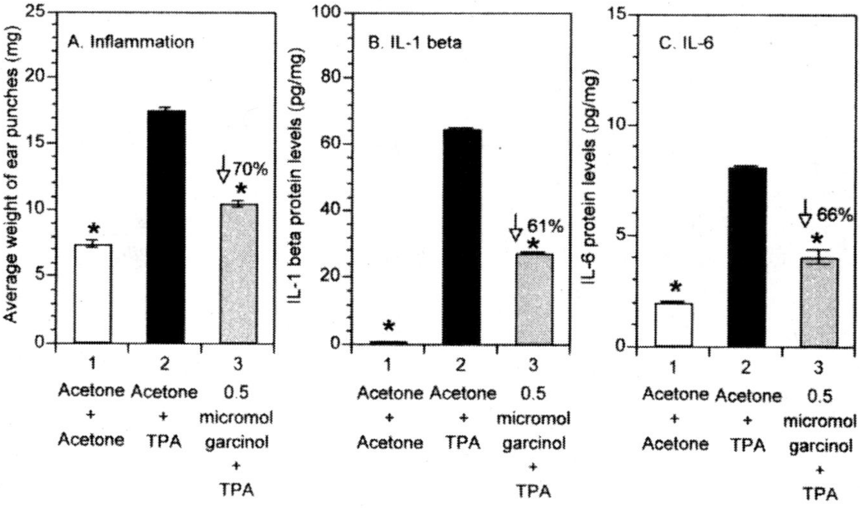

*Figure 2. Effects of topical application of garcinol on TPA-induced inflammation and up-expression of pro-inflammatory cytokine, IL-1 beta and IL-6 protein levels in mouse ears. Both ears of female CD-1 male (5 weeks old; 5 mice per group) were treated topically with acetone or garcinol (0.5 μmol) in acetone 10 min prior to application of acetone or TPA (1.5 nmol) in acetone. The mice were sacrificed 6 h after TPA treatment. Ear punches 96-mm in diameter) were taken and weighed. Data are mean±SE. *Statistically different from the group 2 TPA alone (P< 0.05) as determined by the Student' t test.*

Topical application of 0.5 µmol of garcinol 10 min prior to each TPA treatment once a day for 4 days significantly inhibited TPA-induced ear inflammation by 70% (Figure 2A), inhibited TPA-induced up-expression of IL-1 beta protein levels in ears by 61% (Figure 2B), and inhibited TPA-induced up-expression of IL-6 protein levels in ears by 66% (Figure 2C) , respectively. The results showed that topical application of garcinol strongly suppressed TPA-induced persistent inflammation, up-expression of IL-1 beta and IL-6 protein levels in ears of mice.

Inhibitory Effect of Oral Administration of Garcinol on UVB-induced Ear Inflammation and Up-expression of Pro-inflammatory Cytokine IL-1 beta and IL-6 Protein Levels in Ears.

Female CD-1 mice (5 weeks old; 6 mice per group) irradiation with UVB (180 mJ/cm^2) once a day for 4 days resulted in induction of ear inflammation by increase in average weight of ear punch form 8.6 mg/punch (non-UVB-treated group) to 20.5 mg/punch (UVB-treated group). Oral gavages of 5 mg of garcinol in 0.25 mL corn oil to mice 20 min prior to each UVB (180 mJ/cm^2) irradiation once a day for 4 days inhibited UVB-induced ear inflammation by 48% (Figure 3A). The mice irradiation with UVB (180 mJ/cm^2) once a day for 4 days resulted in increase in expression of IL-1 beta from 1.5 pg/mg to 26.5 pg/mg (non-UVB-treatment vs. UVB-treatment), and up-expression of IL-6 protein levels in ears from 1.7 pg/mg to 7.1 pg/mg of wet tissue weight (non-UVB-treatment vs. UVB-treatment). Oral gavages of garcinol (5 mg) in 0.25 mL corn oil to mice 20 min prior to each UVB (180 mJ/cm^2) irradiation once a day for 4 days inhibited UVB-induced up-expression of IL-1 beta protein levels in ears by 14% and inhibited UVB-induced up-expression of IL-6 protein levels in ears by 12%. All these inhibitions are statistically different from UVB-treated group.

Inhibition of Skin Carcinogenesis by Garcinol

Inhibitory effect of topical application of garcinol on TPA-induced skin tumor promotion in mice previously initiated with DMBA.

Chemicals that are able to inhibit TPA-induced ear inflammation inhibit TPA-induced skin tumor promotion in mice previously initiated with 7,12-dimethyl-benz[a]anthracene (DMBA) (*10-11*). Since garcinol is a potent inhibitor of TPA-induced ear inflammation. We have evaluated the effect of garcinol on TPA-induced skin tumor promoting activity in mice previously initiated with DMBA (Figure 4). Female CD-1 mice (8-9 weeks old; 30 mice per group) were initiated with a single dose of DMBA. A week later, the mice promotion with 5 nmol of TPA twice weekly for 10 weeks or 14 weeks resulted in formation of the average of 9.2 or 16.5 skin tumors per mouse, and 87 or 97% of mice had skin tumors. Topical application of 2 or 10 µmol of garcinol 10 min prior to each TPA (5 nmol) treatment resulted in production of an average number of skin tumors per mouse by 2.4 or 1.6 after 10 weeks of TPA promotion and 3.6 or 2.9 skin tumors per mouse after 14 weeks of TPA promotion (see Figure 4). The results demonstrated that topical application of 2 or 10 µmol of garcinol 10 min prior to each TPA (5 nmol) treatment resulted in inhibiting TPA-induced an average numbers of skin tumors per mouse by 74%, or 83%, respectively, after 10 weeks of TPA treatment. Topical application of 2 or 10 µmol of garcinol 10 min prior to each TPA (5 nmol) treatment resulted in inhibiting TPA-induced an average numbers of skin tumors per mouse by 61%, or 82%, respectively, after 14 weeks of TPA treatment.

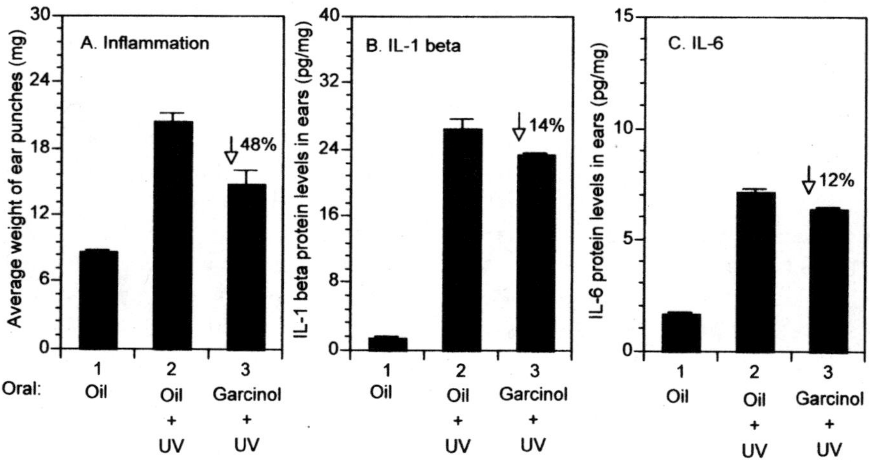

Figure 3. Effects of oral administration of garcinol on UVB-induced edema, and up-expression of pro-inflammatory cytokine IL-1 beta and IL-6 protein levels in ears. Female CD-1 mice (5 weeks old; 5 mice per group) were orally intubated with 0.25 mL corn oil or gracinol (5 mg) in 0.25 mL corn oil 20 min prior to each UVB (180 mJ/cm^2) irradiation once a day for 4 days. The mice were sacrificed 6 h after the last dose UVB irradiation. Ear punches (6-mm in diameter) were taken and weighed. Ear samples were pooled together and homogenized in a phosphate buffer for cytokine ELISA assays.

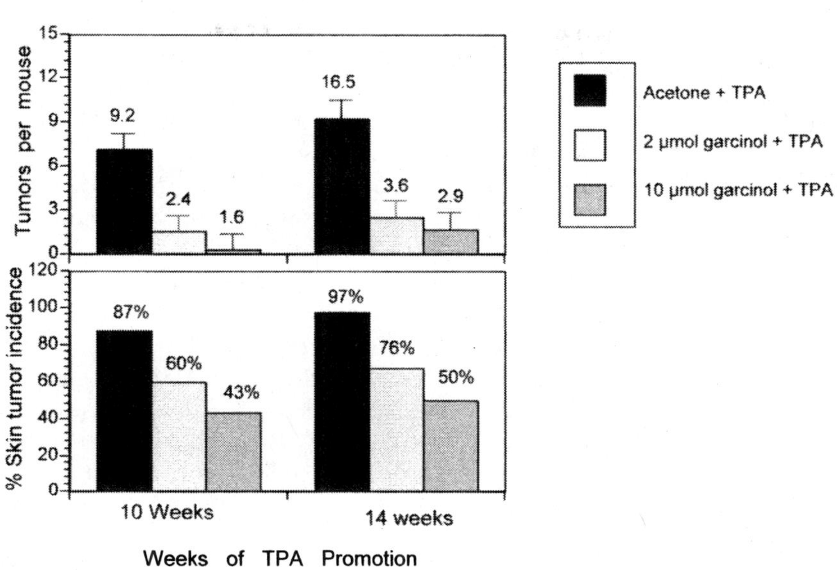

Figure 4. The effects of topical application of garcinol on TPA-induced skin tumor promotion in mice previously initiated with DMBA. Female CD-1 mice (8-9 weeks old; 30 mice per group) were treated topically with 100 μl acetone or a single dose of 7,12-dimethylbenz[a]anthracene (200 nmol) in acetone. One week later, the mice were promoted with 5 nmol TPA in acetone twice weekly for 14 weeks. The mice were treated topically with 100 μL acetone or 2 μmol or 10 μmol of garcinol in acetone 10 min each TPA treatment twice a week for 14 weeks. Skin tumors in diameter greater than 1 mm were counted and recorded.

Discussion

In the present study, we have shown that a single topical application of TPA on ears of CD-1 mice induced edema, and increase in weight of ear punches. The degree of ear inflammation is proportional to increasing in weight of ear punches. Topical application of garcinol and curcumin strongly inhibited TPA-induced ear inflammation in a dose-dependent manner (Table I). Multi-application of TPA (once a day for 4 days) induced persistent ear inflammation and up-expression of pro-inflammatory cytokine IL-1 beta and IL-6 protein levels in ears of CD-1 mice. Topical application of garcinol 20 min prior to each TPA treatment once a day for 4 days strongly inhibited TPA-induced persistent inflammation and up-expression of IL-1 beta and Il-6 protein levels in ears

(Figure 2). We also demonstrated that oral gavages of garcinol to CD-1 mice inhibited TPA-induced ear inflammation in a dose-dependent manner. Oral gavages of garcinol to mice also inhibited UVB-induced inflammation and up-expression of IL-1 beta and Il-6 protein levels in ears. Our results clearly suggest that garcinol has good bioavailability when it is given by orally.

Since garcinol has a novel chemical structure and it contains 1 phenolic catechol group and 3 beta diketone groups. Garcinol possesses high antioxidant and anti-inflammatory activities. We have shown that both topical application and oral administration of garcinol to mice strongly inhibited chemically- and UVB-induced inflammation, up-expression pro-inflammatory cytokine IL-1 beta and IL-6 protein levels in ears (Table I, Figures 2 and 3).

We also demonstrated that topical application of garcinol strongly inhibited TPA-induced development of average numbers of skin tumors per mouse as well as inhibited skin tumor incidence. Tanaka *et al* (*12*) showed that feeding rats with garcinol in the diets inhibited azoxymethane (AOM)-induced aberrant crypt foci (ACF) in colon of male F344 rats. Dietary garcinol (500 ppm) significantly decreased the incidence and multiplicity of 4-nitroquinoline 1-oxide (4-NQO)-induced tongue neoplasms and/or pre-neoplasms as compared to the control diet (*13*). Yoshida *et al* (*13*) suggested that dietary administration of garcinol inhibited 4-NQO-induced tongue carcinogenesis through suppression of increased cell proliferation activity in the target tissues and/or COX-2 expression in the tongue lesions.

Molecular mechanisms of garcinol's actions have been reported in some detail. Mattigel analysis showed that exposure of HT-29 cells to 10 μM garcinol inhibited cell invasion, and decreased the dose-dependent tyrosine phosphorylation of focal adhesion kinase (FAK) (*14*). Garcinol modulates tyrosine phosphorylation of FAK and subsequently induces apoptosis through down-regulation of Src, ERK, and AKt survival signaling in human colon cancer cells (*14*). Garcinol modulates arachidonic metabolism by blocking the phosphorylation of $cPLA_2$ and decreases iNOS protein levels by inhibiting STAT-1 activation (*11*). Garcinol inhibited NF-κB activation and COX-2 expression of ERK1/2 phosphorylation (*11*). Garcinol is reported to be a natural histone acetyltransferase inhibitor, represses chromatin transcription and alters global gene expression (*15)*. Garcinol has been proved to be as a potent antioxidant and anti-inflammatory agent as well as a natural potent cancer chemopreventive agent.

References

1. Krishnamurthy, N; Lewis, Y.S.; Ravindranath, B. *Tetrahedron* **1981**, *22*, 793-796.

2. Krishnamurthy, N.; Ravindranath, B. *Tetrahedron* **1982**, *23*, 2233-2236.
3. Bakana, P.; Claeys, M.; Totte, J.; Pieters, L.A.C.; Van Hoof, L.; Tamba Vemba, D.A.; Vlie-Tink, A.J. *J. Ethnopharmacol.* **1987**, *21*, 75-84.
4. Salhu, A.; Das, B.; Chatterjee, A. *Phytochemistry* **1989**, *28*, 1233-1235.
5. Yamaguchi, F.; Ariga, T.; Yoshimura, Y.; Nakazawa, H. *J. Agric. Food Chem.* **2000**, *48*, 180-185.
6. Yamaguchi, F.; Saito, M.; Ariga, T.; Yosimura, Y.; Nakazawa, H. *J. Agric. Food Chem.* **2000**, *48*, 2320-2325.
7. Liao, C.H.; Ho, C.-T.; Lin, J.K. *Biochem. Biophys. Res. Commun.* **2005**, *329*, 1306-1314,.
8. Huang, M.T.; Smart, R.C.; Wong, C.Q.; Conney, A.H. *Cancer Res.* **1988**, *48*, 5941-5946.
9. Smart, R.C.; Huang, M.T.; Monteiro-Riviere, N.A.; Wong, C.Q.; Mills, K.J.; Conney, A.H. *Carcinogenesis* **1988**, *9*, 2221-2226.
10. Liao, C.H.; Sang, S.; Liang, Y.C.; Ho, C.-T.; Lin, J.K. *Molecular Carcibnogenesis* **2004**, *41*, 140-149.
11. Hong, J.; Sang, S.; Park, H.J.; Kwon, S.J.; Suh, N.; Huang, M.T.; Yang, C.S. *Carcinogenesis* **2005**, *27*, 278-286.
12. Tanaka, T.; Kohno, H.; Shimada, R.; Kagami, S.; Yamaguchi, F.; Kayaoka, S.; Ariga, T.; Murakami, A.; Koshimizu, K.; Ohigashi, H. *Carcinogenesis* **2000**, *21*, 1183-1189.
13. Yoshida, K.; Tanaka, T.; Hirose, Y.; Yamaguchi, F.; Kohno, H.; Toida, M.; Hara, A.; Sugie, S.; Shibata, T.; Mori, H. *Cancer Lett.* **2005**, *221*, 29-39.
14. Liao, C.H.; Sang, S.; Ho, C.-T.; Lin, J.K. *J. Cell. Biochem.* **2005**, *96*, 155-169.
15. Balasubramanyam, K.; Altaf, M.; Varier, R.A.; Swaminathan, V.; Ravindran, A.; Sadhale, P.P.; Kundu, T.K. *J. Biol. Chem.* **2004**, *279*, 33716-33726.

Chapter 21

Inhibition of Inflammation, Expression of Pro-inflammatory Cytokines, Formation of Leukotriene B_4 and Tumor Promotion in Mouse Skin by *Boswellia serrata* Extracts

Mou-Tuan Huang[1], Yue Liu[1], Vladimir Badmaev[2], and Chi-Tang Ho[3]

[1]Department of Chemical Biology, Susan Lehman Cullman Laboratory for Cancer Research, School of Pharmacy, Rutgers University, Piscataway, NJ 08854-8020
[2]Sabinsa Corporation, Piscataway, NJ 08854
[3]Department of Food Science, Rutgers University, New Brunswick, NJ 08901

The *Boswellia serrata* extract (BE) that contains β-boswellic acids and related compounds has been used as an herbal medicine (Ayurvedic system) for the treatment of arthritis and other inflammation related diseases in India and China. In this manuscript, we report that a preparation of *Boswellia* extract (BE) that contained about 35-40% β-boswellic acid and its derivatives inhibited 12-*O*-tetradecanoylphorbol-13-acetate (TPA)-induced increases in inflammation, up-expression of pro-inflammatory cytokine protein levels, and decreased formation of arachidonic acid metabolite, leukotriene B_4 in mouse ears as well as inhibited TPA-induced increases in epidermal proliferation and tumor promotion in mouse epidermis previously initiated with 7,12-dimethylbenz[a]anthrance (DMBA) in CD-1 mice. A single topical application of TPA to ears of CD-1 mice induced a time and dose-dependent increases in acute inflammation (edema). A single topical application of 0.06 - 0,24 mg of BE 20 min prior to 1.0 nmol of TPA treatment inhibited TPA-induced acute inflammation by 40-84% in a dose-dependent manner.

Application of BE to mouse ears 20 min prior to each TPA application once a day for 4 days inhibited TPA-induced persistent inflammation as well as inhibited TPA-induced increases in up-expression of IL-1β and Il-6 protein levels. BE also inhibited TPA-induced increases in formation of arachidonic acid metabolite. Topical application of 1.2-3.6 mg of BE 10 min prior to each TPA (5 nmol) treatment once a day for 2 days strongly inhibited TPA-induced increases in incorporation of bromodeoxyuridine (BrdUr) into epidermal DNA by 71-81%. Topical application of 1.2-3.6 mg of BE 10 min prior to each TPA (nmol) twice a week for 18 weeks treatment to initiated with DMBA mice inhibited the number of skin tumors per mouse by 82-98% and skin tumor incidence was inhibited by 48-90%. The ability of BE to inhibit TPA-induced increases in persistent inflammation, up-expression of pro-inflammatory cytokine proteins as well as to inhibit formation of leukotriene (LTB_4) in mouse skin that may play a role on its inhibitory effect on TPA-induced tumor promotion.

Introduction

The gum resin of plant *Boswellia serrata* grows in the dry part of China and India. The gum resin exudates of B *serrata* is collected from the stem of the tree *B. serrata* and is used as a commercial source for the herbal medicine (*1*). The anti-inflammatory activity of Boswellic acids is due to their ability to inhibit 5-lipoxygenase activity (*1-5*). β-Boswellic acids, the pentacyclic triterpenic acids (Figure 1) are the main constituents in the gum resin of the plant (*5*). The gum resin exudates of B serrata has also been reported to have anti-hyperlipidemic and anti-atherosclerotic activities (*6,7*), and anti-colorectal carcinogenic effects on mouse model (*8,9*). β-Bosweelic acid and its derivatives also have been reported to inhibit the growth of several cancer cells in cultures (*10-13*). At present study, we report that topical application of BE to ears of CD-1 mice inhibited TPA-induced acute and persistent inflammation and inhibited TPA-induced increases in expression of pro-inflammatory cytokine IL-1β and IL-6 protein levels in ears of mice, and decreased formation of arachidonic acid metabolite, leukotriene B_4 in mouse ears as well as inhibited TPA-induced increases in epidermal proliferation and tumor promotion in mouse epidermis previously initiated with DMBA in CD-1 mice.

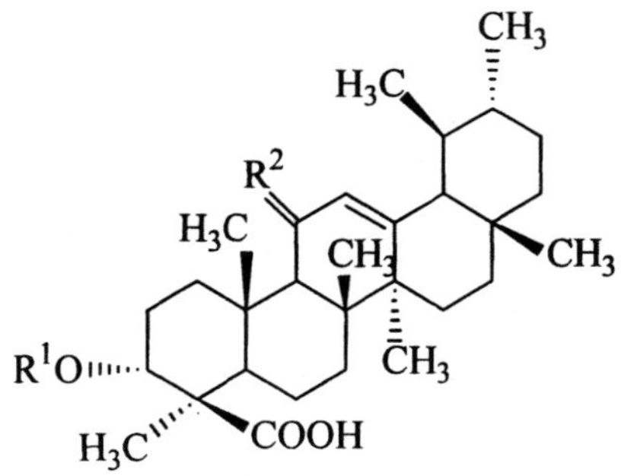

R^1 = H, R^2 = H_2	β-boswellic acid
R^1 = Ac, R^2 = H_2	3-*O*-acetyl-β-boswellic acid
R^1 = H, R^2 = O	11-keto-β-boswellic acid
R^1 = Ac, R^2 = O	3-*O*-acetyl-11-keto-β-boswellic acid

Figure 1. Chemical structures of β-boswellic acid and its derivatives.

Materials and Methods

Preparation of *Boswellin serrata* Extract and Chemical Composition

Oleogum resin exudates of *Boswellia serrata* (100 g) was extracted with methanol (200 mL x 3), the methanol evaporated in vacuum to give 45 g of dried extract. The dried extract (30 g) was dissolved in 2% KOH solution (200 mL) and extracted with ethyl acetyl and the ethyl acetate fraction was discarded. The aqueous solution was neutralized with 2% HCl to pH 6.0, and then extracted again with ethyl acetate (5 x 150 mL). The combined ethyl acetate solution was washed with water, and dried with anhydrous Na_2SO_4 overnight, and then evaporated to dryness to produce 18 g of residue (Boswellin, a mixture of boswellic axcid and its derivatives). Boswellin (BE) contains about 13-18% of β-boswellic acid, 11-17% 3-*O*-acetyl-β-boswellic acid, 6-8% 11-keto-β-boswellic acid and 4-7% 3-*O*-acetyl-11-keto-β-boswellic acid.

Animals

Female CD-1 mice (23-25 days for ear inflammation experiments), or (6-7 weeks old for skin tumor experiments) were purchased from Charles River Laboratories (Kingston, NY). The mice were kept in the Susan Lehman Cullman Laboratory for Cancer Research animal facility for at least 1 week before use. Mice were fed with a Purina Laboratory Chow 5001 diet from Ralston-Purina Co. (St Louis, NO) and water *ad libitum* and kept on a 12 h light/12 h dark cycle. The dorsal region of each mouse was shaved by electric clippers at least 2 days before treatment with TPA or DMBA. Only mice that did not show signs of hair re-growth were used.

Measurement of TPA-induced Acute Inflammation in Ears of CD-1 Mice

Measurement of mouse ear edema was done according to the following procedure. Both ears of female CD-1 mice (23-25 days old; 5-6 mice per group) weren treated topically with 20 μL acetone, 0.5 nmol TPA in acetone or Boswellin extract (BE) together with 0.5 nmol TPA in acetone. Six hours later the mice were sacrificed by cervical dislocation, and 6-mm (diameter) ear punches biopsies were taken and weighed. The increases in weight of ear punches were directly proportional to the degree of inflammation (*15*).

Determination of TPA-induced Persistent Inflammation, Upexpression of Cytokine IL-1 and IL-6 proteine Levels, Formationof PGE2 and LTB4 Levels in Ears of CD-1 Mice

For persistent inflammation, both ears of female CD-1 mice were treated topically once a day for 4 days either 10 μL acetone (vehicle) or test compound in acetone 20 min prior to each application of acetone or 0.8 nmol TPA in acetone. Ears were persistent inflammation during the 4 days of TPA treatment. The mice were sacrificed 6 h after the last TPA treatment. Ear punches (6-mm in diameter) were taken and weighed. Ear samples from each group were pooled and homogenized as described in the next section (*15*)

Preparation of Ear Homogenate for ELISA Assays

Ear tissues were homogenized in a PBS containing 0.4 M NaCl, 0.05% Tween-20, 0.5% BSA, 0.1 mM PMSF, 0.1 mM benzethonium, 10 mM EDTA, and 20 U aprotinin per ml. The supernatant fraction was used for determination of cytokine protein levels. A two-site sandwich ELISA was used to assay for cytokines (*15*).

Determination of Proliferation of Epidermis by Bromodeoxyuridine (BrdUr) Labeling Index

Immunohistochemical staining for BrdUr was performed according to the procedure of the commercial assay kit as described in our previous publication (*9*).

Tumor Studies on Mouse Skin

The dorsal region of female mice CD-1 mice (7-8 weeks old) was shaved with electric clippers. For studies on the inhibitory effect of topical application of BE on TPA-induce tumor promotion, the mice (30 per group) were treated topically with 200 nmol of 7,12-dimethylbenz[a]athracene (DMBA) in 100 μL acetone. After 1 week of the mice were treated topically with 200 μl of acetone, 5 nmol of TPA, or 1.2 or 3.6 mg of BE 10 min prior to 5 nmol TPA twice weekly for 20 weeks. A control group of mice received 200 μl acetone alone. Skin tumors greater than 1 mm in diameter were counted and recorded every 2 weeks. All skin tumors were examined histopathologically (*16*).

Results

Inhibitory Effect of BE on TPA-induced Increases in Acute Inflammation Mouse Ears

The possibility that BE could inhibit TPA-induced edema (acute inflammation) was evaluated by studying the effects of BE on TPA-induced edema of mouse ears. A preparation of *Boswellia* extract (BE) contained about 35-40% of total β-boswellic acids (w/w) was used for our studies. Topical application of 0.06-0.24 mg of BE with 0.5 nmol of TPA to the ears of mice inhibited TPA-induced edema by 40-84% (Figure 2). The inhibitory effect of BE on TPA-induced ear edema was dose-dependent when BE was given together with TPA or 30 min before TPA. Several preparations of BE were evaluated their anti-inflammatory activity. The preparation of BE containing the higher total amount of β-boswellic acids was the higher the anti-inflammatory activity.

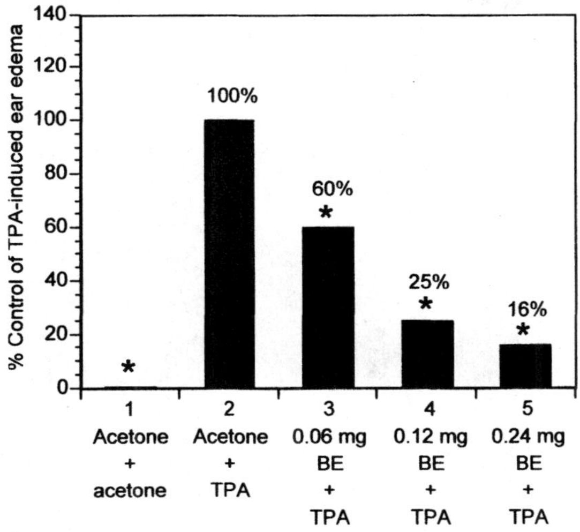

*Figure 2. Effect of BE on TPA-induced acute inflammation in ears of CD-1 mice. Both ears of mice were treated topically with acetone or test compound in acetone at 20 min prior to topical application of acetone or TPA in acetone. The mice were sacrificed at 6 h after TPA treatment. Ear punches were taken and weighed. Data are mean±SE from 10 ear samples per group. *Statistically different fro the group 2 (P<0.05) as determined by the Student's t test.*

Inhibitory Effect of BE on TPA-induced Increases in Persistent Inflammation, Up-expression of Pro-inflammatory Cytokine Protein Levels in Ears of CD-1 Mice

Topical application of 0.8 nmol TPA to ears of CD-1 mice once a day for 4 days resulted persistent inflammation and up-expression of pro-inflammatory cytokine IL-1 ad IL-6 protein levels in ears in a dose- and time dependent manner. Topical application of 0.18-0.36 mg of BE 20 min prior to each TPA treatment inhibited TPA-induced persistent inflammation by 54-74% (Figure 3, upper panel), inhibited TPA-induced up-expression of cytokine IL-1β protein levels in ears by 16-60% (Figure 3, middle panel) and inhibited TPA-induced up-expression of cytokine IL-6 protein levels by 47-78% (Figure 3, lower panel). These results showed that topical application of BE strongly inhibited TPA-induced acute and persistent ear inflammation as well as inhibited TPA-induced increases in up-expression of pro-inflammatory cytokine IL-1 and IL-6 protein levels in ears.

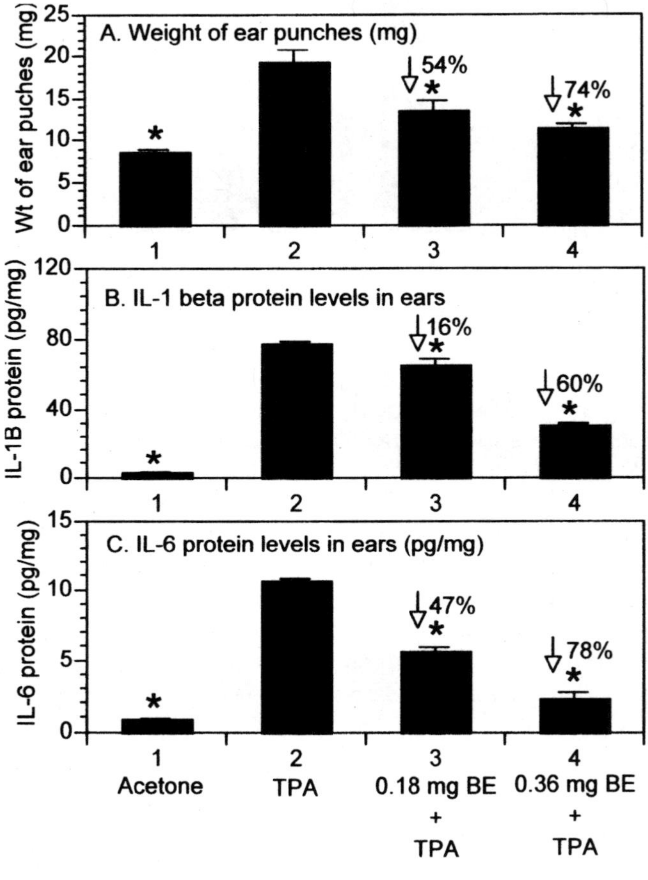

Figure 3. Effect of BE on (A, upper panel), TPA-induced persistent inflammation, (B, middle panel), TPA-induced up-expression of pro-inflammatory cytokine IL-1 beta protein levels, and (C, lower panel), TPA-induced up-expression of pro-inflammatory cytokine IL-6 protein levels in ears of CD-1 mice.

Inhibitory Effect of BE on TPA-induced Increases in Incorporation of BrdUr into Epidermal DNA

BE strongly inhibited TPA-induced proliferation of mouse epidermis as determined by the incorporation of bromodeoxyuridine (BrdUr) into epidermal DNA. Twenty four hours after the last dose of topical application of 5 nmol of TPA once a day for 2 days resulted in a 43% BrdUr labeling index. Topical application of 1.2-3.6 mg of BE 10 min prior to each 5 nmol of TPA treatment once a day for 2 days resulted in a decrease BrdUr labeling to 12.5 or 8.3%. Thus BE inhibited TPA-induced epidermal proliferation by 71 or 81% at1.2 or 3.6 mg, respectively of BE (Figure 4).

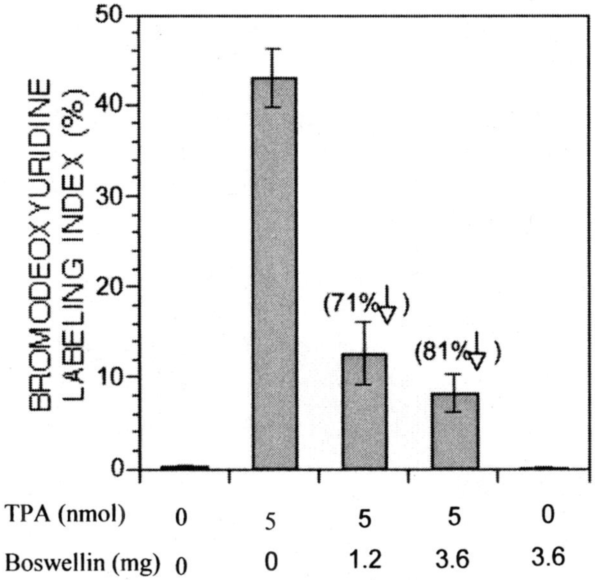

Figure 4. Effect of BE on TPA-induced incorporation of BrdUr into epidermis of CD-1 mice

Inhibitory Effect of BE on TPA-induced Tumor Promotion in the Skin of Mice Previously Initiated with DMBA

Application of 1.2-3.6 mg of BE at 10-30 min before TPA or together with TPA (5 nmol) strongly inhibited TPA-induced skin tumor promotion in female CD-1 mice previously initiated with DMBA. Female CD-1 mice initiated with a single dose of 200 nmol DMBA and promoted with 5 nmol of TPA twice weekly for 16 weeks developed an average of 15.8 skin tumors per mouse and 90% of the mice had skin tumors. Topical application of 1.2 or 3.6 mg of BE with TPA (5 nmol) for 16 weeks reduced significantly the average number of skin tumors by 87 and 99%, respectively. The percentage of the mice with skin tumors was also significantly decreased by 59 or 92%, respectively, when 1.2 mg or 3.6 mg of BE was applied topically with TPA (5 nmol) twice weekly for 16 weeks. BE treatment delayed the appearance of the first skin tumor in comparison to the control DMBA/TPA treatment group. The latency time for the DMBA/TPA positive group was 4 weeks. The latent time for 1.2 mg of BE treatment was 8 weeks and was greater than 8 weeks with 3.6 mg of BE treatment. These results indicate that BE is a potent inhibitor of TPA-induced skin tumor promotion on mouse skin (Figure 5).

Discussion

The results of present study demonstrate that topical application of BE to ears of CD-1 mice strongly inhibited TPA-induced increases in both acute inflammation (edema) and persistent inflammation. Topical application of BE also inhibited TPA-induced increases in up-expression of pro-inflammatory cytokine IL-1β and IL-6 protein levels in ears of the mice. Inflammatory reactions are involved in many acute and chronic diseases (*17*). Abnormal and prolonged over-expression of some inflammation-related genes may lead to certain chronic diseases such as arthritis, diabetes, atherosclerosis and some types of cancer. The ability to detect elevated levels of expression of some inflammation-related genes may be useful as biomarkers for the early detection of certain chronic diseases.

Chronic inflammation produces excessive amounts of reactive oxygen species (ROS) and leads to certain degenerative diseases. Dietary and medicinal plants have been used for the treatment of inflammatory diseases by traditional medicinal plants. Inflammation plays important roles in the pathology of many human diseases. An extract of *Boswellia serrata* (BE) that contains β-boswellicacid and other related pentacyclic triterpene acids (Figure 1) has potential anti-inflammatory activity. BE has been used as an herbal medicine for the treatment of inflammation, inflammatory related diseases and arthritis (*1-5*). In this manuscript we report that BE inhibited TPA-induced increases in inflammation, over-expression of pro-inflammatory cytokine protein levels, and arachidonic acid metabolism in mouse ears as well as inhibited TPA-induced increases in epidermal proliferation and tumor promotion in mouse epidermis previously initiated with 7,12-dimethylbenz[a]anthrance (DMBA).

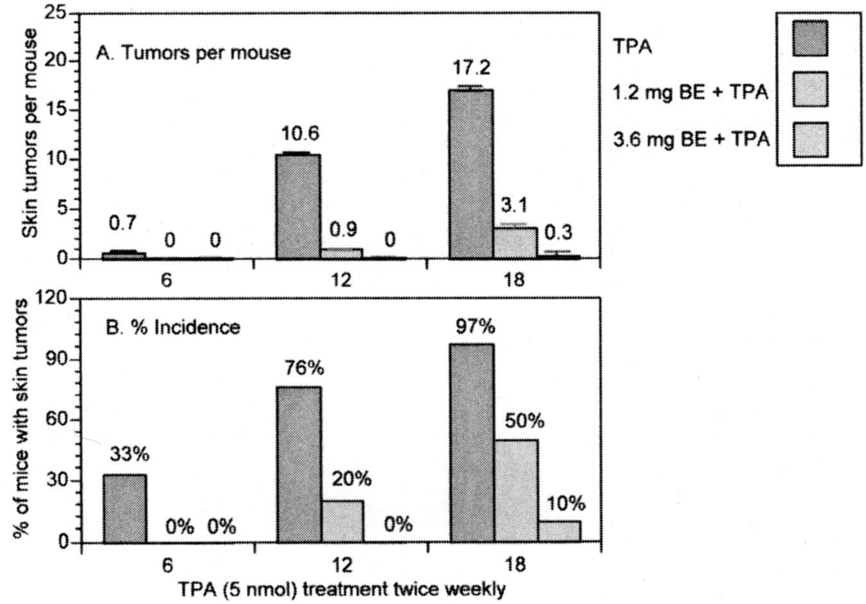

Figure 5. Inhibitory effect of topical application of BE on TPA-induced tumor promotion in skin of CD-1 mice. Famale CD-1 mice (7-8 weeks old: 30 mice per group) were shaved with electric clippers 2 days before treated with of 7,12-dimethylbenz[a]thracene (DMBA) or TPA. The mice were treated topically with a single dose of 200 nmol of DMBA. A week later, the mice were treated topically with 200 µl acetone, 1.8 mg or 3.6 mg of BE in acetone at 10 min prior to each application of acetone or TPA (5 nmol) in acetone twice a week for 18 weeks. Skin tumors greater than 1 mm in diameter were counted and recorded every 2 weeks.

References

1. Sharma, M. L.; Bani, S.; Singh, G.B. *Int. J. Immunopharmacol.* **1989**, *11*, 647-652.
2. Safayhi, H.; Mack, T.; Sabieraj, J.; Anazodo, M.I.; Subramanian, L.R.; Ammon, H.P. *J. Pharmacol. Exp. Ther.* **1992**, *261*, 1143-1146.
3. Ammon, H.P.; Safayhi, H.; Mack, T.; Sabieraj, J. *J. Ethnopharmacol.* **1993**, *38*, 113-119.
4. Ammon, H.P.; Safayhi, H.; Mack, T.; Sabieraj, J. *J. Ethnopharmacol.* **1993**, *38*, 113-119.
5. Majeed, M.; Badmaev, V.; Gopinathan, S.; Rajendran, R.; Norton, T. *The anti-inflammatory Phytonutrient, Nutriscience*, Publishers, Inc., Piscataway, NJ , 1996.
6. Atal, C.K.; Gupta, O.P.; Singh, G.B. *Br. J. Pharm.* **1981**, *74*, 115-116.
7. Zutshi, U. *Ind. J. Pharm.* **1986**, *18*, 182-183.
8. Anthoni, C.; Laukoetter, M.G.; Rijcken, E.; Vowinkel, T.; Mennigen, R.; Muller, S.; Senninger, N.; Russell, J.; Jauch, J.; Bergmann, J.; Granger, D.N.; Krieglstein, C.F. *Am. J. Physiol. Gastrointest. Liver Physiol.* **2006**, *290*, G1131-1137.
9. Huang, M.T.; Badmaev, V.; Ding, Y.; Liu, Y.; Xie, J.G.; Ho, C.-T. *Biofactors* **2000**, *13*, 225-230.
10. Andersson, D.; Liu, J.J.; Nilsson, A.; Duan, R.D. *Anticancer Res.* **2003**, *23*, 3317-3322.
11. Zhao, W.; Entschladen, F.; Liu, H.; Niggemann, B.; Fang, Q.; Zaenker, K.S.; Han, R. *Cancer Detect. Prev.* **2003**, *27*, 67-75.
12. Jing, Y.; Nakajo, S.; Xia, L.; Nakaya, K.; Fang, Q.; Waxman, S.; Han, R. *Leuk. Res.* **1999**, *23*, 43-50.
13. Shao, Y.; Ho, C.-T.; Chin, C.K.; Badmaev, V.; Ma, W.; Huang, M.T. *Planta Med.* **1998**, *64*, 328-331.
14. Safayhi, H.; Rall, B.; Sailer, E.R.; Ammon, H.P. *J. Pharmacol. Exp. Ther.* **1997**, *281*, 460-463.
15. Huang, M.T.; Liu, Y.; Ramji, D.; Lo, C.Y.; Ghai, G.; Dushenkov, S.; Ho C.-T. *Mol. Nutr. Food Res.* **2006**, *50*, 115-122.
16. Huang, M.T.; Smart, R.C.; Wong, C.Q.; Conney, A.H. *Cancer Res.* **1988**, *48*, 5941-5946.
17. Huang, M.T.; Ghai, G.; Ho, C.-T. *Comp. Rev.Food Sci. Food Safety* **2004**, *3*, 127-139.

Indexes

Author Index

Subject Index

B

C

D

F

G

H

I

K

L

M

N

O

P

Q

R

T